Lectures on

Particles and Fields

# Lectures on Particles and Fields

*Edited by*

## H. H. ALY

*Southern Illinois University*
*Edwardsville*

GORDON AND BREACH Science Publishers

New York      London      Paris

# Preface

To communicate coherently through the vast number of published papers in the field of high energy physics appearing almost daily in professional journals would be a very hard task indeed. This fact has led to summer schools, conferences, and many topical seminars. The outcome of these efforts proved to be of a vital interest to students planning a career in high energy physics.

In this volume we put together a series of review and original articles in the hope of making some diverse aspects of recent developments in particle physics available to graduate students and research workers in the field. Needless to say that this rapidly changing field of physics renders a comprehensively written article so as to cover all related topics and credit all authors a particularly hard job.

No doubt some authors and their works may not have been emphasized sufficiently or even mentioned in these articles. But this is the nature of such a rapidly field as high energy physics. To all these authors a sincere apology is in order.

*Edwardsville, Illinois*
*December, 1969*

H.H.ALY

# Contents

# Photoabsorption Sum Rules and the Nonrelativistic Quark Model

G. BARTON

*University of Sussex, Brighton, England*

## Contents



## 1 INTRODUCTION

Sum rules equating integrals over photoabsorption cross-sections to static parameters of the ground state have long been familiar in atomic [1] and nuclear physics[2]. In particle physics, several new ones have been derived

1

recently from rather general assumptions, which might be expected to make them universally applicable. Here we intend to illustrate simple implications of some of these new rules for spin $\frac{1}{2}$, $i$-spin $\frac{1}{2}$ targets, by applying them in the old context, namely to nonrelativistic bound states, exploiting the experience available in nuclear physics, (though ignoring the complication of exchange currents), and introducing relativistic corrections only where they are unavoidable.

Prima facie, the restriction to such nonrelativistic (NR) systems implies both (a) that the speeds (kinetic energies) of the constituents are NR, and (b) that the binding energy is much smaller than the individual rest masses. Nevertheless we shall apply our discussion to the usual NR quark model[3] of the nucleon, where (a) may hold but (b) certainly does not. This is done partly for definitiveness, partly because the illustration is entertaining, and partly because the points of view which emerge are quite clear-cut, and may prove useful in organizing one's questions and ideas even if they cannot be trusted quantitatively. In other words, our results will be suggestive, not conclusive; the quark model is so little understood in detail that the weight attached to such suggestions must be a matter of taste.

In this first section, we introduce the most important sum rules, and recall some generalities. In section 2, we outline the relevant features of the quark model, (2A); discuss the chief simplifications for making photoabsorption by composite systems tractable, (2B); and write down the conventional NR coupling to the radiation field, (2C), which will have to be revised in section 7. Sections 3 and 4 consider the Cabibbo–Radicati[4] (CR) and Beg–Kawarabayashi–Suzuki[5] (BKS) sum rules; both can be understood straightforwardly in a conventional NR framework. The principal implications of marrying them with the quark model are given in (3.8), (3.15), (3.16), and (4.5). In section 5 we discuss very briefly the Pagels–Harari sum rules[6], which are readily understood as adaptations of the ancestral Thomas–Reiche–Kuhn sum rule to the quark model. In section 6, we consider a rule proposed by Gottfried[7] which appears to be contradicted by the model. Finally in section 7 we consider the Drell–Hearn–Gerasimov (DHG) sum rule[8]; it turns out that this can be understood only by making essentially relativistic modifications to the conventional NR radiative coupling, and that consequently it does not lend itself to interpreting experimental data in the light of the NR quark model.

Our approach will be to assume that a rule holds for the constituents, and to ask whether we can then show by ordinary quantum mechanics either that it does hold, or that it does not need to hold, for the bound state; recall that both in nuclei, and in the quark model, the constituents themselves have spin and $i$-spin $\frac{1}{2}$. (The only exception is Gottfried's rule, which is not applicable to the quarks themselves, and where we proceed differently.) Should the answer be negative, we would then have a possibly hypothetical but logically admissible counter example, and the generality of the assumptions underlying the sum rule would have to be restricted. In fact, all but the Gottfried rule will pass the test; the chief gain from the exercise is insight into the detailed mechanisms that validate the original rules and some additional new rules which arise when the original ones are applied to the quark model.

The reader is reminded, briefly, of the basic ingredients of all sum rules: (i) The dispersion representation of the forward Compton amplitude[9] to order $e^2$, either for physical, or separately for isoscalar and isovector photons. (ii) The optical theorems to the required order $e^2$, which express the spectral functions in the dispersion integrals in terms of total absorption cross-sections for various combinations of photon and target polarizations in ordinary and in isotopic space. (iii) A threshold theorem prescribing the value of an amplitude, or linear combination of amplitudes, at zero photon frequency. Some such theorems have been proved by Low and by Gell–Mann and Goldberger[10], assuming only gauge and Lorentz covariance; others need the equal-time commutators of the conserved isospin currents, which for brevity we call "current algebra"[11]. The commutators depend on whether the electromagnetic coupling is minimal in a sense to be discussed in Section 3.4. (iv) The assumption that the amplitude in question vanishes at infinite photon energy. The successive application of these ingredients will be sketched for a particular example, after quoting the more important sum rules.

Let $\sigma(\omega)$ be a total photon absorption cross-section, and define the integrals

$$J = \int_0^\infty d\omega \, \sigma(\omega)/\omega; \quad K = \int_0^\infty d\omega \, \sigma(\omega). \tag{1.1}$$

Since we work to order $e^2$, and since there are no zero-mass hadrons, the true lower limits in (1.1) are the inelastic (photoproduction) thresholds. $J$ is called the Bremsstrahlung-weighted integral.

Introduce suffices and superfixes, used interchangeably, on $\sigma$, and correspondingly on $J$ and $K$, as follows. $\sigma_{P,A}$: absorption of circularly polarized photons with their spins parallel (antiparallel) to that of the (spin $\frac{1}{2}$) target;

$\sigma^{V,S}$: absorption of isovector (isoscalar) photons; $\sigma^V_{1/2,3/2}$: absorption of iso-vector photons (on $i$-spin $\frac{1}{2}$ target) leading to final states with total $i$-spin $\frac{1}{2}$ ($\frac{3}{2}$). Let $eQ$, $e\mu$, $M$ be the total charge, magnetic moment, and mass of the target, distinguishing nucleons and quarks by suffices $N$ and $q$ where necessary; let $\varkappa$ be the dimensionless Pauli moment, so that

$$\mu = (Q + \varkappa)/2M. \tag{1.2}$$

$F_{1,2}$ are the Dirac and Pauli, $G_{E,M}$ the electric and magnetic form factors, satisfying

$$F_1(0) = Q, \quad F_2(0) = \varkappa; \tag{1.3}$$

$$G_E(q^2) = F_1(q^2) - q^2 F_2(q^2)/4M^2,$$

$$G_M(q^2) = \{F_1(q^2) + F_2(q^2)\}/2M, \tag{1.4}$$

whence

$$G_E(0) = Q, \quad G_M(0) = \mu. \tag{1.5}$$

Isovector and isoscalar moments are defined by

$$\mu_{V,S} = \mu\,(I_3 = +\tfrac{1}{2}) \mp \mu\,(I_3 = -\tfrac{1}{2}),$$

and similarly for charges and all form factors. The fine structure constant is denoted by $\alpha$, and we use natural units: $\hbar = 1 = c$.

Lastly, we need the expressions for the mean squared radii[12] defined by

$$\langle R^2 \rangle = \int d\mathbf{x}\, \varrho(\mathbf{x})\, \mathbf{x}^2, \tag{1.6}$$

where $\varrho(\mathbf{x})$ is the three-dimensional charge density in units of $e$, given by[13]

$$\varrho(\mathbf{x}) = \int d\mathbf{q}\, \exp(i\mathbf{q}\mathbf{x})\, \tilde{\varrho}(\mathbf{q}),$$

$$\tilde{\varrho}(\mathbf{q}) = G_E(\mathbf{q}^2)\,(1 + \mathbf{q}^2/4M^2)^{-1/2}. \tag{1.7}$$

Noting that $q^2$ is positive when spacelike, we have

$$\langle R^2 \rangle = -6\tilde{\varrho}'(0) = 6\,\{-G'_E(0) + Q/8M^2\}. \tag{1.8}$$

We are now in a position to quote the sum rules for the nucleon. First, the $CR$ rule[4], based on minimal coupling[14] and current algebra, reads

$$\{2J^V_{1/2} - J^V_{3/2}\} = -4\pi^2\alpha F_1^{V'}(0) - 2\pi^2\alpha\varkappa_V^2/4M^2. \tag{1.9}$$

It can be recast, first in terms of $G'_E$ and $\mu$, and then in terms of $\langle R^2 \rangle$ and $\mu$:

$$\{2J^V_{1/2} - J^V_{3/2}\} = -4\pi^2\alpha G_E^{V'}(0) - 2\pi^2\alpha\mu_V^2 + 2\pi^2\alpha/4M^2$$

$$= \frac{2\pi^2\alpha}{3}\langle R^2 \rangle^V - 2\pi^2\alpha\mu_V^2. \tag{1.10}$$

The final form of (1.10) will prove important because in NR bound states it is $\langle R^2 \rangle$ and $\mu$ that have simple additive properties.

Next, the BKS rule[5], based on similar assumptions, reads

$$(2K_{1/2} - K_{3/2})_P^V - (2K_{1/2} - K_{3/2})_A^V = 4\pi^2 \alpha \mu_V. \tag{1.11}$$

Lastly, the *DHG* rule[8], derived without current algebra, reads:

$$J_P - J_A = 2\pi^2 \alpha \varkappa^2 / M^2. \tag{1.12}$$

By additional appeal to current algebra[5,15] it could be separated into isoscalar and isovector parts, but we will not consider this.

We sketch the typical steps (i) to (iv) for the DHG rule (1.12). Steps (i) an (ii): the forward Compton amplitude may be written as

$$f(\omega) = \varepsilon'^* \cdot \varepsilon f_1(\omega) + i\omega \varepsilon'^* \times \varepsilon \cdot \sigma f_2(\omega), \tag{1.13}$$

where $\varepsilon$ and $\varepsilon'$ are the initial and final photon polarization vectors. With a provisional assumption about convergence, the dispersion representations for the $f_i$ are as follows:

$$f_1(\omega) = f_1(\infty) + \frac{1}{4\pi^2} \int d\omega' \frac{\omega'^2}{\omega'^2 - \omega^2} [\sigma_P(\omega') + \sigma_A(\omega')], \tag{1.14}$$

$$f_2(\omega) = f_2(\infty) - \frac{1}{4\pi^2} \int d\omega' \frac{\omega'}{\omega'^2 - \omega^2} [\sigma_P(\omega') - \sigma_A(\omega')]. \tag{1.15}$$

Step (iii): the threshold theorems[10] yield

$$f_1(0) = -\alpha/M, \quad f_2(0) = -\alpha \varkappa^2/2M^2. \tag{1.16}$$

Setting $\omega = 0$ in (1.14) and (1.15), and substituting from (1.16), we get

$$-\alpha/M = f_1(\infty) + \frac{1}{4\pi^2} \int d\omega \, [\sigma_P(\omega) + \sigma_A(\omega)], \tag{1.17}$$

$$-\alpha \varkappa^2/2M^2 = f_2(\infty) - \frac{1}{4\pi^2} \int d\omega \, [\sigma_P(\omega) - \sigma_A(\omega)]. \tag{1.18}$$

Step (iv): Assume $f_2(\infty) = 0$ (in addition to the convergence of the integral in (1.18)); this yields the DHG rule.

Note that we cannot assume $f_1(\infty) = 0$, since the right hand side of (1.17) would then be positive definite, while the left is negative.

It is apparent already that the rules (1.10) to (1.12) are, in that order, increasingly likely to need a relativistic interpretation. This is indicated by the fact that the leading terms on their right are of increasingly higher order in $M^{-1}$. (For lightly bound states, $M^{-1}$ is comparable to the Compton wavelengths of the constituents.) Thus, for the CR rule the leading term, $\langle R^2 \rangle$, is of zero order in $M^{-1}$, and measures the spatial extension of the bound state wavefunction, while the $\mu^2$ term is a correction formally of order $M^{-2}$. But by the uncertainty principle,

$$\mu^2/\langle R^2 \rangle \approx 1/\langle R^2 \rangle M^2 \approx \langle T \rangle/M \ll 1, \qquad (1.19)$$

where $\langle T \rangle$ estimates the internal kinetic energy. By contrast, the right hand side of the BKS rule is of order $M^{-1}$, and that of the DHG rule of order $M^{-2}$. As already noted, relativistic considerations will indeed be crucial in understanding the DHG rule; the other two will turn out to be "relativistic" only to the extent that all magnetic relative to electric effects are of order $v/c$. Of course, in the quark model the extremely high numerical value of $\varkappa$ (see section 2.1) invalidates (1.19), and for any quantitative purpose we shall take this into account.

## 2  BOUND STATES

### 2.1  The quark model

Our quark model is the standard one[3] originally inspired by full $SU(6)$ symmetry; for our purposes rather less is needed, and we summarize the relevant features and assumptions.

The quarks have spin $\frac{1}{2}$ and baryon number $\frac{1}{3}$; the two non-strange ones $\pi$ and $v$, form an isodoublet, with charges $2e/3$ and $-e/3$ respectively. The proton, $p$, contains $\pi\pi v$, and the neutron, $n$, contains $\pi v v$. The nucleon is in an $S$ state, totally symmetric in spin and $i$-spin, and either totally antisymmetric in the orbital variables (if the quarks obey Fermi statistics), or totally symmetric (if they obey parastatistics of an appropriate kind). In particular, this implies that the two identical quarks are always coupled to total spin 1, which simplifies calculations. The quarks move nonrelativistically both in the nucleon, and in the isobars reached by photoexcitation; isobars are formed by exciting the quark motion, and not, for instance, by creating extra pions or quark-antiquark pairs. In other words the only relevant degrees of freedom are those of the three nonrelativistic quarks originally present. Note that we are not assuming that all the isobars can be fitted into $SU(6)$ representations.

Assuming only that the electromagnetic current transforms like an octet member under $SU(3)$, all quark electromagnetic form factors, magnetic moments, charge densities, etc., are proportional to the charges, which can be written as

$$Q_q = \tfrac{1}{2}\tau_0(q) + \tfrac{1}{6}, \tag{2.1}$$

whence

$$Q_q^V \equiv Q_\pi - Q_v = 1 = Q_P - Q_n \equiv Q_N^V.$$

It is convenient to define, by reference to (1.6)–(1.8), an intrinsic mean square radius for the quarks by

$$\langle R^2 \rangle_\pi = \tfrac{2}{3}\langle R^2 \rangle_q = -2\langle R^2 \rangle_v. \tag{2.2}$$

With this definition, the usual addition of the squared radii for a composite system implies

$$\langle R^2 \rangle_P = \langle 2Q_\pi\,(\xi_\pi^2 + \langle R^2 \rangle_q) + Q_v\,(\xi_v^2 + \langle R^2 \rangle_q) \rangle,$$

where $\xi_q$ is the coordinate of quark $q$ relative to the mass centre. Because of the orbital equivalence of all three quarks in the nucleon,

$$\langle \xi_\pi^2 \rangle = \langle \xi_v^2 \rangle \equiv \langle R^2 \rangle_\psi, \tag{2.3}$$

and we have

$$\langle R^2 \rangle_P = \langle R^2 \rangle_\psi + \langle R^2 \rangle_q, \tag{2.4}$$

where $\langle R^2 \rangle_\psi$ directly measures the spread of the wavefunction. A similar argument shows that the neutron's charge density, and therefore $G_{E,n}$ is zero, so that

$$\langle R^2 \rangle_N^V = \langle R^2 \rangle_P.$$

For the magnetic moments, simple vector addition, using $\mu_\pi = -2\mu_v$, gives

$$\mu_P = \tfrac{3}{2}\mu_\pi = -\tfrac{3}{2}\mu_n,$$

whence

$$\mu_\pi - \mu_v \equiv \mu_q^V = \mu_P,$$

$$\mu_P - \mu_n \equiv \mu_N^V = 5\mu_P/3. \tag{2.5}$$

This shows that even though the quark mass is supposedly very large, $(M_q/M_N \gg 1)$, nonetheless the total quark and nucleon magnetic moments are comparable. Therefore it is an unavoidable peculiarity of the quark model, compared with normal bound states like nuclei, that electromagnetic cross-sections governed by $\mu^2$ need not be negligible compared to those governed by the charges, even though we would normally reckon $\mu^2$ to be of order $M^{-2}$, and therefore of higher order relativistically.

## 2.2  Additivity

A bound state can absorb photons by two basically distinct mechanisms: either the system as a whole absorbs, or each individual constituent absorbs roughly as if it were free, the others acting as mere spectators. In a nucleus, the first mechanism dominates at low energies, and leads to photoexcitation (including disintegration) without pion production; the second would govern pion photoproduction at high energies. In general the two mechanisms co-exist and interfere; there might be a reasonable chance of disentangling them in practice in a special case like the deuteron, which is so loosely bound that the disintegration cross-section is already very small at the pion production threshold, and where the constituents are so well separated, on any relevant length scale, that the production process in indeed likely to be additive to a good approximation. Additivity in production is plausible at energies high enough for the Glauber theory[16] to apply, but not so high that one must consider diffraction-dissociation of the photon into neutral vector mesons[17], or the propagation of individually excited constituents inside the bound system (e.g. a nucleon propagating as an $N^*$ inside, the nucleus). In the Glauber region, additivity is favoured automatically by the mere fact that the cross-sections are of order $\alpha$, and therefore small compared to the geometric size of the system; and one can reasonably hope that the overall contribution to the integrals $J$ and $K$ from the diffraction dissociation region will be small.

Thus, even in the quark model, we shall assume for simplicity that as far as their contributions to $J$ and $K$ are concerned, the photoexcitation of isobars and photoproduction from quasi-free individual quarks can be treated independently, and moreover, that the latter process is simply additive amongst the quarks. In conventional terms the former process is responsible for "resonance formation", and the latter for "background".

With these simplifications, each integral $J$, say, can be decomposed into an "excitation" or "resonance" part $J(1)$, and an "intrinsic" or "background" part $J(2)$;

$$J = J(1) + J(2); \qquad\qquad (2.6)$$

and similarly for the $K$'s. In true loosely bound states, like the deuteron, but not of course in the quark model, this separation may correspond to that between low-frequency disintegration and high frequency photoproduction contributions. We shall see that for the reasonable sum rules, the intrinsic contributions are given directly by the assumption that the constituents themselves obey the rule; but the excitation contributions will have to be calculated, as far as possible, by ordinary NR quantum mechanics and closure. It is these calculations that reveal the possibly interesting new applications in genuinely NR contexts.

A counterpart of additivity for the cross-sections is the assumption that at high energies the forward Compton amplitude $f_1(\omega)$ itself becomes additive. This assumption, when combined with the quark model, suggests the Pagels–Harari sum rule which we will consider briefly in section 5.

## 2.3 Radiative coupling

To calculate the excitation integrals for the CR and BKS rules, we shall need only the standard electric dipole ($E1$) and magnetic dipole ($M1$) couplings of the constituents to the radiation field[2]. We can ignore the dependence of the fields on the positions of the individual particles because, with our approximations, retardation can be neglected, i.e. we can put $\exp(i\mathbf{k}\cdot\mathbf{r}_i) = 1$, where $\mathbf{k}$ is the photon momentum. To see this, note that by energy conservation the relevant values of $|\mathbf{k}| = \omega$ are of the order of a typical excitation energy, expected to be comparable to the internal kinetic energies. Thus, in an obvious notation, $\omega \sim \langle p^2\rangle/2M$, whence, by the uncertainty principle,

$$\mathbf{k}\cdot\mathbf{r} \sim \omega\,|\mathbf{r}| \sim \omega/|\mathbf{p}| \sim |\mathbf{p}|/M \ll 1. \tag{2.7}$$

Therefore the interaction Hamiltonian is given by

$$\mathscr{H} = -e\mathbf{E}\cdot\mathbf{D} - e\mathbf{H}\cdot\mathscr{M}. \tag{2.8}$$

The $M1$ operator is

$$\mathscr{M} = \Sigma_i \left\{\mu_i\sigma_i + \frac{Q_i}{2M_i}\,\mathbf{l}_i\right\}, \tag{2.9}$$

where $\mathbf{l} = \mathbf{r}\times\mathbf{p}$. It is convenient to refer the $E1$ operator $\mathbf{D}$ to the mass centre, because in matrix elements linking the ground and excited states, only relative coordinate are effective, but not the coordinate $\mathbf{R}$ of the mass centre itself. Thus we can write

$$\mathbf{D} = \Sigma_i Q_i\xi_i, \tag{2.10}$$

where

$$\xi_i = \mathbf{r}_i - \mathbf{R} = \mathbf{r}_i - A^{-1}\Sigma_j\mathbf{r}_j. \tag{2.11}$$

The sums in (2.9)–(2.11) run over all the (equal mass) particles present, whose number is $A$.

When considering the DHG rule, additional (spin-orbit type) couplings will be needed in $\mathcal{H}$, and we shall have to modify the NR definition of relative and mass-centre co-ordinates implicit in (2.11).

By substituting (2.1) into (2.10), we recover the well-known result that the $\tau_0$-independent terms cancel, so that we have

$$\mathbf{D} = \mathbf{D}^0 = \Sigma \, \tfrac{1}{2}\tau_0(i)\, \xi_i; \tag{2.12}$$

in other words $\mathbf{D}$ is the thirdcomponent of an isovector. It will be crucial that, in consequence, only isovector photons are absorbed in $E1$ transitions.

It must be stressed that we shall rely throughout on the no-retardation approximation justified by (2.7). Significant departures from this would necessitate a reappraisal of any quantitative expectations built on our results. (It can be verified straightforwardly that the only potentially dangerous retardation corrections, the cross-terms between $E1$ and once-retarded $M1$ amplitudes, and between $M1$ and once-retarded $E1$ amplitudes, vanish in all our applications.)


## 3   THE CABIBBO—RADICATI RULE

### 3.1   Additivity

We consider the rule (1.10) based on minimal coupling, and relegate the non-minimal case to section 3.4. A realistic application to the $A = 3$ nuclei has been discussed elsewhere[18].

For free quarks, the CR rule reads

$$(2J_{1/2} - J_{3/2})_q^V = \frac{2\pi^2\alpha}{3} \, \langle R^2 \rangle_q - 2\pi^2\alpha\mu_p^2, \tag{3.1}$$

where (2.2) and (2.5) have been used.

In fact, the free-quark combination on the left is precisely equal to the intrinsic integral for the nucleon, for the following reason. The combination $(2J_{1/2}^V - J_{3/2}^V)$ is merely another way of expressing the (parallel) – (anti-parallel) combination in isospace; in other words, $(2\sigma_{1/2}^V(\gamma N) - \sigma_{3/2}^V(\gamma N))$ is proportional to $(\sigma(\gamma^+ p) - \sigma(\gamma^- p))$, where the $\gamma^\pm$ are the charged isovector "photons". This will emerge in detail from the discussion following (3.5) below. But absorption of $\gamma^+$ by a $\pi$, (or $\gamma^-$ by $av$), contributes the isotopic "parallel" cross-section, while similar absorption of a $\gamma^-$ contributes the "antiparallel" amount. Therefore in view of their quark content the "intrinsic" part of the proton-neutron difference equals the $\pi - \nu$ difference, as stated. The same argument applies to polarizations in ordinary rather than isospace, and will be needed again in sections 4 and 7.

Thus, for the nucleon excitation integrals we have the consistency condition,

$$(2J_{1/2}(1) - J_{3/2}(1))_N^V = \frac{2\pi^2\alpha}{3} (\langle R^2\rangle_N^V - \langle R^2\rangle_q) - 2\pi^2\alpha\,(\mu_N^{V2} - \mu_p^2)$$

$$= \frac{2\pi^2\alpha}{3} (\langle R^2\rangle_N^V - \langle R^2\rangle_q) - 2\pi^2\alpha \left(\frac{25}{9}\mu_p^2 - \mu_p^2\right),$$

$$(2J_{1/2}(1) - J_{3/2}(1))_N^V = \frac{2\pi^2\alpha}{3} \langle R^2\rangle_\psi - \frac{32\pi^2\alpha}{9}\mu_p^2, \tag{3.2}$$

where several results from section 2A have been used.

We show that $E1$ transitions account for the first term on the right of (3.2), and $M1$ transitions for the second.


## 3.2  Electric dipole transitions

In this subsection, all cross-sections are of the excitation type, are $E1$, and consequently isovector; these labels will generally be omitted. We want to show

$$2J_{1/2} - J_{3/2} = \frac{2\pi^2\alpha}{3} \langle R^2\rangle_\psi^V. \tag{3.3}$$

(In the quark model, $\langle R^2\rangle_\psi^V = \langle R^2\rangle_\psi$ because of the total orbital equivalence, but this is not needed in proving (3.3)).

We start from the Golden Rule for photo-absorption into a final state $|f\rangle$,

$$\sigma_f(\omega) = 4\pi^2\alpha\omega\,|\langle f|\boldsymbol{\varepsilon}\cdot\mathbf{D}|\,0\rangle|^2\,\delta\,(E_0 + \omega - E_f). \tag{3.4}$$

Here, $\boldsymbol{\varepsilon}$ is the unit polarization vector of the photon; in this section $\sigma_P$ and $\sigma_A$ are added, so that circular polarizations are not involved, and we can take $\boldsymbol{\varepsilon}$ as real.

Evidently,

$$(2J_{1/2} - J_{3/2}) = 4\pi^2\alpha \left\{2\sum_{I_f=1/2} - \sum_{I_f=3/2}\right\} |\langle f|\,\boldsymbol{\varepsilon}\cdot\mathbf{D}^0|\,0\rangle|^2, \tag{3.5}$$

where $\mathbf{D}^0$ is taken from (2.12), and its other isotopic components will be written $\mathbf{D}^\pm$. For states with $I_f = \frac{1}{2}$ we can anticipate the subsums over $I_z$ by using the Wigner–Eckart theorem to write

$$2 \langle 0 | \boldsymbol{\varepsilon} \cdot \mathbf{D}^0 | f \rangle \langle f | \boldsymbol{\varepsilon} \cdot \mathbf{D}^0 | 0 \rangle$$

$$= \{ \langle 0 | \boldsymbol{\varepsilon} \cdot \mathbf{D}^+ | f \rangle \langle f | \boldsymbol{\varepsilon} \cdot \mathbf{D}^- | 0 \rangle - \langle 0 | \boldsymbol{\varepsilon} \cdot \mathbf{D}^- | f \rangle \langle f | \boldsymbol{\varepsilon} \cdot \mathbf{D}^+ | 0 \rangle \}; \quad (3.6)$$

for $I_f = \frac{3}{2}$ a similar relation holds but with the factor 2 prefacing the left hand side replaced by $-1$. Thus (3.5) becomes

$$(2J_{1/2} - J_{3/2}) = 4\pi^2 \alpha \, \Sigma_f \, \{ \langle 0 | \boldsymbol{\varepsilon} \cdot \mathbf{D}^+ | f \rangle \langle f | \boldsymbol{\varepsilon} \cdot \mathbf{D}^- | 0 \rangle$$

$$- \langle 0 | \boldsymbol{\varepsilon} \cdot \mathbf{D}^- | f \rangle \langle f | \boldsymbol{\varepsilon} \cdot \mathbf{D}^+ | 0 \rangle \},$$

where the sum now extends over all states, since the matrix elements vanish if $I_f > \frac{3}{2}$, and since $\langle 0 | \mathbf{D} | 0 \rangle$ vanishes by reflection invariance. Then closure gives

$$(2J_{1/2} - J_{3/2}) = 4\pi^2 \alpha \, \langle 0 | \, [\boldsymbol{\varepsilon} \cdot \mathbf{D}^+, \boldsymbol{\varepsilon} \cdot \mathbf{D}^-] \, | 0 \rangle.$$

For definiteness take $\boldsymbol{\varepsilon} = \hat{\mathbf{x}}$; use (2.3), (2.12), the commutation rules for the Pauli matrices $\vec{\tau}$, and note that

$$\langle R_x^2 \rangle = \tfrac{1}{3} \langle \mathbf{R}^2 \rangle_\psi.$$

In general this holds because all spins are summed over; in our special case it holds already because $|0\rangle$ is an $S$ state. Thus we find

$$(2J_{1/2} - J_{3/2}) = 4\pi^2 \alpha \left\langle 0 \left| \sum_{i,j} \left[ \frac{1}{2} \tau^+(i) \, \xi_x(i), \, \frac{1}{2} \tau^-(j) \, \xi_x(j) \right] \right| 0 \right\rangle$$

$$= 4\pi^2 \alpha \left\langle 0 \left| \Sigma_i \, \frac{1}{2} \tau^0(i) \, \xi_x^2(i) \right| 0 \right\rangle = \frac{2\alpha\pi^2}{3} \langle R^2 \rangle_\psi,$$

as required by (3.3).

A remarkable new result follows if (3.3) is joined to the standard Bremsstrahlung-weighted sum rule for the total $E1$ cross-section. This is got by writing

$$J_T = 4\pi^2 \alpha \sum_f |\langle f | \boldsymbol{\varepsilon} \cdot \mathbf{D} | 0 \rangle|^2 = 4\alpha\pi^2 \, \langle 0 | D_x^2 | 0 \rangle,$$

where the sum runs over all states. But by exploiting the total orbital equivalence of the constituents in $|0\rangle$, Foldy[19] showed that the two-particle terms in $D_x^2$ can be eliminated, and found

$$J_T = J_{1/2} + J_{3/2} = \frac{4\pi^2 \alpha}{3} \langle R^2 \rangle_\psi. \quad (3.7)$$

But (3.3) and (3.7) together imply

$$J_{1/2} = J_{3/2} = \frac{2\alpha\pi^2}{3} \langle R^2 \rangle_\psi. \qquad (3.8)$$

The new rules (3.8) are peculiar to the quark model, and rely, moreover, on the orbital equivalence of the quarks. They constrain, separately, the $E1$ photo-excitation of $I = \tfrac{1}{2}$ and $I = \tfrac{3}{2}$ states. In principle, they could be used to obtain an empirical estimate of $\langle R^2 \rangle_\psi$ by disentangling the resonance from the background contributions to the $J_N(E1)$.

### 3.3  Magnetic dipole transitions

The $M1$ operator is given in (2.9). If, as expected, $M_q/M_N \gg 1$, while nevertheless $\mu_q \sim \mu_N$, then the second, orbital part is negligible compared to the first. Moreover, if in the ground state $|0\rangle$ the individual orbital momenta $l_i^2$ are zero, then the orbital part of $M1$ annihilates $|0\rangle$ and cannot induce absorption. (This applies to the $A = 3$ isodoublet, and also to the nucleon if the quarks obey parastatistics). But in any case such orbital contributions to $J$ will be comparable to other retardation corrections which have already been neglected. Therefore we retain only the spin part and write, using (2.5),

$$\mathcal{M} = \mu_p \Sigma_i \, \boldsymbol{\sigma}(i) \, [\tfrac{1}{2}\tau_0(i) + \tfrac{1}{6}].$$

But $\tfrac{1}{2}\sum \boldsymbol{\sigma}(i)$ is simply the total angular momentum operator; $|C\rangle$ is one of its eigenfunctions, and it cannot therefore connect $|0\rangle$ to any excited states. Hence the isoscalar part of $\mathcal{M}$ can also be dropped; *the $M1$ as well as the $E1$ transitions are pure isovector*, and we have effectively

$$\mathcal{M} = \mathcal{M}^0 \equiv \tfrac{1}{2}\mu_p \Sigma_i \, \boldsymbol{\sigma}(i) \, \tau_0(i). \qquad (3.9)$$

For the rest of this subsection we mostly omit the labels $M1$ and $V$. The method already used for $E1$ yields, for the $M1$ contribution to $J(1)$:

$$(2J_{1/2} - J_{3/2}) = 4\pi^2 \alpha \mu_p^2 \, \{\langle 0|\tfrac{1}{2} \Sigma \, \tau_0(i)|0\rangle - 2 \, |\langle 0|\tfrac{1}{2} \Sigma \, \tau_0(i) \, \sigma_0(i)|0\rangle|^2\}.$$

$$(3.10)$$

where $|0\rangle$ is now a proton state with spin up. By direct calculation,

$$\langle 0 \, |\Sigma \, \tau_0(i)| \, 0\rangle = 1, \quad |\langle 0 \, |\Sigma \, \tau_0(i) \, \sigma_0(i)| \, 0\rangle|^2 = 25/9; \qquad (3.11)$$

hence

$$2J_{1/2} \, (M1) - J_{3/2} \, (M1) = -32\pi^2 \alpha \mu_p^2 /9, \qquad (3.12)$$

as required by the $M1$ part of the consistency condition (3.2).

A sum rule for $J_T(M1)$, analogous to (3.7), has been derived by Gerasimov[20]:

$$J_T(M1) = J_{1/2}(M1) + J_{3/2}(M1) = 32\pi^2\alpha\mu_p^2/9. \qquad (3.13)$$

Comparison of (3.12) and (3.13) shows that

$$J_{1/2}(M1) = 0, \qquad (3.14)$$

$$J_{3/2}(M1) = 32\pi^2\alpha\mu_p^2/9. \qquad (3.15)$$

Summarizing, with our approximations the quark model leads to the following result:

*Resonant* M1 *absorption is pure isovector: isobars with* $I = \frac{1}{2}$ *are*
*not excited, and the sum rule* (3.15) *applies.*   (3.16)

In exact $SU(6)$ these results are trivial because then there exists only a single $M1$ transition, leading to the $(3, 3)$ resonance, and this amplitude is predicted in terms of $\mu_p$ by the symmetry. But here we have shown that (3.16) depends only on the assumed properties of the nucleon ground state, and applies even if the excited isobars do not have $SU(6)$-symmetric wavefunctions.

(Gerasimov has also shown[20] that the orbital part of $\mathcal{M}$, if retained, contributes to $J_T(M1)$ an amount

$$\pi^2\alpha \langle \mathbf{l}_i^2 \rangle/3M_q^2 ,$$

where $\mathbf{l}_i$ is the orbital momentum of any one of the three quarks).

The isobar most likely to contradict (3.16) is the Roper resonance, which shares all the quantum numbers of the nucleon but appears nevertheless to be photoexcited quite strongly[21], necessarily by an $M1$ transition. Presumably this implies that to an appreciable extent it is a genuine meson-nucleon resonance rather than a three-quark isobar.

### 3.4   Nonminimal coupling

The CR rule, as discussed so far, depends on commutation rules which can be abstracted from an isospin current density of the form

$$\vec{s}_\lambda(x) = \bar{\psi}(x)\,\tfrac{1}{2}\vec{\tau}\gamma_\lambda\psi(x), \qquad (3.17)$$

where $\psi$ is the quark field. Naturally the Dirac form of $\vec{s}_\lambda(x)$ does not imply that the quarks have no Pauli moment. On the contrary, their strong couplings will endow them both with Pauli moments and with a varying Dirac form factor $F_1(q)$; and it is these same strong couplings that allow photoproduction to order $e^2$ from free or quasi-free quarks.

It is instructive to consider a different extreme case, where the quarks are point particles, not admitting photoproduction when free (so that all "intrinsic" integrals vanish), and where their "observed" Pauli moments indicated by (2.5) are due to a nonminimal Pauli term written directly into $\vec{s}_\lambda$. Correspondingly we must take $F_1$ and $F_2$ as constant. Then (1.4) and (1.8) imply

$$\langle R^2 \rangle_q = 3\mu_p/M_q - 3/4M_q^2, \tag{3.18}$$

and (3.17) is replaced by

$$\vec{s}_\lambda = \bar{\psi}\, \tfrac{1}{2}\vec{\tau}\gamma_\lambda\psi - \frac{\varkappa_q^V}{2M_q}\, \partial^\mu (\bar{\psi}\sigma_{\lambda\mu}\tfrac{1}{2}\vec{\tau}\psi). \tag{3.19}$$

Kawarabayashi and Suzuki[14] have shown that because of the consequent change in the commutators, the CR rule (1.9) for a particle $i$ must then be amended to read

$$(2J_{1/2}^V - J_{3/2}^V)_i = -4\pi^2\alpha F_{i1}^{V1}(0) - 2\pi^2\alpha\varkappa_i^{V2}/4M_i^2 + 2\pi^2\alpha\varkappa_V^{q2}/4M_q^2. \tag{3.20}$$

Thus, for free quarks, of the kind defined above, i.e. for $i = q$, the $J$'s and $F_1^{V1}(0)$ vanish, and the new term, the third, exactly cancels the second, correctly reducing (3.20) to the identity $0 = 0$. What interests us is the application to nucleons, when the addend of course remains the same as in (3.20), so that one gets (since the $J(2)$'s all vanish),

$$(2J_{1/2}(1) - J_{3/2}(1))_N^V = 4\pi^2\alpha F_{N1}^{V1}(0) - 2\pi^2\alpha\varkappa_N^{V2}/4M_N^2 + 2\pi^2\alpha\varkappa_q^{V2}/4M_q^2. \tag{3.21}$$

As in section 2, this can be rewritten:

$$(2J_{1/2}(1) - J_{3/2}(1))_N^V = \frac{2\pi^2\alpha}{3}\langle R^2 \rangle_N^V - 2\pi^2\alpha\mu_V^{N2} + 2\pi^2\alpha\varkappa_q^{V2}/4M_q^2. \tag{3.22}$$

By using (3.3) and (3.12) for the excitation integrals, (2.4) and (3.18) for the charge radii, and by doing some algebra, one can verify the modified consistency condition (3.22). Note that the simplifications underlying it are just those that one would naturally make in an application to nuclear physics, if one wished to be consistent in neglecting pion production while retaining the Pauli moments of the nucleons.

## 4  THE BEG–KAWARABAYASHI–SUZUKI SUM RULE

Consider the intrinsic contribution to the integrals on the left of (1.11). Extending the argument in section 3.1 to both isotopic and ordinary polarizations, we see that the contribution of each quark $i$ equals in magnitude the free-quark value of the RHS, with the sign of $\tau_0(i)\,\sigma_0(i)$ in the spin-up proton wavefunction $|0\rangle$. In other words

$$\{2K^V_{1/2}(2) - K^V_{3/2}(2)\}^{P-A}_N = 4\pi^2\alpha\,\langle 0|\,\mu^V_q\,\Sigma\,\tau_0(i)\,\sigma_0(i)\,|0\rangle. \qquad (4.1)$$

But the operator in the matrix element is simply the spin-dependent part of the total isovector $M1$ operator; recall the discussion in section 3.3. In our case of an $S$ state, the expectation value of the orbital part would vanish even if the individual quark moments $\langle \mathbf{l}_i^2 \rangle$ do not vanish. Hence the right hand side of (4.1) precisely equals the right hand side of the full BKS rule, and the consistency condition becomes

$$\{2K^V_{1/2}(1) - K^V_{3/2}(1)\}^{P-A}_N = 0. \qquad (4.2)$$

We interpret this to mean that the combination of excitation integrals on the left of (4.2) must be much smaller than $4\pi^2\alpha\mu^V_N$.

To see that in this sense (4.2) is satisfied to the order to which we are working, note first that the $E1$ operator (2.10), (2.11) is independent of spins; hence, $E1$ transitions starting from the $S$ ground state contribute equally to $\sigma_P$ and $\sigma_A$, and cancel from the difference. (Brennan[22] has shown that the $E1$ contribution to (4.2) is

$$4\pi^2\alpha\left\langle 0\left|\frac{1}{2M_q}\,\Sigma\,\tau_0(i)\,l_0(i)\right|0\right\rangle, \qquad (4.3)$$

which provides precisely the orbital part of the isovector magnetic moment required by the sum rule in the general case, and evidently vanishes for an $S$ state. In fact (4.3) is just a decomposition of the Thomas-Reiche-Kuhn sum rule by isospin and polarization, in the same sense in which the CR rule (3.3) is an isotopic decomposition of the Bremsstrahlung-weighted rule $J_T = 4\pi^2\alpha\,\langle 0|\,D_x^2\,|0\rangle$.)

It is less easy to be convinced that the $M1$ contributions can be regarded as satisfying (4.2). Formally, they ought to be negligible since the $M1$ cross-sections are of order $M_q^{-2}$ while $\mu^V_q$ is of order $M_q^{-1}$; but this argument is invalidated quantitatively by the large numerical value of $\mu_q$, as discussed at the end of section 1. Instead, we argue that the $M1$ contribution to $K$-integrals are of order $\mu^2\Delta$, where $\Delta$ is some mean excitation; it would be legitimate to neglect this if one has $\mu^2\Delta \ll \mu$, i.e. if

$$\mu\Delta \ll 1. \tag{4.4}$$

Taking a typical $\mu$ as $(2M_N)^{-1}$, and a typical $\Delta$ as roughly the excitation energy of the (3, 3) resonance, we estimate

$$\mu\Delta \approx 0.16.$$

To the extent that this does satisfy (4.4), we can accept (4.1) and (4.2), and conclude the following:

> *The contributions to the integrals K which validate the BKS rule*
> *are expected to be those from the nonresonant background rather*
> *than from isobar formation.*    (4.5)

## 5  THE PAGELS–HARARI SUM RULES

Section 1 showed that for the spin-averaged Compton amplitude $f_1$, the assumption $f_1(\infty) = 0$ leads to an absurdity, precluding any corresponding sum rules for individual particles. But if scattering from bound states satisfies the relation

$$\lim_{\omega \to \infty} \{f_1(\omega) - \Sigma_i f_{i1}(\omega)\} = 0, \tag{5.1}$$

where the sum runs over all constituents, then one can set $\omega = 0$ and use the Thomson limit (1.16), to obtain[9]

$$-\alpha Q^2/M + \sum_i \alpha Q_i^2/M_i = \frac{1}{2\pi^2} \int_0^\infty d\omega \left[ \sigma(\omega) - \sum_i \sigma_i(\omega) \right], \tag{5.2}$$

which is the relativistic form of the Thomas-Reiche-Kuhn (TRK) sum rule used in nuclear physics[2], with $\sigma = \sigma_T = \frac{1}{2}(\sigma_P + \sigma_A)$.

No consistency problems arise here: the TRK rule for $E1$ transitions is the sum rule par excellence; moreover, Gerasimov[23] has shown that retardation and higher multipole corrections to the leading $E1$ terms cancel exactly.

The quark model allows one to select many sets of particles, from linear combinations of which the sums appearing in (5.2) cancel. The resultant sum rules have been written down by Pagels and by Harari[6]; they depend only on the assumed quark content and on additivity in the form (5.1), but not on the more detailed additivity between intrinsic and excitation contributions that we had to rely on in previous sections. Here we quote only a single example, for pions. (Do not confuse the charged and neutral pions, $\pi^+$ and $\pi^0$, with the quark $\pi$.) The quark content of $\pi^+$ and $\pi^0$ is $(\pi\bar{v})$ and $(\pi\bar{\pi} - v\bar{v})/\sqrt{2}$, respectively. But by charge conjugation invariance photo-cross-sections from particle and antiparticle are equal, so that we have for $\pi^+$:

$$-\alpha M_\pi^{-1} + \alpha M_q^{-1}\,(5/9) = \frac{1}{2\pi^2} \int d\omega\,[\sigma_{\pi^+}(\omega) - \sigma_\pi(\omega) - \sigma_v(\omega)],$$

and for $\pi^0$,

$$\alpha M_q^{-1}\,(5/9) = \frac{1}{2\pi^2} \int d\omega\,[\sigma_{\pi^0}(\omega) - \sigma_\pi(\omega) - \sigma_v(\omega)].$$

Subtracting, we find

$$\alpha M_\pi^{-1} = \frac{1}{2\pi^2} \int d\omega\,[\sigma_{\pi^0}(\omega) - \sigma_{\pi^+}(\omega)]. \tag{5.3}$$

At first sight it may perhaps seem paradoxical that the neutral particle absorbs more strongly than the charged. However, this is typical of a bound state model, as can be seen by considering $E1$ absorption only, and realizing that large electric dipole moments are favoured by the presence of oppositely rather than similarly charged constituents.

## 6  THE GOTTFRIED SUM RULE

By combining fully relativistic methods with some features of the quark model, Gottfried[7] has given a plausibility argument for, and suggested the validity of, the following sum rule for the proton:

$$J_T = \frac{4\pi^2\alpha}{3}\,\langle R^2 \rangle_p - 4\pi^2\alpha\mu_p^2. \tag{6.1}$$

(This is Gottfried's equation (11) rewritten with our definition of $\mu_p$ and with our expression (1.8) for $\langle R^2 \rangle$.) The property of the quark model exploited by Gottfried is the absence of charge correlations, which, at least for point quarks, follows from the identity

$$\sum_{i \neq j} Q_i Q_j = 0, \tag{6.2}$$

the sums running over the quarks in the proton.

We ask whether, or under what additional approximations, (6.1) is consistent with the simple quark model as used in the present paper; (6.1) is most vulnerable through being a rule for the total cross-section rather than for some differences.

In the now familiar way we split $J_T$ into an intrinsic part $J(2)$ and an excitation part $J(1)$, which splits further in $J(E1)$ and $J(M1)$. We use (3.7) for $J(E1)$, (3.13) for $J(M1)$, and express $\langle R^2 \rangle_p$ as in (2.4). This transforms (6.1) into

$$ J(2) + \frac{4\pi^2\alpha}{3} \langle R^2 \rangle_\psi + \frac{32\pi^2\alpha}{9} \mu_p^2 = \frac{4\pi^2\alpha}{3} (\langle R^2 \rangle_\psi + \langle R^2 \rangle_q) - 4\pi^2\alpha\mu_p^2, $$

which reduces to a rule for the intrinsic part alone:

$$ J_p(2) = 2J_\pi + J_v, $$

$$ J_p(2) = \frac{4\pi^2\alpha}{3} \langle R^2 \rangle_q - \frac{68\pi^2\alpha}{9} \mu_p^2. \tag{6.3} $$

If one replaces $\langle R^2 \rangle_q$ by $\langle R^2 \rangle_p$, clearly a very liberal upper limit, then the second term on the right is roughly two-thirds the first, still allowing the resultant to be positive. However, there seems to be no fundamental reason for an evidently nontrivial rule like (6.3) to apply to quarks. Indeed, if the quarks approximate to point particles in the sense that $\langle R^2 \rangle_q \ll \langle R^2 \rangle_p$, then the right hand side of (6.3) becomes negative, which would be absurd. Therefore from this viewpoint Gottfried's sum rule does not appear to be plausible. Evidently, the detailed objection uncovered by (6.3) is that the derivation neglects charge correlations within the individual quarks.


## 7　THE DRELL–HEARN–GERASIMOV SUM RULE

In contrast to all the preceding examples, the DHG rule can be verified in detail only by essential augmenting the conventional NR radiation coupling which we have used hitherto. Moreover, a *simple* verification proves possible only in cases where the binding energy is negligible compared to the rest masses. Hence the application to the quark model is purely academic; nevertheless we give the argument as an illustration for this, rather than break the continuity by discussing a nuclear or atomic application. For this section only, we therefore pretend that

$$ M_N = 3M_q. \tag{7.1} $$

To begin with, we must add to the interaction Hamiltonian (2.8) the terms of order $M^{-2}$ which emerge from the Foldy-Wouthuysen transformation[24] of the relativistic Dirac and Pauli couplings. For a single spin $\frac{1}{2}$ particle of charge $e$, mass $M$, and Pauli moment $\varkappa$, the new term is

$$-\frac{e}{4M^2}\,(1+2\varkappa)\,\boldsymbol{\sigma}\cdot\mathbf{E}x\,(\mathbf{p}-e\mathbf{A}) = -\frac{e}{2M}\left(2\mu - \frac{1}{2M}\right)\boldsymbol{\sigma}\cdot\mathbf{E}x\,(\mathbf{p}-e\mathbf{A}). \quad (7.2)$$

Note the disposition of the factors 2 connected with the Thomas precession. Here we need only the parts linear in $e$; then (2.8) must be replaced by

$$\mathscr{H} = -e\,\{\mathbf{E}\cdot\mathbf{D} + \mathbf{H}\cdot\mathscr{M} + \mathbf{E}\cdot\mathscr{N}\}, \quad (7.3)$$

where

$$\mathscr{N} = \frac{e}{4M_q^2}\,(1+2\varkappa_q)\,\Sigma_i\,Q_i\mathbf{p}(i)\,x\boldsymbol{\sigma}\,(i). \quad (7.4)$$

We have defined

$$\mu_q \equiv eQ_q\,(1+\varkappa_q)/2M_q = eQ_q\mu_p. \quad (7.5)$$

But the crux of the matter is that a further modification is also essential[25]. It can be made by two different methods which are equivalent, at least to our approximation. Consider first a two-body bound state, with constituents $a$ and $b$. Brodsky and Primack[26], starting from the Bethe-Salpeter equation, show that one can continue to use precisely the conventional NR methods, including in particular our former definition (2.10, 2.11) of the $E1$ operator $\mathbf{D}$, if one adds to (7.3) yet another term, namely,

$$\frac{e}{4\,(M_a+M_b)}\left\{\frac{\boldsymbol{\sigma}(a)}{M_a} - \frac{\boldsymbol{\sigma}(b)}{M_b}\right\}\{Q_b\mathbf{E}_b\times(\mathbf{p}_a - eQ_a\mathbf{A}_a) - Q_a\mathbf{E}_a\times(\mathbf{p}_b - eQ_b\mathbf{A}_b)\},$$

$$(7.5)$$

The suffices on $\mathbf{E}$ and $\mathbf{A}$ indicate the positions where the fields are to be evaluated, though in the nonretarded approximation these become irrelevant. Brodsky and Primack emphasize the radically new feature that, in following their method, $\mathscr{H}$ evidently ceases to be simply the sum of the NR couplings of the two particles.

Here, we shall adopt a different approach, due to Osborn[27]. In our approximation, Osborn retains the coupling (7.3) (without the addend (7.5)), but recalls that in a relativistic system of two particles with spin, the definition of the canonical centre-of-mass and relative coordinates and momenta differs from the conventional nonrelativistic form[28]. The differences are relevant because they change the definition of the coordinates $\xi$ of the particles relative to the mass centre, and because these changes in turn affect the expression (2.10) for the $E1$ operator $\mathbf{D}$. It turns out that for calculating $J_P - J_A$, only the leading spin-dependent corrections to the $\xi$'s are important. Osborn shows[29] that for a two-particle system these lead to the following expressions:

$$\xi_a = \frac{M_b}{M_T}\mathbf{r} - \frac{1}{2M_TM_a}\mathbf{q}\times\mathbf{S}(a) + \frac{1}{2M_TM_b}\mathbf{q}\times\mathbf{S}(b), \qquad (7.6)$$

where

$$M_T = M_a + M_b, \qquad (7.7)$$

is the total mass, $\mathbf{S}(a)$ and $\mathbf{S}(b)$ are the spin operators of the particles, and where $\mathbf{r}$ and $\mathbf{q}$ are the canonically conjugate relative coordinate and momentum, which commute with those of the mass centre. $\xi_b$ is given by a similar expression with $a$ and $b$ interchanged and with an overall minus sign. Hence $\mathbf{D}$ becomes

$$\mathbf{D} = Q_a\xi_a + Q_b\xi_b,$$

$$\mathbf{D} = \mathbf{r}\left\{\frac{Q_aM_b - Q_bM_a}{M_T}\right\} - \left\{\frac{1}{M_a}\mathbf{q}\times\mathbf{S}(a)\right.$$

$$\left. - \frac{1}{M_b}\mathbf{q}\times\mathbf{S}(b)\right\}\left\{\frac{Q_a + Q_b}{2M_T}\right\} \equiv \mathbf{D}_1 + \mathbf{D}_2. \qquad (7.8)$$

We are now in a position to verify the consistency of the DHG rule, say for the proton. For brevity, define

$$J_i \equiv J_i^P - J_i^A, \qquad (7.9)$$

and let $|0\rangle$ denote a proton with spin up. It is convenient to think of $|0\rangle$ as a $(\pi\pi)$ diquark, $\delta$, of spin 1, coupled to $\nu$:

$$|0\rangle = \sqrt{\tfrac{2}{3}}\,|m_\delta = 1, m_\nu = -\tfrac{1}{2}\rangle - \sqrt{\tfrac{1}{3}}\,|m_\delta = 0, m_\nu = \tfrac{1}{2}\rangle, \qquad (7.10)$$

where $m$ denotes the third component of spin. It is clear from (7.10) that the intrinsic part $J_p(2)$ of $J_p = J_p(1) + J_p(2)$ is given by

$$J_p(2) = \tfrac{2}{3}(2J_\pi - J_\nu) + \tfrac{1}{3}J_\nu = \tfrac{4}{3}J_\pi - \tfrac{1}{3}J_\nu;$$

this is merely the argument of section 3A used again. But by the assumption that the quarks obey the DHG rule,

$$J_\pi = 4J_\nu = \frac{2\pi^2\alpha\varkappa_q^2}{M_q^2}\frac{4}{9},$$

whence

$$J_p(2) = \frac{2\pi^2 \alpha \varkappa_q^2}{M_q^2} \frac{5}{9}.\tag{7.11}$$

Contributions to $J_p(1)$ will come from (i) $M1$ transitions, $J_p(M1)$; (ii) interference terms between ordinary $E1$ transitions induced by the conventional part $\mathbf{D}_1$ of $\mathbf{D}$, and the spin-orbit coupling $\mathcal{N}$, $J_p(SL)$; and (iii) interference terms between ordinary $E1$ transitions, and these induced by the spin-dependent part $\mathbf{D}_2$ of $\mathbf{D}$, $J_p(E1)$.

From the Golden Rule we have for $M1$ transitions

$$J_p^P(M1) = 4\pi^2 \alpha \, \Sigma_f' \, |\langle f|\boldsymbol{\varepsilon}^+ \cdot \mathcal{M}|\, 0\rangle|^2,\tag{7.12}$$

where the prime excludes the (two orientations of the) proton ground state, and where we write

$$\boldsymbol{\varepsilon}^\pm = \mp \frac{1}{\sqrt{2}}\,(\hat{\mathbf{x}} + i\hat{\mathbf{y}}),\tag{7.13}$$

for the photon polarization vector when parallel (antiparallel) to the proton spin $|0\rangle$ is the proton with spin up. Evidently

$$J_p(M1) = 4\pi^2 \alpha \Sigma_f' \left\{ |\langle f|\boldsymbol{\varepsilon}^+ \cdot \mathcal{M}|\, 0\rangle|^2 - |\langle f|\boldsymbol{\varepsilon}^- \cdot \mathcal{M}|\, 0\rangle|^2 \right\}$$

$$= 4\pi^2 \alpha \left\{ \langle 0|\boldsymbol{\varepsilon}^+ \cdot \mathcal{M}\boldsymbol{\varepsilon}^{+*} \cdot \mathcal{M} - \boldsymbol{\varepsilon}^- \cdot \mathcal{M}\boldsymbol{\varepsilon}^{-*} \cdot \mathcal{M}|\, 0\rangle \right.$$

$$\left. - \sum_m \left( |\langle m|\boldsymbol{\varepsilon}^+ \cdot \mathcal{M}|\, 0\rangle|^2 - |\langle m|\boldsymbol{\varepsilon}^- \cdot \mathcal{M}|\, 0\rangle|^2 \right) \right\},\tag{7.14}$$

where $\Sigma_m$ runs over the two spin orientations of the proton (so that $|0\rangle = |m = \tfrac{1}{2}\rangle$).

By using the identity

$$\boldsymbol{\varepsilon}_+^* \cdot \mathbf{A}\boldsymbol{\varepsilon}_+ \cdot \mathbf{B} - \boldsymbol{\varepsilon}_-^* \cdot \mathbf{A}\boldsymbol{\varepsilon}_- \cdot \mathbf{B} = i\,(A_x B_y - A_y B_x),\tag{7.15}$$

the expression (2.9) for $\mathcal{M}$, the commutation rules, and the Wigner–Eckart theorem for the correction terms in (7.14), we find straight-forwardly

$$J_p(M1) = 4\pi^2 \alpha \left(\frac{1+\varkappa_q}{2M_q}\right)^2 \left\{ -2 \left\langle 0 \left| \left[ \frac{4}{9}(\sigma_0(1)+\sigma_0(2)) + \frac{1}{9}\sigma_0(3) \right] \right| 0 \right\rangle + 2 \right\},\tag{7.16}$$

where the quarks $(\pi, \pi, \nu)$ have been labelled $(1, 2, 3)$. But by (7.10), the expectation value equals $5/9$, whence

$$J_p(M1) = 8\pi^2 \alpha\,(1 + \varkappa_q)^2/9M_q^2.\tag{7.17}$$

The calculation of $J_p(SL)$ and $J_p(E1)$ is simplified by observing that the $(\pi\pi)$ subsystem $\delta$ can have no internal $E1$ transitions, so that as regard such transitions $|0\rangle$ can be treated like a composite of the two particles $\delta$ and $\nu$. Then the formula (7.8) becomes applicable, with the identification

$$\mathbf{S}_a = \mathbf{S}(1) + \mathbf{S}(2), \quad Q_a = \tfrac{4}{3}, \quad M_a = 2M_q,$$
$$\mathbf{S}_b = \mathbf{S}(3), \quad Q_b = -\tfrac{1}{3}, \quad M_b = M_q. \tag{7.18}$$

Thence the calculations exactly parallel that of $J_p(M1)$, with only the difference that we need the commutator between $\mathbf{r}$ and $\mathbf{q}$ instead of those between the angular momentum operators in $\mathcal{M}$; and because of the different parity there are no correction terms like the sums in (7.14). A straightforward calculation yields the following results:

$$J_p(SL) = 8\pi^2\alpha i \langle 0 \,|[D_x, \mathcal{N}_y]|\, 0\rangle$$

$$= 8\pi^2\alpha i \left(\frac{1 + 2\varkappa_q}{4M_q^2}\right) \left\langle 0 \left| \frac{2i}{9} \left[\sigma_0(1) + \sigma_0(2) + \sigma_0(3)\right] \right| 0 \right\rangle,$$

$$J_p(SL) = 4\pi^2\alpha\,(1 + 2\varkappa_q)/9\mu_q^2. \tag{7.19}$$

Similarly,

$$J_p(E1) = -\frac{4\pi^2\alpha i}{9\mu_q^2} \langle 0\,|[r_x, (\mathbf{q}x\,\{\mathbf{S}(1) + \mathbf{S}(2) - 2\mathbf{S}(3)\})_y]|\, 0\rangle$$

$$= -\frac{4\pi^2\alpha}{9M_q^2} \langle 0\,|[S_0(1) + S_0(2) - 2S_0(3)]|\, 0\rangle.$$

But (7.10) shows the expectation value to be unity, whence

$$J_p(E1) = -4\pi^2\alpha/9\mu_q^2. \tag{7.20}$$

Finally we need the other side of the DHG rule, expressed in terms of the quark parameters; the crucial point is that the partition of a given total moment $\mu$ into Dirac and Pauli parts depends on the total mass and charge of the system. Thus, using (2.5) and (7.1), we get

$$\frac{2\pi^2\alpha\varkappa_p^2}{M_p^2} = 8\pi^2\alpha \left(\mu_p - \frac{1}{2M_p}\right)^2$$

$$= 8\pi^2\alpha \left[\frac{1 + \varkappa_q}{2M_q} - \frac{1}{6M_q}\right]^2 = \frac{2\pi^2\alpha}{9\mu_q^2}\,(2 + 3\varkappa_q)^2. \tag{7.21}$$

The consistency condition for the DHG rule can be written as

$$\frac{2\pi^2\alpha}{9M_q^2}\,(2 + 3\varkappa_q)^2 = J_p(2) + J_p(M1) + J_p(SL) + J_p(E1);$$

by (7.11), (7.17), (7.19) and (7.20), this reduces to

$$\frac{2\pi^2\alpha}{9M_q^2}\,(2 + 3\varkappa_q)^2 = \frac{2\pi^2\alpha}{9M_q^2}\,\{5\varkappa_q^2 + 4\,(1 + \varkappa_q)^2 + 2\,(1 + 2\varkappa_q) - 2\}, \qquad (7.22)$$

which is, indeed, an identity.

This explicit verification of the DHG rule is interesting chiefly because it evidences the need for either the Brodsky–Primack or the Osborn-type modification in radiative calculations carried to order $M^{-2}$. As regards the quark model, we see that however slowly the quarks may move inside the nucleon, the mere fact that they are tightly bound unbalances (7.22), because departures from (7.1) change the relation (7.21) between $\varkappa_q$ and $\varkappa_p$. Therefore the transitions which may validate the DHG rule experimentally[30] certainly cannot all be understood in terms of the simple NR classification into $M1$, $E1$, and $SL$ type transitions. In other words, as we have pointed out earlier, the DHG rule is essentially relativistic.

(For completeness, we mention that the methods of the present section can be used to show that there are no $0(M^{-2})$ contributions to the CR rule, other than the $M1$ contributions already taken into account. This explains why the $M1$ contributions by themselves obey the CR rule; as verified in section 3C. In fact $0(M^{-2})$ contributions other than $M1$ cancel not only from the CR difference, but also from the total cross-sections.)

## Acknowledgments

It is a pleasure to thank the Rutherford Laboratory, where this work was started, for its hospitality; Professors A. Clegg, S. D. Drell and S. B. Gerasimov for correspondence; and Drs. J. S. Bell, N. Dombey, G. Moorhouse, H. Osborn and C. Wilkin for discussions.

## Notes and references

1. H. A. Bethe and E. E. Salpeter, *Quantum Mechanics of One- and Two-Electron Atoms*, Springer Verlag, 1957, section 61.
2. J. S. Levinger, *Nuclear photodisintegration*, Oxford University Press, New York, 1960; M. Danos, University of Maryland Technical Report No. 221, 1961 (unpublished).
3. For reviews and references, see R. H. Dalitz, in *Proceedings of the 1965 Oxford Inter-*

*national Conference on Elementary Particles*, Rutherford High Energy Laboratory, 1966, in *Proceedings of the 13th International Conference on High Energy Physics, Berkeley, 1966*, and in *Panel Discussion Section of the Topical Conference on $\pi N$ Scattering at Irvine, California* Oxford preprint, 1967; also H.J.Lipkin, in *Proceedings of the Heidelberg International Conference on Elementary Particles*, 1967 North Holland, 1968.

4. N.Cabibbo and L.Radicati, *Phys. Letters*, **19**, 697 (1966).

5. M.A.B.Beg, *Phys. Rev. Letters*, **17**, 333 (1966); *Phys. Rev.*, **150**, 1276 (1966); K.Kawarabayashi and M.Suzuki, *ibid.*, **150**, 1181 (1966).

6. H.Pagels, *Phys. Rev. Letters*, **18**, 316 (1967); H.Harari, *ibid.*, **18**, 319 (1967).

7. K.Gottfried, *Phys. Rev. Letters*, **18**, 1174 (1967).

8. S.B.Gerasimov, *J. Nucl. Phys. (USSR)*, **2**, 598 (1965); S.D.Drell and A.C.Hearn, *Phys. Rev. Letters*, **16**, 908 (1966). See also L.I.Lapidus and Chou Kuang-Chao, *JETP*, **41**, 1546 (1961); English translation: *Soviet Physics, JETP*, **14**, 1102 (1962); M.Hosoda and K.Yamamoto, *Prog. Theor. Physics, (Kyoto)*, **30**, 425 (1966).

9. M.Gell-Mann, M.L.Goldberger, and W.E.Thirring, *Phys. Rev.*, **95**, 1612 (1954); for a correction, see Drell and Hearn, reference (8).

10. F.E.Low, *Phys. Rev.*, **96**, 1428 (1954); beware of misprints. M.Gell-Mann and M.L.Goldberger, *ibid.*, **96**, 1433 (1954).

11. See S.L.Adler and R.F.Dashen, *Current Algebras*, W.A.Benjamin, Inc., New York, (1968).

12. Note that $\langle R \rangle^2$ can be negative if $\varrho$ is.

13. G.Barton, *Introduction to Dispersion Techniques in Field Theory*, W.A.Benjamin, Inc., New York, 1965; sections 7.4 and 13.1. M.Gourdin, *Diffusion des Electrons de Haute Energie*, Masson et Cie, Paris, 1966, section II.3.

14. K.Kawarabayashi and M.Suzuki, *Phys. Rev.*, **152**, 1383 (1966).

15. K.Kawarabayashi and W.W.Wada, *Phys. Rev.*, **152**, 1286 (1966).

16. R.J.Glauber, in *Lectures on Theoretical Physics*, Colorado Summer School 1958; Interscience, New York, 1959; and in *Proceedings of the Second International Conference on High Energy Physics and Nuclear Structure*, Rehovoth Israel, 1967.

17. For a review and reference, see J.S.Bell, *Weak Interactions in the nuclear shadow*. Cern report, TH.887 (1968), see the article by J.S.Bell appearing in the present volume; and R.H.Bassel and C.Wilkin, *High Energy proton scattering and the structure of light nuclei, Brookhaven preprint BNL 12430 (1967)*.

18. G.Barton, *Nuclear Physics*, **A104**, 189 (1967); L.M.Delves and A.C.Phillips, M.I.T. preprint (1968), to appear in *Rev. Mod. Phys.*; section V.D; F.Scheck and L.Schülke, *Physics Letters*, **25B**, 526 (1967).

19. L.L.Foldy, *Phys. Rev.*, **107**, 1303 (1957).

20. S.B.Gerasimov, Dubna preprint P-2439 (1965).

21. Y.C.Chau, N.Dombey and R.G.Moorhouse, *Phys. Rev.*, **163**, 1632 (1967).

22. J.A.Brennan, Sussex thesis, 1968 (unpublished).

23. S.B.Gerasimov, *Phys. Letters*, **13**, 240 (1964).

24. L.L.Foldy and S.A.Wouthuysen, *Phys. Rev.*, **78**, 29 (1950); H.Neuer and P.Urban, *Acta Phys. Austriaca*, **15**, 380 (1962).

25. This necessary modification was ignored in a previous paper, which therefore concluded, incorrectly, that the DHG rule could not generally hold both for a lightly bound composite system and for its constituents. See G.Barton and N.Dombey, *Phys. Rev.*,

**162**, 1520 (1967). In fact, omission of this modification also leads to an apparent contradiction with the threshold theorem of reference (10); see references (26) and (27) below.

26. S.J.Brodsky and J.Primack, Stanford preprint, Slac-Pub-440, (1968), to appear in Phys. Rev.; and *"The Electromagnetic Interactions of Composite Systems"*, to be published.

27. H.Osborn: *Relativistic centre of mass variables for two particle systems with spin*, and *Relativistic corrections to nonrelativistic two-particle dynamical calculations*, Sussex preprints (1968); to appear in Phys. Rev.

28. See for instance B.Bakamjian and L.H.Thomas, *Phys. Rev.*, **92**, 1300 (1953); R.Fong and J.Sucher. *J. Math. Phys.*, **5**, 456 (1964).

29. See the first paper in reference (27), final equation of section IV.

30. Y.C.Chau, N.Dombey, and R.G.Moorhouse, to be published; Drell and Hearn, reference (8); H.R.Pagels, *Phys. Rev.*, **158**, 1566 (1967).

# Strong Interaction Sum Rules—
# An Introduction to Recent Developments

G. COSTA

*Università Padova, Italy*
*Istituto Nazionale di Fisica Nucleare, Sezione di Padova, Italy*

and

GRAHAM SHAW

*Columbia University, New York, U.S.A.*

Contents

---

## 1  INTRODUCTION

Since the publication of the famous paper of De Alfaro et al.[1] on super-convergence relations, a vast amount of work has been carried out exploiting sum rules based on the analyticity and assumed high energy behaviour of strong interaction amplitudes. In fact (see section 2) many of the sum rule types which have aroused such interest have been known for many years. What has happened has been a realisation of the power and usefulness of these ideas in a wide variety of contexts. It is the purpose of this article to serve as an introduction to the ideas and approximations used in these many applications. We stress that we do not aim at a comprehensive survey of the very large number of applications published, but at an account of the types of approaches used, illustrated by a few simple examples. We make no pretense at completeness.

One of the principal reasons for this revival of interest in sum rules is simply the substantial advance in our experimental knowledge. Phase shift analyses and the increased knowledge of the resonance spectra gained in recent years has enabled us to evaluate the low energy contributions to sum rules with some confidence, and the analysis of high energy data in terms of Regge pole parametric forms has led to greater confidence in the description of high energy scattering. Further, the bonus yielded by the unexpected success of low cut-off calculations—which will be discussed in detail in sections 4 and 5—has been a major factor in the usefulness of these methods.

In principle the sum rules, which have the characteristic property of relating the high and low energy behaviour of the scattering amplitudes, can be used to test the rather general principles on which they rest (section 2). For example, the Cottingam sum rules[2], described in § 3, provide an interesting test of whether the mass splittings within isospin multiplets are of electromagnetic origin, nor not. More usually, because of this very generality, they are not very fruitful unless combined with other more specific ideas or models, on which they impose severe restrictions. They have, used in this spirit, proved very useful tools in the analysis of experimental data. Thus, in cases in which the high energy contribution is known, or is negligible (superconvergence relations[1]) they reduce to consistency conditions on the parameters of low energy scattering, and can be used to determine some of these if others are known (section 3). On the other hand, if the low energy amplitude is known, then the sum rules can provide powerful restrictions on the parameters of Regge pole models at high energy (section 4). Here the application of continuous moment sum rules[3] (see section 2), and the use of low cut-offs (the energy above which the Regge parameterisation is used) have made this an especially useful technique. The reasons why such low cut-offs are successful are closely related to the theoretical models of the scattering amplitude discussed in section 5, where the important idea of "duality"[4] is introduced. In particular, the model in which the amplitude is dominated by resonances lying on Regge trajectories in both the direct and crossed channels leads to a new kind of bootstrap[5,6], also discussed in section 5. After this, the last sort of application that we consider is concerned with asymptotic symmetries. If a symmetry is expected to hold asymptoically, we can take advantage of this to choose combinations of amplitudes which will obey rapidly convergent sum rules[7,8] (section 2). The most interesting applications of this type are the spectral function sum rules (section 6), first obtained by Weimberg under rather different assumptions[9].

Finally, in order to completely define what we have included, we mention some topics we have left out. Firstly, if the substraction terms occurring in some of the sum rules in section 2 are fixed by soft pion theorems, using current algebra and PCAC, we arrive at many of the famous current algebra sum rules. Since these have been discussed so thoroughly elsewhere (see e.g. ref. 10), we do not consider them here. Nor do we describe the results of numerical evaluation of superconvergence relations and finite energy sum rules [4,5,11] in the approximation of resonance saturation—these have been reviewed by Soloviev[12] and Gilman and Harari[13]. Further, the latter authors[13] have given a very clear discussion of the relationship between superconvergence and $SU(6)$, showing which sum rules, when saturated by an $SU(6)$ multiplet, will lead to symmetry results, and which will not. We refer the interested reader to the above article for a discussion of this topic.

## 2  TYPES OF SUM RULE USED

### 2.1  Superconvergence and finite energy sum rules

Consider a crossing anti-symmetric amplitude $F(v, t)$ which satisfies an unsubtracted dispersion relation*

$$F(v, t) = \frac{2v}{\pi} \int_0^\infty dv' \, \frac{\operatorname{Im} F(v', t)}{v'^2 - v^2} \,. \tag{2.1}$$

If $F(v, t)$ decreases faster than $v^{-1}$ as $v \to \infty$, taking the limit $v \to \infty$ in (2.1) immediately leads to the *superconvergence* relation**

$$0 = \int_0^\infty \operatorname{Im} F(v, t) \, dv \,. \tag{2.2}$$

Let us now suppose that the amplitude does not decrease rapidly enough at high energies for the above relations to convergence, but that above some energy $v_L$, it can be represented by a sum powers in $v$:

$$F(v, t) = \sum_i \frac{\beta_i(t)\,(1 - e^{-i\pi\alpha_i(t)})}{\Gamma(\alpha_i + 1)\sin\pi\alpha_i(t)} \, v^{\alpha_i(t)}, \quad v > N. \tag{2.3}$$

Each term correspond to a Regge pole with trajectory function $\alpha_i(t)$ and residue $\beta_i(t)$; the phase factor follows directly from analyticity and crossing symmetry (see e.g. ref. 16). Following a trick first used by Igi[17], if we subtract off the terms for which $\alpha_i > -1$, the resultant amplitude will obey a superconvergence relation

$$\int_0^\infty dv \left\{ \operatorname{Im} F(v, t) - \sum_{\alpha_i > -1} \frac{\beta_i v^{\alpha_i}}{\Gamma(\alpha_i + 1)} \right\} = 0. \tag{2.4}$$

---

* Here and in the following, the variable $v$ is defined by $v = \dfrac{s - u}{4m}$ ; $s, u, t$ are the usual Mandelstam variables, and $m$ is the mass of the target particle. For processes in which the particles in the final and initial states are the same, $v$ is simply related to the lab. energy of the incident particle, $v = \omega_L + t/4m$.

** This type of sum rule has been exploited in specific instances for some time; see for example refs 14, 15. However, their general importance was first stressed by De Alfaro et al[1], who obtained such sum rules from differences of charge algebra sum rules, and who also emphasized the usefulness of Regge pole theory in deciding which amplitudes converge rapidly enough to satisfy relations such as (2.2).

Splitting the range of integration into two parts, $v > v_L$, $v < v_L$, we have:

$$\int_0^{v_L} dv\, \mathrm{Im}\, F(v, t) - \int_0^{v_L} dv \sum_{\alpha_i > -1} \frac{\beta_i v^{\alpha_i}}{\Gamma(\alpha_i + 1)} + \int_{v_L}^{\infty} dv \sum_{\alpha_i < -1} \frac{\beta_i v^{\alpha_i}}{\Gamma(\alpha_i + 1)} = 0,$$

so that we finally arrive at*

$$\int_0^{v_L} dv\, \mathrm{Im}\, F(v, t) = \sum_{\alpha_i} \frac{\beta_i v_L^{\alpha_i + 1}}{\Gamma(\alpha_i + 2)}. \tag{2.5}$$

These relations are the finite energy sum rules, and were first derived by several authors[4,5,11]; we here follow closely the discussion of Horn and Schmid[4]. An important point about (2.5) is that all Regge poles enter in exactly the same way, irrespective of their value. This result follows immediately from a more direct and elegant derivation originally given by Kadyshevsky et al.[18], and by Gatto[19].

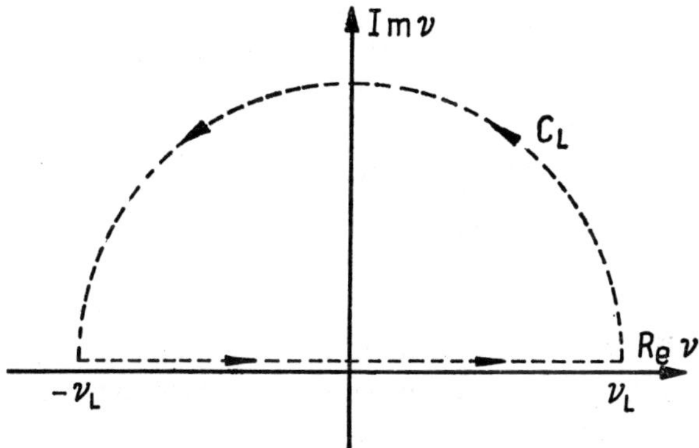


**Figure 1**   Contour of integration for eq. (2.6).

Consider Cauchy's theorem for $F(v, t)$ along the contour shown in Fig. 1

$$\int_{-v_L}^{v_L} dv\, \mathrm{Im}\, F(v, t) + \int_{c_L} F(v, t)\, dv = 0. \tag{2.6}$$

Then, evaluating explicitly the integral around the arc, using the expression (2.3) for $F(v, t)$ gives (2.5) directly. The equivalence of the two assumptions is guaranteed by the Phragman–Lindelöff theorem[20].

---

* Eq. (2.5) holds also if poles at $-1$ are present; in this case, their residues are included in the sum on the r.h.s. of eq. (2.5) and would appear also in the r.h.s. of eq. (2.4).

The equations (2.5) are the focal point of this article. They are of interest because they relate the parameters of high and low energy scattering, and hence in conventional theory, of direct and crossed processes. In particular, the superconvergence relations can be obtained in appropriate cases by letting $\gamma \to \infty$ in (5). However, since even these are normally evaluated only up to some cut-off $\nu_L$, they are in practice an evaluation of (5) on the assumption that the right hand side is negligible.

So far we have derived sum rules only on the amplitude $F(\nu, t)$ itself. Clearly, the derivation will hold for $\nu^n F(\nu, t)$, where $n$ is an even integer for crossing odd $F$, and an odd integer for crossing even $F$. We then arrive at positive and negative *moment sum rules*

$$n \geqq 0 \qquad \int_0^{\nu_L} \nu^n \, \mathrm{Im}\, F(\nu, t)\, d\nu = \sum_i \frac{\beta_i(t)\, \nu_L^{\alpha_i(t)+n+1}}{(\alpha_i + n + 1)\, \Gamma(\alpha_i + 1)} \qquad (2.7)$$

$$n > 0 \qquad \int_0^{\nu_L} \frac{\mathrm{Im}\, F(\nu, t)\, d\nu}{\nu^{n+1}} - \frac{\pi F^n(0, t)}{n!} = \sum_i \frac{\beta_i \nu_L^{\alpha_i - n}}{(\alpha_i - n)\, \Gamma(\alpha_i + 1)}. \qquad (2.8)$$

### 2.2   Schwarz sum rules

The derivation of eq. (2.5) holds only for odd functions, since the integral

$$\int_{-\nu_L}^{\nu_L} d\nu \, \mathrm{Im}\, F^{(+)}(\nu, t)$$

vanishes identically, for even functions $F^{(+)}(\nu, t)$, by crossing symmetry. In the absence of fixed poles in the $J$-plane however, Schwarz[21] has shown that the sum rules hold also for $F^{(+)}(\nu, t)$:

$$0 = \int_0^{\nu_L} d\nu \, \mathrm{Im}\, F^{(+)}(\nu, t) - \sum_i \frac{\beta_i \nu_L^{\alpha_i + 1}}{\Gamma(\alpha_i + 2)}. \qquad (2.9)$$

In general, it has been shown[22] that such poles can occur in strong interaction amplitudes, at wrong signature nonsense points, and the left hand side of eq. (2.9) is then given by the residue of the fixed pole

$$\gamma(t) = \int_0^{\nu_L} d\nu \, \mathrm{Im}\, F^{(+)}(\nu, t) - \sum_i \frac{\beta_i \nu_L^{\alpha_i + 1}}{\Gamma(\alpha_i + 2)}. \qquad (2.9')$$

On inspection, the Schwarz sum rules (2.9) have been found to be very badly violated[21],[23], and the residues of the fixed poles are large in some cases at least. Even if such fixed poles do not contribute to the asymptotic behaviour of the physical amplitudes, they can have physical consequence in leading to singular Regge residues and invalidating the usual interpretation of dips in differential cross sections.

For a review of this topic, which will not be discussed in the following, we refer to the paper by Roskies[24] and references therein.

### 2.3 Continuous moment sum rules

It was pointed out by Gilbert[25] that if the function $F(v, t)$ obeyed the relation (2.1), then an "inverse" dispersion relation could be obtained by defining the function

$$G(v, t) = \frac{iF(v, t)}{(v^2 - v_0^2)^{1/2}},$$

(2.10)

where the square root is defined to be positive (negative) for real $v > v_0$ ($v < -v_0$), and positive imaginary for $-v_0 < v < v_0$. The dispersion relation for $G(v, t)$ then gives

$$G(v, t) = \frac{2v}{\pi} \int_{v_0}^{\infty} \frac{dv'}{v'^2 - v^2} \frac{\operatorname{Re} F(v', t)}{(v'^2 - v_0^2)^{1/2}}.$$

(2.11)

Liu and Okubo[3] generalised this, and derived the corresponding superconvergence relations. Introducing the function

$$C(v, t, \gamma) = \frac{e^{i\pi\gamma} F(v, t)}{(v^2 - v_0^2)^{\gamma}} \quad \gamma < 1$$

(2.12)

we have

$$C(v, t, \gamma) - \frac{2v}{\pi} \int_{v_0}^{\infty} \frac{dv'}{(v'^2 - v^2)} \left\{ \frac{\cos \pi\gamma \operatorname{Im} F(v', t) + \sin \pi\gamma \operatorname{Re} F(v', t)}{(v'^2 - v^2)^{\gamma}} \right\}$$

(2.13)

so that, if $F(v, t)$ decreases faster than $v^{2\gamma - 1}$, letting $v \to \infty$ we arrive at the continuous family of superconvergence relations

$$0 = \int_{v_0}^{\infty} dv \left\{ \frac{\cos \pi\gamma \operatorname{Im} F(v, t) + \sin \pi\gamma \operatorname{Re} F(v, t)}{(v^2 - v_0^2)^{\gamma}} \right\}.$$

(2.14)

As before, we can go on to derive, in this case, a continuous set of FESR— *the continuous moment sum rules* (CMSR)[26]. Parametrising the high energy behaviour in the form*

$$F(v, t) = \sum_i \frac{\beta_i(t)\,(1 - e^{-i\pi\alpha_i(t)})}{\Gamma(\alpha_i + 1)\sin\pi\alpha_i}\, v\,(v^2 - v_0^2)^{1/2(\alpha_i - 1)} \qquad v > v_L \qquad (2.15)$$

this takes the form

$$\int_{v_0}^{v_L} dv \left\{ \frac{\cos\pi\gamma\,\operatorname{Im} F(v, t) + \sin\pi\gamma\,\operatorname{Re} F(v, t)}{(v^2 - v_0^2)^\gamma} \right\}$$

$$= \sum_i \frac{\beta_i(t)}{\Gamma(\alpha_i + 1)} \cdot \frac{(v_L^2 - v_0^2)}{(\alpha_i + 1 - 2\gamma)} \cdot \frac{\cos\frac{1}{2}\pi\,(\alpha_i - 2\gamma)}{\cos\frac{1}{2}\pi\alpha_i}\,. \qquad (2.16)$$

If $F(v, t)$ is an even amplitude, then assuming

$$F(v, t) = -\sum_i \frac{\beta_i(t)\,(1 + e^{-i\pi\alpha_i(t)})}{\Gamma(\alpha_i + 1)\sin\pi\alpha_i}\,(v^2 - v_0^2)^{1/2\alpha_i} \qquad v > v_L \qquad (2.17)$$

and repeating the argument for $vF(v, t)$ we obtain

$$\int_{v_0}^{v_L} dv \left\{ \frac{v\cos\pi\gamma\,\operatorname{Im} F(v, t) + v\sin\pi\gamma\,\operatorname{Re} F(v, t)}{(v^2 - v_0^2)^\gamma} \right\}$$

$$= \sum_i \frac{\beta_i(t)}{\Gamma(\alpha_i + 1)} \frac{(v_L^2 - v_0^2)^{1/2(\alpha_i + 2 - 2\gamma)}}{(\alpha_i + 2 - 2\gamma)} \frac{\sin\frac{1}{2}\pi\,(\alpha_i - 2\gamma)}{\sin\frac{1}{2}\pi\alpha_i}\,. \qquad (2.18)$$

Let us stress again that all the relations derived so far are consequences of Regge high energy behaviour and the fixed $t$-dispersion relation. But, although (2.11) adds nothing to (2.1), equations (2.14) and (2.16) are powerful generalisations of (2.7), and as we shall see in section 4, prove an especially useful tool for the determination of the parameters of high energy scattering. Finally, we note that although in deriving the explicit formulae we have assumed that the high energy scattering amplitude can be described by a sum of Regge pole terms, the arguments used can be trivially extended to any assumed form of the high energy scattering amplitude.

------

* In this paragraph, we have implied that $v_0$ is the beginning of the right hand cut, whereas in the previous paragraph we have taken $v_0 = 0$, in which case taking integer $\gamma$ in (2.16) exactly reproduces (2.7). The distinction between these two forms is of no consequence for reasonable values of the cut-off $v_L$.

## 2.4  Sum rules from asymptotic symmetries

In the previous Sections we have discussed sum rules which relate the low and high energy behaviour of the scattering amplitude, where the latter have been described by a sum of Regge pole terms. For amplitudes for which these are believed to be negligible, the sum rules reduce to consistency relations between the low energy parameters themselves—the so-called super-convergence relations. It was pointed out by Costa and Zimerman[7] that, if symmetries were believed to hold asymptotically between amplitudes then this could be exploited to form combinations of amplitudes which would superconverge. These authors exploited universal coupling of the rho trajectory in charge exchange processes. Others have used the asymptotic predictions of the quark model[27]. However, the most exciting applications of asymptotic symmetries have been to the propagators and form factors of the weak vector and axial currents, where, as shown by Das, Mathur and Okubo[28], application of the chiral symmetry group $SU(2) \times SU(2)$ leads to the Weinberg sum rules[9] and other interesting results. These will be discussed in section 6.


## 3  LOW ENERGY PHENOMENOLOGY

A very useful feature of the sum rules described in section 2 is the possibility of obtaining information about low energy parameters in some cases in which a more direct determination is rather difficult. What is required is the knowledge of the high energy behaviour of the scattering amplitude, and then the sum rules reduce to consistency reltaions among low energy parameters.

We shall consider in the following two typical examples of such use: the prediction of the KNY coupling constants, for which previous determinations suffer from rather large uncertainties, and the evaluation of electromagnetic mass differences of hadrons.

### 3.1  Determination of the KNY coupling constants

Previous determinations of the $KN\Lambda$ and $KN\Sigma$ coupling constants, $g_\Lambda^2$ and $g_\Sigma^2$, obtained from $K^\pm p$ and $K^\pm n$ forward dispersion relations[29], are affected by the uncertainty connected with the extrapolation of the $\overline{K}N$ scattering lengths into the unphysical region. This extrapolation is strongly dependent on the parametrization used[30],[31].

Determinations of the $g_\Lambda^2$ and $g_\Sigma^2$ coupling constants, which are not so sensitive to the extrapolation to the unphysical region, have been provided by Martin and Ross[32], using FESR, and by Chan and Meiere[33]. We shall discuss here the determination of the latter authors, which is based on the use of CMSR.

The sum rule is written in form of a dispersion relation for the amplitude

$$F_\gamma(\nu) = A'\,(\nu, t = 0)\,(\nu - m_K)^{\gamma - 1}\,(\nu - \nu_c)^{-\gamma}, \tag{3.1}$$

where $\nu$ is the laboratory energy of the $K$-meson, $\nu_c$ is the lowest $\pi Y$ threshold, and $\gamma$ is an arbitrary real parameter, $0 < \gamma < 1$. The factor multiplying the usual invariant forward amplitude $A'(\nu)$ is taken to be negative for real $\nu < \nu_c$, and positive for real $\nu > m_K$. Two sum rules are written for the two values of the iso-spin, $I = 0$ and 1 in the $t$-channel. We give explicitly, as an example, the sum rule corresponding to $I = 0$:

$$\frac{2g_\Lambda^2 X(\Lambda)}{(\nu_\Lambda + m_K)(m_K - \nu_\Lambda)^{1-\gamma}(\nu_c - \nu_\Lambda)^\gamma} = \frac{1}{4\pi^2} \int_0^\infty \frac{d\nu}{\nu} \left(\frac{\nu + m_K}{\nu + \nu_c}\right)^\gamma [2\sigma_{K+p}(\nu) - \sigma_{K+n}(\nu)]$$

$$- \frac{1}{4\pi^2} \int_0^\infty \frac{d\nu}{\nu} \left(\frac{\nu - m_K}{\nu - \nu_c}\right)^\gamma [2\sigma_{K-p}(\nu) - \sigma_{K-n}(\nu)]$$

$$+ \frac{1}{\pi} \int_{\nu_c}^{m_K} \frac{d\nu}{\nu + m_K} \frac{\cos \pi\gamma \, \mathrm{Im}\, A'(\nu) + \sin \pi\gamma \, \mathrm{Re}\, A'(\nu)}{(m_K - \nu)^{1-\gamma}(\nu - \nu_c)^\gamma}$$

$$- \frac{A'(-m_K)}{(2m_K)^{1-\gamma}(\nu_c + m_K)^\gamma}, \tag{3.2}$$

where:

$$X(\Lambda) = [m_K^2 - (m_\Lambda - m_N)^2]/4m_N^2.$$

The advantags of the above choice for the amplitude are evident from eq. (3.2). In fact, in the physical region, $\mathrm{Im}\, F_\gamma(\nu)$ is proportional to $\mathrm{Im}\, A'(\nu)$ and it is expressed in terms of total cross sections, while in the unphysical region, $\nu_c < \nu < m_K$, it is given by

$$\mathrm{Im}\, F_\gamma(\nu) = \cos \pi\gamma \, \mathrm{Im}\, A'(\nu) + \sin \pi\gamma \, \mathrm{Re}\, A'(\nu). \tag{3.3}$$

Then, by varying $\gamma$ one can give different weights to the imaginary and real parts of $A'(\nu)$, that is different weigths to the contribution from the unphysical cut. This is evaluated making use of the multichannel effective-range parametrisation given by Kim[31]. The contributions in the physical region are evaluated in terms of effective-range parameters[31],[34] and total cross sections up to 18 BeV/c[35]. The asymptotic behaviour above this point is approximated by $\sigma_{\bar{K}N} - \sigma_{KN} \sim k^{-1/2}$; it is estimated to contribute very little to $g_A^2$ and $g_\Sigma^2$. All these pieces of information seem to provide a self-consistent determination of the coupling constants, since this determination is stable varying the parameter $\gamma$ in the range from 0.1 to 0.9. The results are summarized by $g_A^2 = 13 \pm 3$, $g_\Sigma^2 = 0 \pm 1$, which are consistent with $SU(3)$ invariance. It is interesting to note that the same procedure, making use of constant scattering length parametrization[30], give inconsistent results for different values of $\gamma$, although this is to some extent trivial, since the latter parametrization does not incorporate the correct threshold behaviour at $v = v_c$. This will clearly be important for values of $\gamma$ close to one.

## 3.2   Sum rules for electromagnetic mass differences

Another very interesting example is the evaluation of electromagnetic mass differences of hadrons, which can be performed in those cases in which the high energy behaviour of the relevant Compton scattering is known.

It is well known that the electromagnetic self-energy of a hadron, to first order in $e^2$, can be written in terms of the forward Compton scattering amplitude $T_{\mu v}(q^2, v)$ as follows:

$$\Delta M = \frac{1}{8\pi^2} \int_{-\infty}^{\infty} \frac{T_{\mu v}(q^2, v)\, g^{\mu v}}{q^2 - i\varepsilon}\, d^4q, \qquad (3.4)$$

where $q$ is the photon four-momentum, and $v = q_0$. An apparent difficulty with eq. (3.4) is that the amplitude in the integral corresponds to virtual time-like photons (of mass $q^2$). This difficulty was eliminated by Cottingham[36], who noticed that one could safely rotate the contour in the integral in eq. (3.4) from the real to the imaginary axis in the complex $v$-plane. In this way, the self-mass $\Delta M$ can be expressed in term of a space-like photon amplitude $T_{\mu v}(q^2, iv)$, and, in principle, it can be computed making use of the experimental electron scattering data. Replacing $v$ by $iv$, and integrating over the angular variables, eq. (3.4) can be re-written as:

$$\Delta M = -\frac{1}{4\pi} \int_0^{\infty} \frac{dq^2}{q^2} \int_{-q}^{+q} dv\, \sqrt{q^2 - v^2}\, \{3q^2 t_1(q^2, iv) - (q^2 + 2v^2)\, t_2(q^2, iv)\},$$

$$(3.5)$$

where $t_1$, $t_2$ are the two invariant amplitudes defined through ($p$ is the hadron four-momentum):

$$T_{\mu\nu}(q^2, i\nu) = t_1(q^2, i\nu)[q^2 g_{\mu\nu} - q_\mu q_\nu]$$

$$+ t_2(q^2, i\nu)\left[-\nu^2 g_{\mu\nu} + \frac{q^2}{M^2} p_\mu p_\nu + \frac{i\nu}{M}(p_\mu q_\nu + p_\nu q_\mu)\right]. \quad (3.6)$$

The two amplitudes $t_1(q^2, i\nu)$, $t_2(q^2, i\nu)$ could be determined in terms of their absorptive parts, by means of fixed $q^2$ dispersion relation. There is, however, the problem of possible subtractions.

A very important contribution in this connection has been given by Harari[37]. He noticed that for a given isospin multiplet one can define, to lowest order in $e^2$, two terms in the electromagnetic mass differences, $\Delta M^{(1)}$ and $\Delta M^{(2)}$, which transform according to $\Delta I = 1$ and $\Delta I = 2$. Then he assumed that the asymptotic dependence of the forward non-spin-flip $I$ in the $t$-channel is dictated by the Regge pole model. The high energy behaviour is then given by $\nu^{\alpha_I(0)}$ where $\alpha_I(0)$ is the intercept of the relevant Regge trajectory. Assuming $\alpha_2(0) < 0$, as suggested by the absence of low-lying $I = 2$ boson resonances and used extensively in superconvergence sum rules, one can use unsubtracted dispersion relations for both $t_1$ and $t_2$. Using the approximation that the absorptive parts are dominated by the lower states, Harari evaluated the mass differences $\Delta M_\pi^{(2)} = (5 \pm 1)$ MeV and $\Delta M_\Sigma^{(2)} = (1.5 \pm 0.5)$ MeV, which compare rather well with the experimental values. This indicates also that the mass splitting in an iso-spin multiplet can be attributed entirely to *EM* interactions.

In the case of $\Delta I = 1$ mass difference, as in the proton-neutron case, while the amplitude $t_2$ still satisfies an unsubtracted dispersion relation, a subtraction is needed for $t_1$. This is simply related to the fact that now the contribution from an $I = 1$ Regge pole ($A_2$-trajectory with $\alpha(0) \sim 0.4$) is not negligible. However, one can estimate this contribution, and then compute $\Delta M^{(1)}$. We shall not discuss this problem here, but refer to the relevant papers[38].


## 4  HIGH ENERGY PHENOMENOLOGY

The earliest, and perhaps most fruitful use to which the sum rules described in II have been put is the determination of the parameters of high energy scattering in those cases where there is abundant low energy data available.

This technique has proved especially useful recently, because of the introduction of CMSR. In practice, the high energy parameters studied have been the parameters of pure Regge pole models. However, as stressed previously (section 2), the arguments can clearly be extended to any high energy form, and in fact have already been applied to Regge models with absorptive corrections by Michael[39], where essentially a prescription for the inclusion of cuts is given. For the rest of this section, however, we shall only consider the explicit Regge pole forms given in section 2.

Before going on to illustrate the usefulness of this approach, primarily by a discussion of vacuum exchanges in $\pi N$ scattering, let us discuss briefly the sort of approximation we shall make. The idea is simply to evaluate the integrals in (2.7), (2.8) or more generally (2.16), (2.18) from a given phase shift solution, thus determining the combination of Regge parameters on the right hand side. Clearly, to obtain useful information, the value of $n$ in (2.7) say must not be so large that the integral is dominated by a short region at the upper end of the integration range. Nor must it be so large in (2.8) that the Regge terms drop out altogether, and one obtains effectively a super-convergence relation. The most useful sum rules are those in which the integrand is fairly evenly weighted. Equally clear is the major problem of this approach; if the energy $\nu_L$ is the upper limit of the region in which we have phase shift analysis or other partial wave data, it is in general well below the asymptotic region in which the approximation (2.3) is valid. It would seem in practice that this problem is not so serious as might at first be supposed. Let us consider why this might be so for the simplest sum rule we derived (2.5)*

$$\int_0^{\nu_L} d\nu \, \mathrm{Im}\, F\,(\nu, t) = \sum_i \frac{\beta_i \nu_L^{\alpha_i + 1}}{\Gamma\,(\alpha_i + 2)}, \tag{4.1}$$

in the case where all $\alpha_i > -1$. If the form (2.3) were *exact* for $\nu > \nu_L$, then it would also be exact for $\nu < \nu_L$. Equation (4.1) asserts that if (2.3) is a good approximation for $\nu > \nu_L$, then it holds on the average over the range $0 < \nu < \nu_L$ since the right hand side is merely the integral over this range of the Regge form (2.3). Although it does not necessarily follow, it is not implausible that this averaging feature holds more locally

$$\int_{\nu_1}^{\nu_2} d\nu \, \mathrm{Im}\, F\,(\nu, t) = \sum_i \frac{\beta_i}{\Gamma\,(\alpha_i + 2)}\,(\nu_2^{\alpha_i + 1} - \nu_1^{\alpha_i + 1}) \tag{4.2}$$

---

* The discussion is easily extended to other cases.

where the range $v_1$, $v_2$ spans several wiggles, which might be of the order of the widths of the resonances present. In this case, the approach to Regge asymptotics would be as is schematically shown in Fig. 2, and it is clear that one can use a much lower cut-off in the sum rule, where one is concerned with the area under the curve, than when a direct Regge fit to the data is attempted. The sort of behaviour illustrated by (4.2) and Fig. 2 was

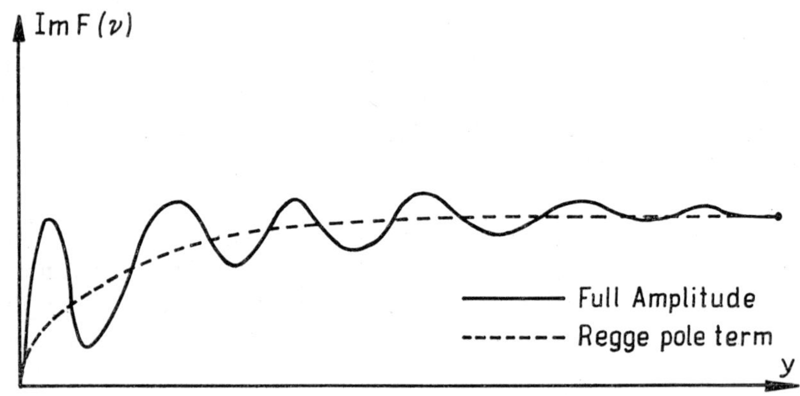


**Figure 2**   Qualitative behaviour of a full scattering amplitude compared with a Regge pole term.

first noted in a resonance model for $\pi N$ charge exchange by Dolen, Horn and Schmid[4],[5] and led directly on to "duality" and the "generative" model to be discussed in the following sections. However, in the present context the justification for low cut-offs is empirical—if the sum rules with a given cut-off predict correctly those parameters which are well known from fits to high energy data, it is reasonable to assume they also predict correctly those parameters which are not known.

Let us now illustrate the use of the various sum rules of section 2 in determining the characteristics of the vacuum exchanges in $\pi N$ scattering.

## 4.1   The Igi sum Rule

The first attempt in this direction, and the first use of what was in effect a FESR, was due to Igi [17], who considered the crossing even forward pion-nucleon scattering amplitude

$$f^+(v) = \frac{1}{4\pi} A'^{(+)}(v, t = 0) = \frac{1}{4\pi}\{A^{(+)}(v, t = 0) + vB^{(+)}(v, t = 0)\} \quad (4.3)$$

where $A^{(+)}$, $B^{(+)}$ are the conventional invariant amplitudes of Chew et al.[40], and addressed himself to the question of whether there were other trajectories than the Pomeron present with $0 < \text{Re}\,\alpha\,(t = 0) < 1$. In the absence of such trajectories, one can subtract the Pomeron contribution from the forward amplitude, and write an unsubtracted dispersion relation for the remainder. Using the optical theorem, this relation, evaluated at threshold is

$$\left(\frac{m+1}{m}\right) a^{(+)} = -\frac{f^2}{m}\frac{1}{1 - 1/4m^2} + \frac{1}{2\pi^2}\int_0^\infty dk'\,[\sigma_t^{(+)}(k') - \sigma_t^{(+)}(\infty)] \quad (4.4)$$

where $a^{(+)}$ is the scattering length, $\sigma_t$ the total cross-section and $k$ the pion lab momentum. From 20 GeV to infinity, the integral was evaluated using the contemporary Regge fits of Udgaonkar[41], and if this and the integral over $\sigma_t(\infty)$ is done explicitly we have a sum rule of the type (2.8) with $n = 0$. The integral up to 20 GeV was then evaluated from the total cross-section data; and the other terms from the low energy parameters of Hamilton and Woolcock[42]. The sum rule (4.4) was found to be badly violated, and on these grounds Igi predicted the existence of a second vacuum trajectory $P'$ with $\alpha_{P'}(0) > 0$.

## 4.2  Application of CMSR to vacuum exchange in $\pi N$ scattering

Although the need for a second trajectory was confirmed by fits to the high energy data, this data could by no means unambiguously determine the parameters of the two trajectories. For example, values of the Pomeron coupling $\sigma_t(\infty)$ have varied between 14.5 mb[43] and 22.1 mb[44], and the spin-flip amplitude $B^{(+)}(\nu, t)$ was very poorly determined (see below). This situation has been decisively improved by the application of the CMSR (2.16), (2.18), which were first used to investigate high energy scattering parameters by Olsson[26]. In this case, one exploits the knowledge of both the real and imaginary parts of the amplitude at low energies, and this is especially useful since for any Regge pole term, the phase and energy dependence are uniquely related through the signature factor[16].

The CMSR were first applied to this process by Barger and Phillips[45] who studied both the spin-flip amplitude $B^{(+)}(\nu, t)$ and the non spin-flip amplitude

$$A'^{(+)}(\nu, t) = A^{(+)}(\nu, t) + \frac{\nu + t/4m}{1 - t/4m^2} B^{(+)}(\nu, t) \quad (4.5)$$

over the range of four momentum transfer $0 > t > -1\cdot 0$ GeV.

Introducing the functions

$$a'^{(+)}(v, t, \gamma) = vA'^{(+)}(v, t)\left(\frac{\lambda^2}{v^2 - v_0^2}\right)^{\gamma} e^{i\pi\gamma} \tag{4.6}$$

and

$$b^{(+)}(v, t, \gamma) = B^{(+)}(v, t)\left(\frac{\lambda^2}{v^2 - v_0^2}\right)^{\gamma} e^{i\pi\gamma} \tag{4.7}$$

where $\lambda = 1$ GeV is a scale constant. They then write the sum rules

$$\int_{v_0}^{v_L} dv \,\, \mathrm{Im}\,\, a'^{(+)}(v, t, \gamma) = \sum_{i=P, P'} \frac{\beta_i'\xi_i}{\alpha_i - 2\gamma + 2}\left(\frac{v_L^2 v_0^2}{\lambda^2}\right) \tag{4.8}$$

and

$$\int_{v_0}^{v_L} dv \,\, \mathrm{Im}\,\, b^{(+)}(v, t, \gamma) = \sum_{i=P, P'} \frac{\beta_i\xi_i}{\alpha_i - 2\gamma} \tag{4.9}$$

where

$$\xi_i = \frac{\sin\frac{1}{2}\pi(\alpha_i - 2\gamma)}{\sin\frac{1}{2}\pi\alpha_i}\left(\frac{v_L^2 - v_0^2}{\lambda^2}\right)^{1/2(\alpha_i - 2\gamma)}. \tag{4.10}$$

The above sum rules (4.8), (4.9) are simply (2.18), (2.16) respectively re-written with a redefinition of the residue functions $\beta(t)$, $\beta'(t)$. They then evaluate the low energy integrals for $1 > \gamma > -\frac{3}{2}$ and $0 > t > -0.8$, using the results of the CERN phase shift analysis[46] and a cut-off corresponding to a centre of mass energy $W = 2.19$ GeV. On comparing the results with the values of the right hand sides of (4.8) (4.9) obtained from $P + P'$ Regge fits to high energy data, they find good agreement with several fits for the non spin-flip amplitude $A'^{(+)}$, but bad agreement with previous assumptions for the spin-flip amplitude; for example, compared to the results of reference (43), the sum rules demand smaller size and opposite sign. The sum rules in fact suggest a $B^{(+)}$ amplitude of the order $A'^{(+)}/v$, and the authors subsequently re-analysed the high energy data imposing this condition, to obtain new values for the residues of these trajectories. These new parameters are shown to be in good agreement with the $A'^{(+)}$ sum rules, and reasonably good, although not perfect, agreement with the $B^{(+)}$ sum rules. Finally, in a subsequent paper[47] the same authors have performed a simultaneous fit to the sum rules and all available high energy data. The results agree in general features with the previous results, but a striking improvement of the fits occurs if a third vacuum pole, $P''$ is added with intercept below zero. The parameters of this third pole depend on the assumptions made on the properties of the $P'$, but the parameter mentioned previously $\sigma_t(\infty)$ is not determined to within about 1 mb (21 mb) and we stress again the improved knowledge of the spin-flip amplitude.

In fact, the necessity of adding a third pole, perhaps to simulate background or cuts, had been previously noted by Olsson[48] and Della Selva et al.[49] from studies of the CMSR for the forward amplitude $A'^{(+)}(\nu, t = 0)$. These authors take the imaginary part from the experimental total cross-section, the real part from the forward relation, and evaluate the CMSR with cut-offs of 5 GeV[48], and 3 and 6 GeV[49]. Both groups find the addition of a $P''$ trajectory with $\alpha(0) < 0$ necessary to fit the sum rules, and both find a value of the coupling $\sigma_t(\infty)$ consistent with Barger and Phillips[45].

*Other Cases*   Before closing this section, we note that the CMSR have been applied in this way in very many cases. A few examples of their application to $\pi N$ charge exchange [45] [50], $KN$ scattering[51] and pion photoproduction[52] being given in the reference.


## 5   DUALITY AND BOOTSTRAPS

At high energies the scattering amplitude can be described by a series of power terms (2.3), (2.17). At low energies we describe the same amplitude by means of the partial wave expansion, and in many processes, many of these partial waves resonate, having the sort of behaviour associated with second sheet poles. These two regimes are related by the FFSR (2.5), where they are divided by the cut-off $\nu_L$. As we have seen in the last section, in many (successful) applications the cut-off $\nu_L$ is taken well below the asymptotic region, so that the transition between the two regions must be smooth, in the sense discussed, to avoid arbitrary cut-off dependence. That this should be so is not surprising in an analytic function, but, as we shall see in this section, the question of the manner in which this occurs has given rise to a number of interesting speculations, and a new type of bootstrap.


### 5.1   The interference model

The first model to be suggested was the interference model, in which the scattering amplitude is assumed to be simply the sum of the resonances and the Regge pole amplitude

$$F(\nu, t) = F_{\text{Res}}(\nu, t) + F_{\text{Regge}}(\nu, t). \tag{5.1}$$

This model was originally proposed[53] as a mean of analysing data in the intermediate energy region, above where phase shift analysis is possible, but below the energy of normal Regge fits ($\gtrsim 5$ GeV). In this energy region, some successful fits have been achieved[53] [54], but whether these will outline more detailed knowledge of the resonance spectrum is not clear. However, the basic idea of the interference model—that the terms which are smoothly varying with energy are coherent in $l$, whereas the rapidly varying structure occurs in only a few partial waves—may well survive the failure of (5.1).

Whereas the validity of (5.1) in the intermediate energy region is uncertain, its failure on extension to all energies[10], [55] is clear[5]. This is illustrated particularly simply[56] by writing a superconvergence relation for the difference of the forward pion nucleon scattering amplitude $f^{(+)}(\nu)$ (4.3) and its Regge contribution

$$\frac{f^2\mu}{2M} + \frac{1}{\pi} \int_0^\infty \{\nu\sigma_{\text{tot}}^{(+)}(\nu) - \operatorname{Im} f_{\text{Regge}}^{(+)}(\nu)\}\, \nu\, d\nu = 0. \qquad (5.2)$$

In the interference model, the subtraction of the Regge contributions leaves only the resonance terms in the total cross-section, and since these are all positive, the relation cannot possibly be satisfied. Another simple example of its failure is the sum rule over the $\pi N$ amplitude $\nu B^{(-)}(\nu)$, where again all the main resonance contributions have the same sign[23].

### 5.2   Duality and the generative model

An alternative possibility (duality) was proposed by Dolen, Horn and Schmid[4], [5], and we have already led up to this in the previous section. There it was suggested that a semi-local averaging feature might hold, and the full amplitude wiggle about the Regge terms in the manner shown in Figure 2, so that, for example

$$\int_{\nu_1}^{\nu_2} d\nu\, \operatorname{Im} F(\nu, t) = \int_{\nu_1}^{\nu_2} d\nu\, \operatorname{Im} F_{\text{Regge}}(\nu, t) \qquad (5.3)$$

where $(\nu_2 - \nu_1)$ is of the order of several wiggles. The above authors attributed the wiggles to resonances, and suggested that the appropriate way to combine resonances and Regge terms was

$$F(\nu, t) = F_{\text{Regge}}(\nu, t) + F_{\text{Res}}(\nu, t) - \langle F_{\text{Res}}(\nu, t)\rangle \qquad (5.4)$$

where $\langle\,\rangle$ denotes a semi-local average in the above sense. If we accept this, we see that the interference model will only work when there are large cancellations among the resonances, so that $\langle F_{\text{Res}}(v, t)\rangle = 0$, and this is in fact the case for those sum rules in which the model has been successfully applied[19],[55]. In the counter-examples we cited the resonant contributions add, so that by omitting the (large) final term in (5.4), the interference model commits what has been called[4],[5] double counting.

Let us consider an amplitude which is dominated by resonances in some region, so that

$$F(v, t) \simeq F_{\text{Res}}(v, t).$$

Then, from (5.5)

$$F_{\text{Regge}}(v, t) \simeq \langle F_{\text{Res}}(v, t)\rangle \tag{5.5}$$

and so in this "generative" model the resonances, in an average sense, build the Regge pole exchanges. This idea is at first sight a little surprising in that the Regge amplitude contains no resonance poles, whereas the resonance terms, *before averaging*, do. As we shall see below, the assumption that the low energy amplitude is dominated by resonances, and that these resonances lie on the same trajectories which in a crossing symmetric case dominate the $t$-channel exchange, leads to a new type of bootstrap. First, however, let us, following Schmid[57], consider the extension of duality to the partial wave amplitude.

From (5.4) we see that $\langle F(v, t)\rangle = F_{\text{Regge}}(v, t)$. If this is approximately true over the physical $t$ range, one would expect it to be also true for the individual partial waves. Schmidt[57] considered this by investigating the following projection of the spin-flip amplitude for pion nucleon charge exchange in the energy region $1.0 < p_L < 2.4$

$$B_l(v) = \frac{1}{2} \int_{-1}^{1} d\cos\theta P_l(\cos\theta) B^{(-)}(v, \cos\theta) \tag{5.6}$$

$$\sim \frac{4\pi}{E - m} \{f_{(l+1)+} - f_{(l+1)-}\}. \tag{5.7}$$

He took the Regge pole parameters from fits to high energy (6–18 GeV) data, continued them down to the region quoted, and performed the above projection. On plotting out the amplitude as a function of energy on an Argand diagram, he found the now famous Schmid circles (e.g. Fig. 3a) somewhat reminescent of he cirles associated with resonances (Fig. 3b). Further, the positions of these circles roughly correspond to the positions of some known $\pi N$ resonances.

The Schmid amplitudes are not in the normal sense resonant; they have no second sheet poles. Hence they do not in general show the characteristic energy variation around the circle of a resonance, nor would one expect them to possess the $s$-channel factorisation properties, characteristic of a pole, which enable one to discuss the production and decay of a resonance in

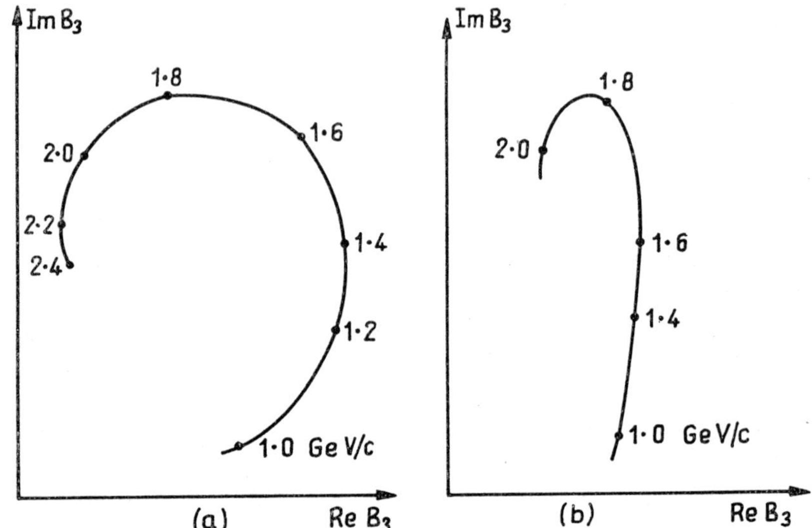


**Figure 3**    (a) Argand plot for the $\pi N$ amplitude $B_{l=3}^{(-)}$ as evaluated by Schmid (the numbers on the curve are the values of the laboratory momentum). (b) Corresponding Argand plot for the experimental $F_{37}$ resonance.

terms of its partial widths for the initial and final states. Nonetheless, if a partial wave is resonant, one would expect its local average to be very like a Schmid circle, and it is not unreasonable to conjecture that when Schmid's analysis leads to a pronounced circle, the physical partial wave resonates. Since the Regge forms are characteristic of $t$-channel exchanges, such an assertion leads to a new kind of bootstrap. However, as we shall see below, this can be discussed without invoking the somewhat controversial conjecture above.

### 5.3   A new kind of bootstrap

Let us return to the possibility of an amplitude which is completely dominated by resonances. Then, the left hand side of a FESR is given by the $s$-channel resonance contributions, whereas the right hand side is characterised by Regge pole terms, which are related to the exchange of particles in the $t$-channel. We have here the ingredients of a new type of bootstrap[5]. However, in order to relate particles in the two channels, it is necessary to have a model which enables the trajectories to be continued from spacelike to timelike $t$, where they are manifested as particles, and/or which allows the sum rules to be evaluated at these $t$-values, where the partial wave expansion will in general diverge. This has been discussed by Mandelstam[6] in the context of models in which the scattering amplitude is dominated by series of narrow resonances lying on infinitely rising linear trajectories, and a number of calculations based on this model have been performed[6],[58]–[62].

It is most easy to illustrate the sort of arguments and approximations used by considering an example. It was pointed out by several authors[58]–[60] that a particularly simple case to study in this context was the reaction

$$\pi(p_1) + \pi(p_2) \to \pi(p_3) + \omega\,(e, q).$$

In this case, there is only one invariant amplitude $A$:

$$T\,(v, t) = \varepsilon_{\mu v \sigma \tau} e^{\mu} p_1^v p_2^\sigma p_3^\tau A\,(v, t).$$

This amplitude is completely crossing symmetric, and selection laws and Bose statistics imply that the intermediate states in all three channels must have $J^{PC} = 1^{--}, 3^{--}, 5^{--} \ldots$ with $I = 1$. Thus we need consider only the rho trajectory, and any others with the same quantum numbers and signature. The above authors consider the first moment sum rule

$$\int_0^v v\,\mathrm{Im}\,A\,(v, t)\,dv = \frac{\beta(t)\,v_L^{\alpha(t)+1}}{\alpha(t) + 1} \tag{5.8}$$

and assume saturation of both sides by the rho trajectory with the cut-off placed between the rho and its first $(3^-)$ Regge recurrence. The first and most striking result of this is immediately obtained—in the narrow width approximation, independent of cut-off, the rho contribution to the integral is proportional to

$$2m_\varrho^2 - m_\omega^2 - 3m_\pi^2 + t$$

which vanishes at

$$t = m_\omega^2 + 3m_\pi^2 - 2m_\varrho^2 \simeq -0.53\ \mathrm{GeV}^2.$$

Hence, $\beta(t)$ must vanish at this point also. The most natural interpretation of this is that it is the zero associated, in the absence of fixed poles, with the wrong signature nonsense point $\alpha(t) = 0$. The famous dip in the differential

cross-section for $\pi^- p \to \pi^0 n$ associated with this same zero occurs experimen
tally at about $t = 0.59$ GeV$^2$, in excellent agreement with the above. One can
then attempt to take the calculation further by assuming a linear trajectory

$$\alpha(t) = at + b \qquad (5.9)$$

and evaluating the sum rule also at the point $\alpha(t) = 1$. Here, the residue
function is given in terms of the rho width, which then cancels from both
sides of the equation. The resultant equation, together with the position of
the zero, enables one to solve for both $a$ and $b$. Unfortunately, although
good agreement with experiment can be obtained with a reasonable cut-off
between $1^-$ and $3^-$ positions, the results are dependent on the cut-off position. Thus one cannot in fact determine both parameters, but given, say, the
rho mass, one can correctly predict the trajectory slope. Having accom-
plished this, it is of course trivial to evaluate the sum rule as a function of $t$,
and obtain an estimate of the residue function.

In the previous paragraph it was argued that the idea of duality suggests
that much lower cut-offs may be used in the sum rules than one would otherwise expect. However, in the above, a very low cut-off has been used, and
there is a certain amount of cut-off dependence in the results. Following

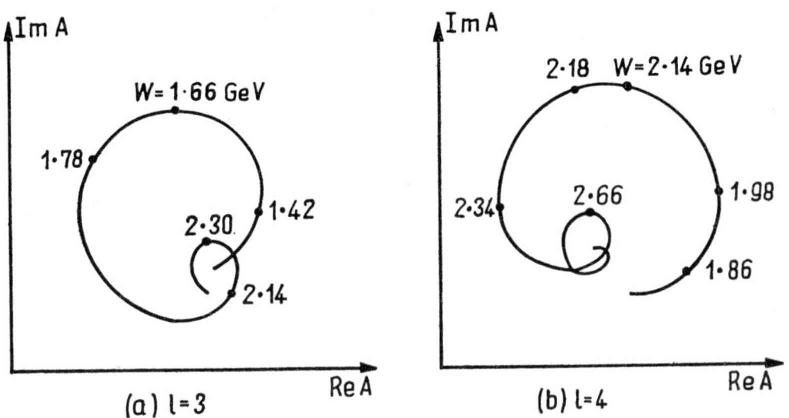


**Figure 4**   Argand plot of the partial wave amplitudes for the process
$\pi\pi \to \pi\omega$: (a) for $l = 3$, (b) for $l = 4$.

Rubinstein et al.[62], let us consider what happens in the above calculation
when higher cut-offs are used, to include $3^-$, $5^-$, ... states. They find that,
retaining only the single trajectory, the equality of the Regge and resonant

sides of (5.8) is still satisfactory at $3^-$ level for a large range of $t$, but when the cut-off is made high enough to include $5^-$, $7^-$ states etc, the resonance side becomes smaller than the Regge side. Since as the cut-off energy is increased,

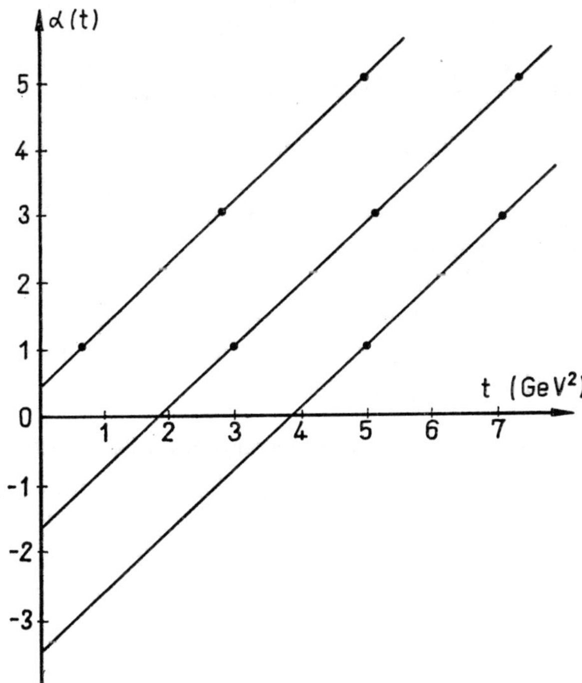


**Figure 5**    Family of trajectories coupled to the $\pi\pi \to \pi\omega$ system.

the rho trajectory should become a better approximation, assuming that the other singularities lie lower, it is reasonable to suppose that extra structure is needed in the resonance spectrum. In order to investigate this, Rubinstein et al.[62] analysed the Regge term into partial waves in the manner suggested by Schmid[57]. They found Argand plots like those shown in Fig.4(a), (b), suggesting the presence of parallel daughter trajectories spaced two units lower than the parent rho trajectory (Fig. 5).

We have tried to illustrate the features of FESR bootstrap calculations by consideration of some simple features of the reaction $\pi\pi \to \pi\omega$. This has led us to a picture in which both direct and crossed channels are dominated by narrow resonances lying on parallel linear Regge trajectories. It is interesting to note that an explicit form for the scattering amplitude has been suggested for this case[63]

$$A\,(s, t, u) = \frac{\bar{\beta}}{\pi}\,[B\,(1 - \alpha(t), 1 - \alpha(s)) + B\,(1 - \alpha(t), 1 - \alpha(u))$$

$$+ B\,(1 - \alpha(s), 1 - \alpha(u))], \qquad\qquad (5.10)$$

where $B\,(x, y) = \dfrac{\Gamma(x)\,\Gamma(y)}{\Gamma(x + y)}$, $\bar{\beta}$ is a constant and $\alpha(y) = ay + b$. As is shown

in reference[63], (5.10) has exact crossing symmetry, contains towers of narrow resonances lying on an infinitely rising Regge trajectory with daughter trajectories lying two units lower, satisfies the FESR, and has (almost) the correct asymptotic behaviour.

Finally, before leaving the question of bootstraps, we note that throughout this section, unitarity has been completely ignored. Relations like (5.8) then only provide linear relations between couplings, so that only ratios, and not magnitudes of these can be determined.

### 5.4   The Pomeron and exchange degeneracy. Harari's conjecture

In considering the circles produced on partial wave analysis of a Regge exchange, Schmid[57] noted that the Pomeron was qualitatively different from other Regge pole exchanges. This arises from the observation, elaborated by Chiu and Kotanski[64], that the circular behaviour originates in the signature factor

$$\frac{\mp 1 - e^{-i\pi\alpha(t)}}{\sin \pi\alpha(t)}.$$

Roughly, the values of $\cos\theta$ in the integration correspond to large values of $t$ as the energy increases, and if $\alpha$ varies as a function of $t$ as it does for most trajectories, the exponential factor leads to a rotation. However, the Pomeron appears to be rather flat with $\alpha(t) \simeq 1$, so that the amplitude is essentially purely imaginary, and there will be no circles. Having noted this, Schmid goes on to show how exchange degeneracy[65] leads to an understanding of the $K^+p$ and $K^-p$ scattering amplitudes. If the odd signature meson trajectories $X\,(\omega, \phi, \varrho,$ etc.) and the even signature trajectories $Y\,(A_2, f, f',$ etc.) are exchange degenerate, then the signature factor in $K^+p$ scattering

$$(1 - e^{-i\pi\alpha})_X - (-1 - e^{-i\pi\alpha})_Y = 2$$

is real, whereas that for $K^-p$

$$(1 - e^{-i\pi\alpha})_X + (-1 - e^{-i\pi\alpha})_Y = -2e^{-i\pi\alpha}$$

is complex. The Schmid analysis therefore predicts that there will be no resonance in $K^+p$, but many in $K^-p$, essentially in accordance with experiment.

Considerations such as the above, and the observation that most successes of resonance saturated FESR were for inelastic processes where the Pomeron did not contribute, led Harari[66] to conjecture that scattering amplitudes could be described by a resonance part from which the sloping trajectories are built; and a non-resonant background from which the Pomeron is built. Reversing the Schmid argument, the absence and presence of resonances in $K^+p$ and $K^-p$ scattering respectively implies exchange degeneracy, and the consequences which follow from this—for example the constancy of the $K^+p\,(NN, \pi^+\Sigma^+,$ etc.) total cross-sections down to low energies. It also guarantees that total cross-sections fall towards their high energy limit, since the resonance contributions can only add to the constant Pomeron contribution, again in agreement with present experimental trends. In a subsequent paper, Gilman, Harari and Zarmi[67] compared the model with $\pi N$ and $KN$ elastic scattering data. They find agreement with the model in the forward direction, and go on to derive the properties of the $P'$ trajectory for $t \neq 0$ from the model. However, this calculation has been criticised by Dance and Shaw[68], who for example use the better estimate of the Pomeron coupling discussed in section 4, and these authors claim that the forward $\pi N$ amplitude is in much better agreement with resonance dominance, than with the Harari prescription. Further, although Harari discussed the model only for imaginary parts, it is reasonable to consider its implications for real parts also. Here its failure is obvious: for example, since the Pomeron is purely imaginary, and there are no resonances, the real part of the $K^+p$ scattering amplitude is predicted to be zero. This is in contradiction to experiment.

### 5.5 Duality and the Deck effect

To close this section we note that Chew and Pignotti[69] have considered the implications of duality for multi-peripheral processes. They suggest that the Deck effect[70], instead of being regarded as an alternative explanation of bumps in, for example, the $\pi\varrho$ mass spectrum in the reaction

$$\pi + N \to \pi + \varrho + N$$

should be regarded as a prediction of resonance formation. We refer to their paper for a discussion of this topic.

## 6  SPECTRAL FUNCTION SUM RULES

As we pointed out in section 2, the most interesting applications of sum rules
from asymptotic symmetries have been to the spectral functions of the vector
and axial vector currents. These spectral function sum rules were first obtained by Weinberg[9], by use of the chiral algebra of currents and specific
hypotheses about the Schwinger terms*. However, here we shall discuss them
following the derivation of Das, Mathur and Okubo[28], which is closer to
the approach to sum rules adopted in the rest of this article.

We first define the propagator functions for the strangeness conserving
vector and axial vector currents as in reference[28]:

$$\Delta^V_{\mu\nu}(q) = \int d^4x \, e^{-iqx} \, \langle 0 \,|T\,\{V^i_{\mu j}(x), V^j_{\nu i}(0)\}|\, 0\rangle \qquad (6.1)$$

$$\Delta^A_{\mu\nu}(q) = \int d^4x \, e^{-iqx} \, \langle 0 \,|T\,\{A^i_{\mu j}(x), A^j_{\nu i}(0)\}|\, 0\rangle \qquad (6.2)$$

where $i$ and $j$ are the $SU(2)$ indices $(i, j = k, 2)$. If chiral $SU(2) \times SU(2)$ were
exact, we would have

$$\Delta^V_{\mu\nu}(q) - \Delta^A_{\mu\nu}(q) = 0 \qquad (6.3)$$

for all values of $q$. We know that in fact the symmetry is broken but it seems
reasonable to assume that (6.3) holds exactly in the limit $q \to \infty$

$$\underset{q \to \infty}{\text{Lt}} \, \{\Delta^V_{\mu\nu}(q) - \Delta^A_{\mu\nu}(q)\} = 0. \qquad (6.4)$$

The difference between the two propagator functions can be written in the
general form

$$\Delta^V_{\mu\nu}(q) - \Delta^A_{\mu\nu}(q) = F(q^2)\,\delta_{\mu\nu} + G\,(q^2)\,q_\mu q_\nu + H\delta_{\mu 4}\delta_{\nu 4} \qquad (6.5)$$

where $H$ is a constant related to the Schwinger terms. The asymptotic condition (6.4) implies that $F(q^2)$ satisfies an unsubtracted dispersion relation,
$G(q^2)$ a superconvergence relation, and that $H$ is zero. The last condition**
is equivalent to the assumption of the identity of the Schwinger terms, and is
of the same kind as the conditions required by Weinberg[9].

---

* Weinberg assumes that the Schwinger terms are either $c$ number, or, if operators,
contain no $\Delta I = 1$ terms. Alternatively, for the derivation of the sum rules one could use
only the algebra of gauge fields[71].

** This condition is implicitly assumed if one requires the superconvergence of $G(q^2)$.

The superconvergence of $G(q^2)$ leads to the first Weinberg sum rule:

$$\int_0^\infty \{\varrho_v'(m^2) - \varrho_A'(m^2)\} \frac{dm^2}{m^2} = f_\pi^2 \qquad (6.6)$$

where $\varrho_v, \varrho_A$ are the spectral functions appearing in the Källen–Lehmann representation of the propagators. That is

$$F(q^2) = -i \int_0^\infty \frac{\varrho_v(m^2) - \varrho_A(m^2)}{q^2 + m^2 - i\varepsilon} dm^2 \qquad (6.7)$$

and*

$$G(q^2) = -i \left[ \int_0^\infty \frac{\varrho_v'(m^2) - \varrho_A'(m^2)}{m^2 (q^2 + m^2 - i\varepsilon)} dm^2 - \frac{f_\pi^2}{q^2 + m_\pi^2 - i\varepsilon} \right]. \qquad (6.8)$$

Assuming a stronger requirement for $F(q^2)$, namely that $F(q^2)$ also super-converges, we obtain the second Weinberg sum rule:

$$\int_0^\infty \{\varrho_v(m^2) - \varrho_A(m^2)\} dm^2 = 0. \qquad (6.9)$$

If we now assume that the vector and axial-vector currents are dominated by the rho and $A_1$ mesons,

$$\varrho_v(m^2) = \varrho_v'(m^2) = g_\varrho^2 \delta (m^2 - m_\varrho^2)$$
$$\varrho_A(m^2) = \varrho_A'(m^2) - g_{A_1}^2 \delta (m^2 - m_{A_1}^2) \qquad (6.10)$$

then (6.6), (6.9) lead to the result

$$\frac{m_{A_1}}{m_\varrho} = \left( 1 - \frac{f_\pi^2 m_\varrho^2}{g_\varrho^2} \right)^{-1/2}.$$

Finally, combining this with the KSRF relation[72]

$$g_\varrho^2 = 2f_\pi^2 m_\varrho^2$$

we arrive at the famous mass relation

$$m_{A_1} = \sqrt{2}\, m_\varrho, \qquad (6.11)$$

in excellent agreement with experiment.

---

* The identity of the spectral functions $\varrho_v(m^2) = \varrho_v'(m^2), \varrho_A(m^2) = \varrho_A'(m^2)$, follows from het conservation of the vector, and partial conservation of the axial current.

Since the derivation of the sum rules (6.6), (6.9) is not based on any assumptions about conserved and partially conserved currents, it can be extended to chiral $SU(3) \times SU(3)$, when it leads to a relation similar to (4.11) between the masses of $K^*(890)$, and a strange axial vector meson[28]. It can also be extended to $SU(3)$ symmetry. For example the asymptotic condition

$$\underset{q \to \infty}{\mathrm{Lt}} \{ \Delta_{\mu\nu}^{(\pi)}(q) - \Delta_{\mu\nu}^{(K)}(q) \} = 0 \tag{6.12}$$

leads to the spectral sum rule

$$\int_0^\infty \frac{dm^2}{m^2} \{ \varrho^{(\pi)}(m^2) - \varrho^{(K)}(m^2) \} = 0 \tag{6.13}$$

and, again assuming vector dominance, a relation between the widths of $\varrho$ and $K^*(890)$ is obtained in good agreement with the experimental data[28]. However, equation (6.13) is of the type of Weinberg's first sum rule, equation (6.6). The corresponding second sum rule (6.9) would be

$$\int_0^\infty dm^2 \{ \varrho^{(\pi)}(m^2) - \varrho^{(K)}(m^2) \} = 0 \tag{6.14}$$

and would give the exact $SU(3)$ results $g_{K^*} = g_\varrho$, $m_{K^*} = m_\varrho$. In fact, the sum rules of the second kind are in general expected to be less reliable than those of the first kind, since they require the symmetry condition to be attained much faster, and Sakurai[73] has shown that, unless vector meson dominance is qualitatively incorrect, the second sum rule for vector current spectral functions must be abandoned.

Several other applications of these ideas can be found in the literature see for example ref.[74] One particularly interesting calculation, which we simply quote here, is the evaluation of the electromagnetic mass difference of pions[75] Using current algebra and $PCAC$, it is possible to express the mass difference in terms of the difference between the spectral functions $\varrho_V(m^2)$, $\varrho_A(m^2)$. Making use of Weinberg sum rules and of approximations similar to those discussed in the previous cases, the final result is*

$$m_{\pi^+}^2 + m_{\pi^0}^2 = \frac{3 \log_e^2}{2\pi} \left( \frac{e^2}{4\pi} \right) m_\varrho^2 . \tag{6.15}$$

---

* Equation (6.15) is obtained, however, in the limit of soft pions. For criticism on its extension to the case of on mass-shell pions we refer to ref.[76]

This gives $m_{\pi^+} - m_{\pi^0} \simeq 5.0$ MeV, to be compared with the experimental number 4.6 MeV.

## Acknowledgements

We would like to thank C.J.Goebels, C.Michael, R.J.N.Phillips and C. A.Savoy for many discussions. We are indebted to R.J.N.Phillips and to C.A.Savoy also for reading part of the manuscript.

## References


1. V. de Alfaro, S.Fubini, G.Furlan and C.Rossetti, *Phys. Lett.*, **21**, 576 (1966).
2. W.N.Cottingham, *Annals of Physics*, *N.Y.* , **25**, 424 (1963).
3. Y.C.Liu and S.Okubo, *Phys. Rev. Letters*, **19**, 190 (1967).
4. D.Horn and C.Schmid, California Institute of Technology Report No. CALT-68-127 (1967), unpublished.
5. R.Dolen, D.Horn and C.Schmid, *Phys. Rev. Lett.*, **19**, 402 (1968).
6. S.Mandelstam, *Phys. Rev.*, **166**, 1539 (1968).
7. G.Costa and A.H.Zimerman, *Nuovo Cimento*, **46A**, 198 (1966).
8. V.A.Matveev, B.V.Struminsky, A.N.Tavkhelidze, *Phys. Lett.*, **23**, 146 (1966).
9. S.Weinberg, *Phys. Rev. Letters*, **18**, 507 (1967).
10. S.Adler and R.Dashen, *Current Algebras*, Benajmin Inc. (1968).
11. K.Igi and S.Matsuda, *Phys. Rev. Lett.*, **18**, 625, 822 (1967); A.A.Logunov, L.D.Soloviev and A.N.Tavkhelidze, *Phys. Letters*, **24B**, 181 (1967).
12. L.D.Soloviev, *Proceedings of the International Conference on Particles and Fields, Rochester* (1967).
13. F.J.Gilman and H.Harari, *Phys. Rev.*, **165**, 1803 (1968).
14. A.Donnachie and J.Hamilton, *Phys. Rev.*, **138**, B678 (1965).
15. L.D.Soloview, *Soviet J. of Nuclear Phys.*, **3**, 131 (1966).
16. R.J.Eden, *High Energy Collisions of Elementary Particles*, p.192, Cambridge University Press (1967).
17. K.Igi, *Phys. Rev. Letters*, **9**, 76 (1962).
18. V.G.Kadyshevsky, R.M.Mir-Kasimov, A.N.Tavkhelidze, JINR Preprint P2 3198 Dubna (1967).
19. R.Gatto, *Phys. Rev. Lett.*, **18**, 803 (1967).
20. E.C.Titchmarsh, *The theory of functions*, Oxford University Press (1939) p.176.
21. J.H.Schwarz, *Phys. Rev.*, **159**, 1269 (1967).
22. C.E.Jones and V.L.Teplitz, *Phys. Rev.*, **159**, 1271 (1967); S.Mandelstam and L. L.Wang, *Phys. Rev.*, **160**, 1490 (1967).
23. R.Dolen, D.Horn and C.Schmid, *Phys. Rev.*, **166**, 1768 (1968).
24. R.Roskies, Weizman Institute of Science preprint (1968).
25. W.Gilbert, *Phys. Rev.*, **108**, 1078 (1957).
26. M.Olsson, *Phys. Letters*, **26B**, 310 (1968).
27. H.Pagels, *Phys. Rev. Lett.*, **18**, 316 (1967); R.Chanda, *Phys. Rev.*, **163**, 1627 (1967).
28. T.Das, V.S.Mathur and S.Okubo, *Phys. Rev. Lett.*, **18**, 761 (1967).

29. Lusignoli *et al.*, *Phys. Letters*, **21**, 229 (1966); G.H.Davies *et al.*, *Nuclear Phys.*, **B3**, (1967); H.P.C.Rood, *Nuovo Cim.*, **50A**, 493 (1967); J.K.Kim, *Phys. Rev. Letters*, **19**, 1079 (1967).

30. J.K.Kim, *Phys. Rev. Lett.*, **14**, 29 (1965).

31. J.K.Kim, *Phys. Rev. Lett.*, **19**, 1074 (1967).

32. A.D.Martin, G.G.Ross, *Phys. Letters*, **26B**, 527 (1968).

33. C.H.Chan and F.T.Meiere, *Phys. Rev. Lett.*, **20**, 568 (1968).

34. S.Goldhaber *et al.*, *Phys. Rev.*, **134**, B1111 (1964).

35. R.J.Abrams *et al.*, *Phys. Rev. Letters*, **19**, 259, 678 (1967); J.D.Davies *et al.*, *Phys. Rev. Letters*, **18**, 62 (1967).

36. W.N.Cottingham, *Annals of Phys. N.Y.*, **25**, 424 (1963).

37. H.Harari, *Phys. Rev. Letters*, **17**, 1303 (1966).

38. Y.Srivastava, *Phys. Rev. Letters*, **20**, 232 (1968); D.J.Gross and H.Pagels, *Phys. Rev.* **172**, 1635 (1968).

39. C.Michael, Rutherford Laboratory preprint (1968).

40. G.F.Chew, M.L.Goldberger, F.E.Low and Y.Nambu, *Phys. Rev.*, **106**, 1337 (1957).

41. B.M.Udgoankar, *Phys. Rev. Letters*, **8**, 142 (1962).

42. J.Hamilton and W.S.Woolcock, *Phys. Rev.*, **118**, 291 (1960).

43. W.Rarita, R.J.Riddel, C.B.Chiu and R.J.N.Phillips, *Phys. Rev.*, **165**, 1615 (1968).

44. K.Z.Foley *et al.*, *Phys. Rev. Letters*, **19**, 330 (1967).

45. V.Barger and R.J.N.Phillips, *Phys. Letters*, **26B**, 730 (1968).

46. A.Donnachie, R.G.Kirsopp and C.Lovelace, *Phys. Lett.*, 26B, 161 (1968).

47. V.Barger and R.J.N.Phillips, Report submitted to the XIV International Conference of High Energy Physics, Vienna (1968).

48. M.G.Olsson, Wisconsin preprint (1968).

49. A.Della Selva, Masperi and R.Odorico, *Nuovo Cimento*, **55A**, 602 (1968).

50. A.Della Selva, L.Masperi and R.Odorico, *Nuovo Cimento*, **54A**, 978 (1968).

51. C.Ferro-Fontan, M.Lusignoli, R.Odorico, R.Restignoli and G.Vislini, *Nuovo Cimento*, **57A**, 442 (1968); P.D.Vecchia, F.Drago and M.L.Paciello, *Phys. Letters*, **26B**, 530 (1968); G.V.Dass and C.Michael, *Phys. Rev. Letters*, **20**, 1066 (1968).

52. A.Bietti, P.Di Vecchia, F.Drago and M.L.Paciello, *Phys. Lett*, **26B**, 457, 736 (1968); P.Di Vecchia, F.Drago, C.Ferro-Fontàn, R.Odorico and M.L.Paciello, *Phys. Lett.*, **27B**, 296 (1968).

53. V.Barger and D.Cline, *Phys. Rev. Letters*, **16**, 913 (1966); *Phys. Rev.*, **155**, 1792 (1967).

54. V.Barger and M.G.Olsson, *Phys. Rev.*, **151**, 1123 (1966).

55. G.V.Dass and C.Michael, *Phys. Rev.*, **162**, 1403 (1967).

56. C.B.Chiu and A.V.Stirling, *Phys. Letters*, **26B**, 236 (1968).

57. C.Schmid, *Phys. Rev. Letters*, **20**, 688 (1968).

58. D.J.Gross, *Phys. Rev. Letters*, **19**, 1303 (1967).

59. M.Ademollo, H.R.Rubinstein, G.Veneziano and H.A.Virasoro, *Phys. Rev. Lett.*, **19**, 1402 (1967).

60. S.Matsuda, *Phys. Rev.*, **169**, 1169 (1968).

61. C.Schmid, *Phys. Rev. Letters*, **20**, 628 (1968).

62. H.R.Rubinstein, A.Schwimmer, G.Veneziano and M.A.Virasoro, *Phys. Rev. Letters*, **21**, 491 (1968).

63. G.Veneziano, *Nuovo Cimento*, **57A**, 190 (1968).

64. C.B.Chiu and Kotanski, *Nuclear Physics*, **B7**, 615 (1968).

65. R.C.Arnold, *Phys. Rev. Letters*, **14**, 657 (1965).

66. H.Harari, *Phys. Rev. Letters*, **20**, 1395 (1968).

67. F.J.Gilman, H.Harari and Y.Zarmi, *Phys. Rev. Letters*, **21**, 323 (1968).

68. D.R.Dance and G.Shaw, Rutherford Laboratory preprint (1968).

69. G.F.Chew and A.Pignotti, *Phys. Rev. Letters*, **20**, 1078 (1968).

70. R.T.Deck, *Phys. Rev. Letters*, **13**, 169 (1964).

71. T.D.Lee, S.Weinberg and B.Zumino, *Phys. Rev. Lett.*, **18**, 1029 (1967).

72. K.Kawarabayashi and H.Suzuki, *Phys. Rev. Lett.*, **16**, 255 (1966); Riazzudin and Fayyazuddin, *Phys. Rev.*, **147**, 1071 (1966).

73. J.J.Sakurai, *Phys. Rev. Letters*, **19**, 803 (1967).

74. S.L.Glashow, H.J.Schmitzer and S.Weinberg, *Phys. Rev. Letters*, **19**, 139 (1967); P.P.Divakaran and L.K.Pandit, *Phys. Rev. Lett.*, **19**, 535 (1967).

75. T.Das, G.S.Guralnik, V.S.Mathur, F.E.Low and J.E.Young, *Phys. Rev. Letters*, **18**, 759 (1967).

76. J.S.Bell, *Proceedings of the 1967 International Conference on Particles and Fields, Rochester (1967)*; W.I.Weisberger, *ibid*.

# Asymptotic Symmetry, Sum Rules, and Renormalization Constants

H. H. ALY

and

P. NARAYANASWAMY
*Southern Illinois University*

## Contents

## 1  INTRODUCTION

Elementary particles which are observed in nature exhibit certain internal symmetries which are however approximate, in so far as their mass differences could be neglected. Scattering amplitudes, propagators and vertex functions involving elementary particles thus admit of symmetries which are violated in nature. It is believed that at infinitely large energies or for large momenta, these symmetries would be valid since the mass differences are negligible. Thus one postulates the validity of symmetries in the asymptotic limit.

In particular the asymptotic symmetry applied to weak scattering amplitudes and weak propagators involving the vector and axial vector currents has a special bearing on the formulation and ideas of the methods of current algebra. It may even be conceivable that the asymptotic symmetry is equivalent to the current algebra approach.

The asymptotic symmetry hypothesis, combined with the powerful technique of dispersion relations leads to several interesting results, some of them being the same as current algebra predictions.

In this review we shall begin with the application of asymptotic $SU(2) \times SU(2)$ (chiral) symmetry to scattering amplitudes and derive the sum rules. This is the content of sections 2 and 3. In section 4, we briefly summarise the application of $SU(2) \times SU(2)$, $SU(3)$ and $SU(3) \times SU(3)$ symmetries to two point functions. In the same section we go on to consider special applications of these symmetries with a view to shed some light on the physical nature of the wave function renormalization constants of the pseudoscalar and vector mesons. Section 5 carries some general remarks and conclusion.

## 2  ASYMPTOTIC $SU(2) \times SU(2)$ SYMMETRY AND ITS APPLICATION TO $\pi N$ AND $\pi \pi$ SCATTERING SUM RULES

The application of current algebras to strong interaction physics consists of the derivation of useful sum rules and relations between the various strong interaction coupling constants. The most celebrated of these sum rules was the calculation of Adler (1) and Weisberger (2), who used the chiral current algebra (3) and the hypothesis of the partial conservation of the axial-vector current (PCAC). These authors derived a relation (AW sum rule) connecting the axial neutron beta decay constant to an integral over the $\pi N$ total cross-sections. This relation is so well in agreement with experiment that it is regarded as a direct evidence for the validity of the chiral current algebra.

In view of the last remark, it is interesting to consider an alternative derivation of the AW sum rule proposed by Fayyazuddin and Hussain (4). This derivation makes no explicit use of the chiral current algebra but is based on the hypothesis of asymptotic symmetry. The concept of asymptotic symmetry was originally suggested by Gell–Mann and Zachariasen (5) to describe the broken symmetry in nature. It is a statement expressing the validity of certain internal symmetries in the limit of infinite energy or infinite momentum where mass differences and other broken symmetry manifestations would be negligible. It is of course not clear as to which of the two is a weaker postulate since it is believed that asymptotic symmetry could be derived if the equal time commutation relations are assumed (6, 7). For our purposes, let us regard the asymptotic symmetry as an alternative or independent hypothesis. This view point is certainly useful as we intend to discuss the derivation of as many consequences of it as possible. We also consider various asymptotic symmetries other than the chiral $SU(2) \times SU(2)$, as we go along. Besides, the FH derivation of the AW sum rule has several interesting and additional features as we will see later.

The basic assumption of this approach is the equality of the weak vector-nucleon and axial-vector nucleon amplitudes in the asymptotic limit when the energy is large. The weak amplitudes in question are defined as

$$M_{\mu\nu}^{(V)\,i,\,j}(\nu, q^2) = i \int d^4x \, e^{-iq \cdot x} \, \theta(x_0) \, \langle N(p) \, |[V_\mu^i(x), V_\nu^j(0)]| \, N(p) \rangle \quad (2.1)$$

$$M_{\mu\nu}^{(A)\,i,\,j}(\nu, q^2) = i \int d^4x \, e^{-iq \cdot x} \, \theta(x_0) \, \langle N(p) \, |[A_\mu^i(x), A_\nu^j(0)]| \, N(p) \rangle \quad (2.2)$$

describing the 'scattering' (see Fig. 1) of a vector (axial-vector) current with momentum $q$ off a nucleon with momentum $p$. We have specialised to the forward direction, the initial and final momenta thus being the same. The variable $\nu$, related to the centre of mass energy (8), is $\nu = \dfrac{-p \cdot q}{m}$ where $m$ is nucleon mass. $i, j = 1, 2, 3$ are iso-spin indices. Here $q^2$ will be regarded as a variable and we will later specialize to the case $q^2 = 0$. Asymptotic $SU(2) \times SU(2)$ symmetry is signified by

$$\operatorname*{Lim}_{\nu \to \infty} [M_{\mu\nu}^{(V)\,i,\,j} - M_{\mu\nu}^{(A)\,i,\,j}] \to 0 \quad (2.3)$$

which implies corresponding asymptotic behaviour for the amplitudes occurring in the invariant decomposition of the difference

$$M_{\mu\nu}^{i,\,j}(\nu, q^2) \equiv M_{\mu\nu}^{(V)\,i,\,j}(\nu, q^2) - M_{\mu\nu}^{(A)\,i,\,j}(\nu, q^2) \quad (2.4)$$

Each of the invariant amplitudes will be assumed to satisfy an unsubtracted dispersion relation. As a consequence, we may be able to write superconvergent dispersion relations for one or more of the invariant amplitudes in the standard manner (9), which can be converted into sum rules by making use of the PCAC hypothesis. Let us briefly review the method of FH.

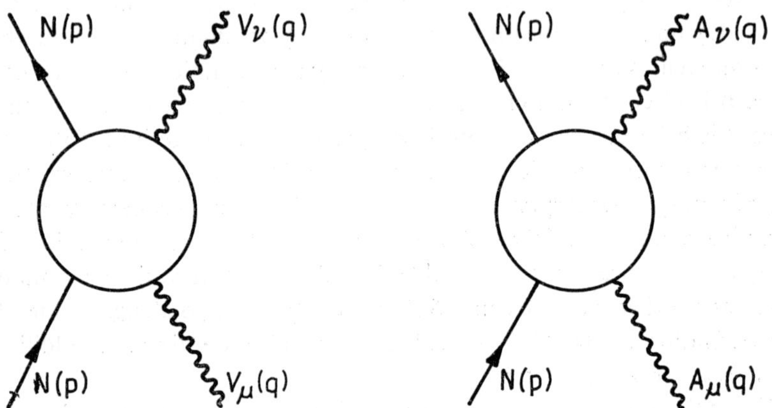


**Figure 1**   Diagrammatic representation of the weak amplitudes $M_{\mu\nu}^{(V)}, M_{\mu\nu}^{(A)}$.

Consider the invariant decomposition

$$M_{\mu\nu}^{i,j} = p_\mu p_\nu A (\nu, q^2) + (p_\mu q_\nu + q_\mu p_\nu) B (\nu, q^2) \tag{2.5}$$

$$+ q_\mu q_\nu C (\nu, q^2) + \delta_{\mu\nu} D (\nu, q^2)$$

where the spins of the nucleons have been summed over. We also have the iso-spin decomposition:

$$M_{\mu\nu}^{i,j} = \delta_{ij} M_{\mu\nu}^{+} + \tfrac{1}{2} [\tau_i, \tau_j] M_{\mu\nu}^{-} \tag{2.6}$$

The asymptotic symmetry assumed in Eq. (2.3) implies

$$\underset{\nu \to \infty}{\mathrm{Lim}} \quad [\nu A (\nu, q^2)] \to 0 \tag{2.7}$$

and it is also observed that $A^- (\nu, q^2)$ is odd under crossing:

$$A^- (\nu, q^2) = - A^- (-\nu, q^2) \tag{2.8}$$

We can thus write the following superconvergent relation:

$$\int \text{Im } A^- (v, q^2) \, dv + \text{poles} = 0. \tag{2.9}$$

Remembering that $M_{\mu v}^{(V)}$ and $M_{\mu v}^{(A)}$ have decompositions similar to Eq. (2.5), we have

$$\int \text{Im } A^{(V)-} (v, q^2) \, dv - \int \text{Im } A^{(A)-} (v, q^2) \, dv + (\text{poles})_V - (\text{poles})_A = 0 \tag{2.10}$$

where the poles are due to the nucleon intermediate states in Eqs. (2.1) and (2.2), because these are precisely the poles which were separated before writing the unsubtracted dispersion relations.

The absorptive part of the vector-nucleon amplitude gives no contributipn in Eq. (2.10). This is due to the conserved vector current hypothesis (CVC) according to which

$$q_\mu \text{ Im } M_{\mu v}^{(V)} = 0 \tag{2.11}$$

which at $q^2 = 0$ gives

$$\text{Im } A^{(V)} = 0 \qquad (v \neq 0) \tag{2.12}$$

PCAC can now be used to relate the absorptive part of axial-vector nucleon amplitude to $\pi N$ cross-sections. PCAC tells us that

$$q_\mu q_v \text{ Im } M_{\mu v}^{(A)} = F_\pi^2 \frac{m_\pi^4}{(q^2 + m_\pi^2)^2} \text{ Im } T_{\pi N} (v, q^2) \tag{2.13}$$

Here, $m_\pi$ is the pion mass, $F_\pi$ is the pion decay constant and PCAC has been used in the form

$$\partial_\mu A_\mu^i(x) = F_\pi m_\pi^2 \phi_\pi^i (x) \tag{2.14}$$

The physical pion-nucleon forward scattering amplitude occurring in Eq. (2.13) is defined by

$$T_{\pi N}^{i,j} = i \int d^4x \, e^{-ip \cdot x} \, \theta(x_0) \langle N(p) \, |[J_\pi^i, J_\pi^j]| \, N(p) \rangle \tag{2.15}$$

and has the iso-spin decomposition

$$T_{\pi N}^{i,j} = \delta_{ij} T_{\pi N}^+ + \tfrac{1}{2} [\tau_i, \tau_j] T_{\pi N}^-. \tag{2.16}$$

We specialize to the case of $q^2 = 0$ which corresponds to zeromass pions in Eq. (2.15). The reason is that the AW sum rule is a relation involving zeromass pions. This should be distinguished from the soft pion for which the four momentum $q$ itself vanishes. The method of FH does not resort to the use of soft pion technique usually used in current algebra derivations. Thus all kinds of ambiguities relating to the soft pion limit are avoided.

Finally we consider the pole contributions. To evaluate these, we need the vertex functions:

$$\langle p' \,|A_\mu^i(0)|\, p\rangle = i\bar{u}(p') \,[\gamma_5\gamma_\mu G_A\,((p'-p)^2)$$

$$+ \gamma_5\,(p'-p)_\mu f\,((p'-p)^2)]\,\tau^i u(p)$$

$$\langle p' \,|V_\mu^i(0)|\, p\rangle = \bar{u}(p') \,[\gamma_\mu F_1\,((p'-p)^2)$$

$$+ i\,(p'-p)_\nu\,\sigma_{\mu\nu}F_2\,((p'-p)^2)]\,\tau^i u(p).$$

$$\tag{2.17}$$

$F_1$ is the nucleon charge form factor and $G_A(0) = G_A$ is the axial $\beta$-decay constant. The other form factors will not contribute to the pole terms at $q^2 = 0$. It is easy to evaluate the pole contributions in the limit $\nu \to \infty$, both for $M_{\mu\nu}^{(V)}$ and $M_{\mu\nu}^{(A)}$. We are interested only in the coefficients of $p_\mu p_\nu$ in the pole terms and we get

$$(\text{pole})_{AV} = \frac{1}{m^3}\,F_1^2(0)$$

$$(\text{pole})_{AA} = \frac{1}{m^3}\,G_A^2.$$

$$\tag{2.18}$$

Combining this with Eqs. (2.13), (2.11) and (2.10), the $AW$ sum rule follows:

$$G_A^2 = F_1^2(0) - \frac{4}{\pi}\,F_\pi^2 \int \frac{\text{Im }T_{\pi N}^-\,(\nu, 0)}{\nu^2}\,d\nu, \tag{2.19}$$

valid for the iso-spin flip $\pi N$ scattering amplitude and for zeromass pions.

At this stage it is worthwhile pointing out some of the interesting features of the FH derivation of the sum rule, Eq. (2.19). These are 1) the sum rule is a consequence of asymptotic $SU(2) \times SU(2)$ symmetry (chiral symmetry) of the weak amplitudes and no use of current algebra is made: 2) the soft pion limit is not needed unlike in the usual current algebra derivations; 3) the sum rule is derived for zeromass pions but the method is not restricted to $q^2 = 0$. In fact the FH calculation discusses the case $q^2 \neq 0$ and leads to a new sum rule (10) for $G_A(q^2)$ and 4) the spin flip sum rule due to Gerstein (11) can also be derived.

We shall now go on to derive the sum rule for pion-pion scattering, originally derived by Adler (1) using current algebra, now by the method of FH with a few modifications.

We begin with the weak pion-vector and pion-axial vector amplitudes, analogues of Eqs. (2.1) and (2.2) where the matrix elements are now taken between pion states of momenta $p$. We shall invoke asymptotic $SU(2) \times SU(2)$ symmetry as before. There are no $\gamma$-matrices involved and the invariant decompositions would be given by Eqs. (2.5) and (2.6). To distinguish it from the $\pi N$ case, let us use different symbols and write

$$M_{\mu\nu}^{i,j} (\nu, q^2) = p_\mu p_\nu A_1 (\nu, q^2) + (p_\mu q_\nu + q_\mu p_\nu) A_2 (\nu, q^2) + q_\mu q_\nu A_3 (\nu, q^2)$$

$$+ \delta_{\mu\nu} A_4 (\nu, q^2). \tag{2.20}$$

We note that the amplitude $A_1^- (\nu, q^2)$ which is relevant for the superconvergence relation to be derived, is odd under crossing ($\nu \to -\nu$).

First we observe that $M_{\mu\nu}^{(V)-}$ has to satisfy the following divergence condition, as a consequence of CVC:

$$q_\mu M_{\mu\nu}^{(V)-} = \frac{1}{m_\pi} p_\nu. \tag{2.21}$$

This result follows when we explicitly assume the equal time commutation relation for the vector currents. Of course we still do not make use of the full chiral algebra. In contrast to the $\pi N$ case, it should be stressed that the Adler–Fubuni sum rule (9, 12) stated in what follows is thus an additional assumption required to derive the $\pi\pi$ scattering sum rule. Eqs. (2.20) and (2.21) together imply the following relation between the iso-spin flip invariant amplitudes

$$A_1^{(V)-} (\nu, q^2) = - \frac{1}{m_\pi^2 \nu} + \frac{q^2}{m_\pi \nu} A_2^{(V)-} (\nu, q^2). \tag{2.22}$$

If $A_2^{(V)} (\nu, q^2)$ is assumed to fall off as $\nu \to \infty$ and $A_1^{(V)} (\nu, q^2)$ satisfies an unsubtracted dispersion relation, we obtain the Adler–Fubini sum rule (9, 12):

$$\frac{2}{\pi} \int \operatorname{Im} A_1^{(V)-} (\nu, q^2) \, d\nu = \frac{1}{m_\pi^2}. \tag{2.23}$$

The asymptotic symmetry which requires the vanishing of $M_{\mu\nu}^{i,j}$ in the limit $\nu \to \infty$ implies a superconvergent relation for $A_1^- (\nu, q^2)$. Together with Eq. (2.23), we thus derive

$$\frac{2}{\pi} \int \operatorname{Im} A^{(A)-} (\nu, q^2) \, d\nu = \frac{1}{m_\pi^2} \tag{2.24}$$

since there are no pole terms. The contributions due to $\varrho$ and $A_1$ intermediate states are included in the continuum. Eq. (2.24) can now be expressed in terms of $\pi\pi$ total cross-sections. By invoking PCAC and relating the absorptive part in Eq. (2.24) to the $\pi\pi$ scattering amplitude

$$T_{\pi\pi}^{i,j}(v, q^2) = i \int d^4x \, e^{-ip \cdot x} \, \theta(x_0) \, \langle \pi(p)| \, [J_\pi^i, J_\pi^j] \, |\pi(p)\rangle \qquad (2.25)$$

we derive the following sum rule for zeromass pions:

$$\frac{2}{\pi} F_\pi^2 \int \frac{\mathrm{Im}\, T_{\pi\pi}^-(v, 0)}{v^2} \, dv = 1 . \qquad (2.26)$$

This is Adler's pion-pion sum rule which can be rewritten in the more familiar form by employing the Goldberger–Trieman relation:

$$\frac{4m_N^2}{g_{\pi N}^2} \frac{1}{\pi} \int \frac{\mathrm{Im}\, T_{\pi\pi}^-(v, 0)}{v^2} \, dv = \frac{2}{G_A^2} . \qquad (2.27)$$

The absorptive part is related to the $\pi\pi$ total cross-section via the optical theorem.

It should be mentioned here that the sum rule in Eq. (2.27) is not in agreement with experiment. Adler (1) saturated the integral by $\varrho$ and $f^0$ contributions and finds the left hand side to be 0.53 whereas the right hand side is 1.43. Unless there is a *large* contribution from a hypothetical $S$-wave resonance, Eq. (2.27) is not supported by experiment (13).

## 3  ASYMPTOTIC SYMMETRY REQUIREMENT
## FOR THE PROPER AMPLITUDES

We shall proceed to examine the consequences of asymptotic symmetry hypothesis for the 'proper' amplitudes, dealing with the same examples as before i.e., $\pi\pi$ and $\pi N$ sum rules. The "proper" amplitudes are the scattering amplitudes involving physical particles and it is thus conceivable that the asymptotic symmetry hypothesis is equally valid for these as for the weak amplitudes defined in terms of matrix elements of vector and axial vector currents. As the weak and the proper amplitudes are simply related through two-point functions, a study of the asymptotic requirement for the proper amplitudes will lead to new sum rules. As we shall see later, this study will also throw more light on the asymptotic behaviour of the weak amplitudes.

It was recently proposed by Barry, Gounaris and Sakurai (14) that the asymptotic symmetry requirement for the proper amplitudes can be used to obtain a modification of the Adler–Fubini sum rule (9, 12). The new sum

rule that follows was taken to be an indication for the validity of this hypothesis. This hypothesis may not necessarily be valid in general, as we shall see
later. Nevertheless it is instructive to examine its consequences.

The proper amplitudes are introduced by the definitions

$$M_{\mu\nu}^{(V)i,j}(\nu, q^2) = \Delta_{\mu\lambda}^{V}(q)\,\Delta_{\nu\sigma}^{V}(q)\,T_{\lambda\sigma}^{(V)i,j}(\nu, q^2)$$

$$\bar{M}_{\mu\lambda}^{(A)i,j}(\nu, q^2) = \Delta_{\mu\lambda}^{A}(q)\,\Delta_{\nu\sigma}^{A}(q)\,T_{\lambda\sigma}^{(A)i,j}(\nu, q^2)$$

(3.1)

which serve to isolate the spin one (vector and axial-vector) poles from the
weak amplitudes $M_{\mu\nu}^{(V),\,(A)}$. Such a separation of poles was used by Schnitzer
and Weinberg in a different context (15) and here we have merely generalized
it. Figure 2 illustrates the procedure. $T_{\mu\nu}^{(V),\,(A)}$ denote the proper amplitudes.

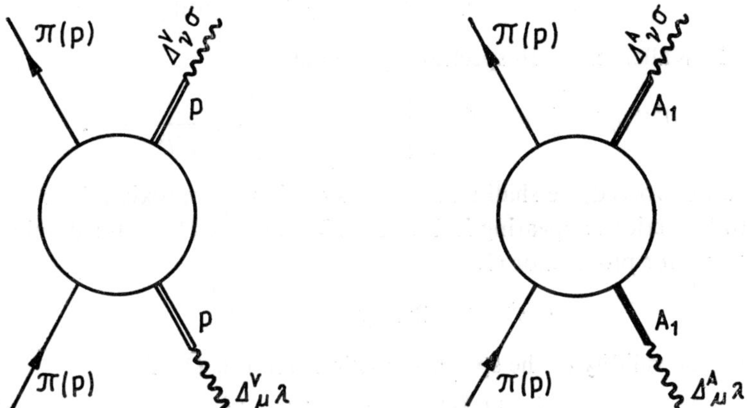


**Figure 2**   Diagram describing the connection between $M_{\mu\nu}^{(V),(A)}$ and $T_{\mu\nu}^{(V),(A)}$.

We shall first take up the $\pi\pi$ case so that $T_{\mu\nu}^{(V)}$ and $T_{\mu\nu}^{(A)}$ stand for the physical $\pi\varrho$ and $\pi A_1$ scattering amplitudes respectively (except for multiplicative
constants $m_\varrho^4$ and $m_{A_1}^4$). By requiring asymptotic $SU(2) \times SU(2)$ for the amplitudes $T_{\mu\nu}$ we shall derive a modified $\pi\pi$ sum rule, a modification of the well
known Adler sum rule. Later we will investigate the $\pi N$ case.

### 3.1   The pion-pion sum rule

In Eq. (3.1), $\bar{M}_{\mu\nu}^{(A)}$ stands for

$$\bar{M}_{\mu\nu}^{(A)i,j}(\nu, q^2) = i \int d^4x\, e^{-iq\cdot x}\theta(x_0)\,\langle\pi(p)|\,[\bar{A}_\mu^i(x), \bar{A}_\nu^j(0)]\,|\pi(p)\rangle \quad (3.2)$$

where

$$\bar{A}_\mu^i(x) = A_\mu^i(x) + iq_\mu m_\pi^{-2}\,(\partial_\lambda A_\lambda^i(x)). \quad (3.3)$$

Note that $\bar{A}_\mu^i(x)$ is a conserved current (by construction) which guarantees that the spin zero part (namely $\partial_\mu A_\mu$) is subtracted and only the conserved part is used in extracting the proper amplitude. It is easy to derive the following useful relation:

$$
\begin{aligned}
M_{\mu\nu}^{(A)i,j} = \; & \bar{M}_{\mu\nu}^{(A)i,j} + iq_\mu q_\nu F_\pi^2 \int d^4x \, e^{-iq\cdot x} \theta(x_0) \, \langle \pi(p)| \, [\phi^i(x), \phi^j(0)] \, |\pi(p)\rangle \\
& + q_\mu F_\pi \int d^4x \, e^{-iq\cdot x} \theta(x_0) \, \langle \pi(p)| \, [\phi^i(x), A_\nu^j(0)] \, |\pi(p)\rangle \\
& + q_\nu F_\pi \int d^4x \, e^{-iq\cdot x} \theta(x_0) \, \langle \pi(p)| \, [A_\mu^i(x), \phi^j(0)] \, |\pi(p)\rangle .
\end{aligned}
\tag{3.4}
$$

The covariant propagators of iso-spin current density occurring in Eq. (3.1)

$$
\Delta_{\mu\nu}^V(q) = \int d^4x \, e^{-iq\cdot X} \langle 0| \, T\,(V_\mu^k(x)\, V_\nu^k(0)) \, |0\rangle
$$

$$
\Delta_{\mu\nu}^A(q) = \int d^4x \, e^{-iq\cdot X} \langle 0| \, T\,(A_\mu^k(x)\, A_\nu^k(0)) \, |0\rangle
\tag{3.5}
$$

have the Källen–Lehman spectral representations

$$
\Delta_{\mu\nu}^{V;A}(q) = \int de^2 \varrho^{V,A}(p^2) \left[ \delta_{\mu\nu} + \frac{q_\mu q_\nu}{l^2} \right] \frac{1}{q^2 + l^2 - i\varepsilon} .
\tag{3.6}
$$

Before we proceed, we shall make some simplifying approximations for the spectral functions appearing in Eq. (3.6). We assume (1) vector dominance for the vector propagator (16)

$$
\varrho^V(l^2) \approx 2m_\varrho^2 F_\pi^2 \delta\,(l^2 - m_\varrho^2)
\tag{3.7}
$$

and (2) the validity of the spectral function sum rule (17–20)

$$
\int dl^2 \, [\varrho^V(l^2) - \varrho^A(l^2)]/l^2 = F_\pi^2 .
\tag{3.8}
$$

We shall need these relations later.

We now invoke asymptotic $SU(2) \times SU(2)$ symmetry on the proper amplitudes expressed by the statement

$$
\lim_{\nu \to \infty} T_{\mu\nu}^{i,j}(\nu, q^2) \to 0
\tag{3.9}
$$

where

$$
T_{\mu\nu}^{i,j}(\nu, q^2) \equiv T_{\mu\nu}^{(V)i,j} - T_{\mu\nu}^{(A)i,j}
\tag{3.10}
$$

and this implies the equality of the physical $\pi\varrho$ and $\pi A_1$ scattering amplitudes as $\nu \to \infty$.

Let us multiply Eq. (3.1) by $q_\mu q_\nu$ and subtract to obtain

$$
\begin{aligned}
q_\mu q_\nu \, [M_{\mu\nu}^{(V)i,j} - M_{\mu\nu}^{(A)i,j}] = \; & q_\mu q_\nu \, \Delta_{\mu\lambda}^V \Delta_{\nu\sigma}^V \, [T_{\lambda\sigma}^{(V)i,j} - T_{\lambda\sigma}^{(A)i,j}] \\
& + F_\pi^2 \, \{ q_\lambda q_\nu \Delta_{\nu\sigma}^V + q_\sigma q_\mu \Delta_{\mu\lambda}^V - F_\pi^2 q_\sigma q_\lambda \} \, T_{\lambda\sigma}^{(A)i,j}
\end{aligned}
\tag{3.11}
$$

This relation has been derived after setting $q^2 = 0$ (which is the case for which the superconvergent relations are going to be used) which allows us to make use of Eqs. (3.7) and (3.8). It should be stressed here that the "bar" sign does not appear on the amplitude $M_{\mu\nu}^{(A)}$ in Eq. (3.1) due to the fact that the following relation is valid at $q^2 = 0$:

$$q_\mu q_\nu \bar{M}_{\mu\nu}^{(A)i,j}(\nu, q^2) = q_\mu q_\nu M_{\mu\nu}^{(A)i,j}(\nu, q^2). \tag{3.12}$$

To prove this, one multiplies Eq. (3.4) by $q_\mu q_\nu$ and carries out integration by parts. Of the several terms that arise, the "$\sigma$"-type equal time commutators cancel each other if we assume (21)

$$\delta(x_0)\, [A_0^i(x), \phi^j(0)] = \sigma^{ij}(x)\, \delta^4(x) \tag{3.13}$$

where $\sigma^{ij} = \sigma^{ji}$, thus arriving at

$$q_\mu q_\nu\, (\bar{M}_{\mu\nu}^{(A)i,j} - M_{\mu\nu}^{(A)i,j})$$
$$= iq^4 F_\pi^2 \int d^4x\, e^{-iq\cdot x}\theta(x_0)\, \langle p|\, [\phi^i(x), \phi^j(0)]\, |p\rangle, \tag{3.14}$$

the right hand side of which is evidently zero since the integrand does not become singular like $1/q^4$ in the limit when $q^2 = 0$. Hence Eq. (3.12) follows.

Let us now proceed to seek a relation between the invariant amplitudes of $M_{\mu\nu}$ and those of $T_{\mu\nu}$, following the method of ref. 14. $T_{\mu\nu}$ can be decomposed as

$$T_{\mu\nu}\,(\nu, q^2) = p_\mu p_\nu a_1\,(\nu, q^2) + (p_\mu q_\nu + q_\mu p_\nu)\, a_2\,(\nu, q^2)$$
$$+ q_\mu q_\nu a_3\,(\nu, q^2) + \delta_{\mu\nu} a_4\,(\nu, q^2), \tag{3.15}$$

to be compared with

$$M_{\mu\nu}\,(\nu, q^2) = p_\mu p_\nu A_1\,(\nu, q^2) + (p_\mu q_\nu + q_\mu p_\nu)\, A_2\,(\nu, q^2)$$
$$+ q_\mu q_\nu A_3\,(\nu, q^2) + \delta_{\mu\nu} A_4\,(\nu, q^2). \tag{3.16}$$

Similar decompositions hold for $T_{\mu\nu}^{(V)}$, $T_{\mu\nu}^{(A)}$ and $M_{\mu\nu}^{(V)}$, $M_{\mu\nu}^{(A)}$ with $a_i^{(V),\,(A)}$ and $A_i^{(V),\,(A)}$. Combining Eqs. (3.11), (3.15) and (3.16) yields, at $q^2 = 0$

$$A_1 = 4F_\pi^4 a_1 + 3F_\pi^4 a_1^{(A)}. \tag{3.17}$$

Our basic assumption, Eq. (3.9) implies a superconvergence relation for the iso-spin flip amplitude $a_i^{(-)}$. We write this as:

$$\frac{2}{\pi} \int [\mathrm{Im}\, a^{(V)-}\,(\nu, q^2) - \mathrm{Im}\, a^{(A)-}(\nu, q^2)]\, d\nu = 0. \tag{3.18}$$

By virtue of Eqs. (3.17) and (2.23) we thus get

$$\frac{2}{\pi} [\mathrm{Im}\, A^{(A)-}(\nu, q^2) + 3F_\pi^4\, \mathrm{Im}\, a^{(A)-}(\nu, q^2)]\, d\nu = \frac{1}{m_\pi^2}. \qquad (3.19)$$

Employing PCAC, we can relate $\mathrm{Im}\, a^{(A)}$ to the absorptive part of $T_{\pi\pi}(\nu, q^2)$ in exact analogy to what was done in Sec. 1.

$$\mathrm{Im}\, a^{(A)-}(\nu, q^2) = \frac{1}{m^2 \nu^2 F_\pi^2}\, \mathrm{Im}\, T_{\pi\pi}^-(\nu, q^2). \qquad (3.20)$$

The sum rule for the iso-spin flip pion-pion scattering follows (one pair of pions have zero mass):

$$\frac{8F_\pi^2}{\pi} \int \frac{\mathrm{Im}\, T_{\pi\pi}^-(\nu, 0)}{\nu^2}\, d\nu = 1. \qquad (3.21)$$

This should be contrasted with Eq. (2.26), previously derived by Adler. As was pointed out earlier, Eq. (2.26) is not in agreement with experiment. On the other hand, if we use Adler's estimate for the integral in Eq. (3.21) the left hand side has the value $\dfrac{4 \times 0.53}{1.43}$, thus bringing it to a better agreement.

This would imply that our basic assumption of asymptotic symmetry for the proper amplitudes is more reasonable than that of asymptotic symmetry for the weak amplitudes.

### 3.2   The pion-nucleon sum rule

The picture which holds for the $\pi\pi$ case does not give any substance in the case of the $\pi N$ sum rule. This is due to the fact that the AW sum rule is in very good agreement with experiment and therefore any modified AW sum rule would be inconsistent. Contrary to the $\pi\pi$ case we will be able to show that in the $\pi N$ case, the asymptotic symmetry hypothesis for the proper amplitudes leads to an inconsistency (22).

The procedure and details of calculation are the same as before. For the sake of completeness we shall outline the essential steps.

The asymptotic $SU(2) \times SU(2)$ for the proper amplitudes (which stand for the physical $N\varrho$ and $NA_1$ scattering amplitudes) implies a superconvergence relation for the invariant amplitude $a^{(-)}(\nu, q^2)$ which is the coefficient of $p_\mu p_\nu$ in the decomposition of $T_{\mu\nu}(\nu, q^2)$:

$$\frac{2}{\pi} \int \mathrm{Im}\, a^-(\nu, 0)\, d\nu + (\text{poles})_a{}^V - (\text{poles})_a{}^A = 0. \qquad (3.22)$$

Following the previous procedure, we arrive at the sum rule

$$G_A^2(0) - F_1^2(0) + 3m^2 F_\pi^4 \, (\text{poles})_{aA} + \frac{16F_\pi^2}{\pi} \int \frac{\text{Im } T_{\pi N}^-(\nu, 0)}{\nu^2} \, d\nu = 0. \quad (3.23)$$

where the pole term can be evaluated to be (nucleon intermediate state):

$$(\text{pole})_{aA} = \frac{1}{m^2 m_{A_1}^4} \, g_{NNA_1}^2. \quad (3.24)$$

It is easy to see that the sum rule, Eq. (3.23) is in contradiction with the well established AW sum rule, Eq. (2.19). By combining Eq. (2.19) and (3.23) we obtain

$$\frac{g_{NNA_1}^2}{4\pi} = \frac{1}{4\pi} \left( \frac{m_{A_1}}{F_\pi} \right)^4 \{G_A(0) - F_1^2(0)\}, \quad (3.25)$$

which can never be satisfied since it would imply too large a coupling constant $g_{NNA_1}$.

Such an inconsistency was realised for a special case by Gerstein (23) who considered the modification of the Adler–Fubini sum rule (9, 12). Referring to Gerstein's analysis, it should be stressed that the Adler-Fubini sum rule has *no* modification if we set $q^2 = 0$ in the sum rule derived in ref. 14. This is indeed the point of view we have taken in our derivation of the pion-pion sum rules Eqs. (2.26) and (3.21) where the Adler–Fubini sum rule plays a crucial role in the calculations. This should be contrasted with the situation in $\pi N$ case where the $AW$ sum rule *does* in fact get modified even at $q^2 = 0$, if the proper amplitudes satisfy asymptotic $SU(2) \times SU(2)$ symmetry. The inconsistency pointed out by Gerstein (23) would thus persist even at $q^2 = q'^2 = 0$.

## 4  APPLICATIONS OF ASYMPTOTIC SYMMETRY TO TWO-POINT FUNCTIONS

In this section we review the application of asymptotic symmetry to two-point functions of the form

$$\Delta_{\mu\nu}^{J,i}(q) = \int d^4x \, e^{-iq \cdot x} \, \langle 0| \, T \, (J_\mu^i(x), J_\nu^i(0)) \, |0\rangle, \quad (4.1)$$

where $J_\mu$ could be a vector or an axial-vector current and $i$ denotes $SU(2)$ or $SU(3)$ indices.

### 4.1  Spectral function sum rules

We shall closely follow the elegant method employed by Das, Mathur and Okubo (DMO) (20) who derived spectral function sum rules (17–20) from the requirement of asymptotic symmetry for the two-point functions. Let us consider the special cases of Eq. (4), namely the weak propagators:

$$\Delta^V_{\mu\nu}(q) = \int d^4x \, e^{-iq\cdot x} \langle 0| \, T\,(V^i_\mu(x),\, V^i_\nu(0))\, |0\rangle \tag{4.2}$$

$$\Delta^A_{\mu\nu}(q) = \int d^4x \, e^{-iq\cdot x} \langle 0| \, T\,(A^i_\mu(x),\, A^i_\nu(0))\, |0\rangle \tag{4.3}$$

where $i = 1, 2, 3$ denotes $SU(2)$ indices (iso-spin). Invoking asymptotic $SU(2)$ on these two-point functions, one expects that in the limit of large $q$:

$$\operatorname*{Lim}_{q\to\infty} [\Delta^V_{\mu\nu}(q) - \Delta^A_{\mu\nu}(q)] \to 0. \tag{4.4}$$

The limiting procedure is usually understood to be $q_0 \to \infty$, keeping $\mathbf{q}$ fixed (Bjorken limit (6)) but has a covariant generalization (24).

Writing the general decomposition

$$[\Delta^V_{\mu\nu}(q) - \Delta^A_{\mu\nu}(q)] = F(q^2)\,\delta_{\mu\nu} + G(q^2)\,q_\mu q_\nu + S\delta_{\mu 4}\delta_{\nu 4}, \tag{4.5}$$

where $S$ is the contribution from the Schwinger terms, it is observed that $F(q^2)$ should satisfy an unsubtracted dispersion relation while $G(q^2)$ should satisfy a superconvergent dispersion relation; the Schwinger terms must be equal for the vector and axial vector propagators.

With the help of Källén–Lehman representation, we can express $F(q^2)$ and $G(q^2)$ as

$$F(q^2) = -i \int_0^\infty \frac{\varrho^V(\mu^2) - \varrho^A(\mu^2)}{q^2 + \mu^2 - i\varepsilon}\, d\mu^2 \tag{4.6}$$

$$G(q^2) = -i \int_0^\infty d\mu^2 \, \frac{[(\varrho^V \mu^2) - \varrho^A(\mu^2)]}{\mu^2\,(q^2 + \mu^2 - i\varepsilon)} + \frac{iF_\pi^2}{q^2 + m_\pi^2 - i\varepsilon} \tag{4.7}$$

where we have explicity separated the contribution due to the pion pole in the axial-vector propagator. These steps immediately lead to the two spectral function sum rules originally derived by Weinberg and others (17–20)

$$\int d\mu^2 \, \frac{[\varrho^V(\mu^2) - \varrho^A(\mu^2)]}{\mu^2_{\;1}} = F_\pi^2 \tag{4.8}$$

$$\int d\mu^2 \, [\varrho^V(\mu^2) - \varrho^A(\mu^2)] = 0. \tag{4.9}$$

As was pointed by DMO (20) one has to know beyond what value of $q$ is Eq. (4.1) valid i.e., how fast does the difference vanish as $q \to \infty$. The second sum rule requires a convergence faster than $1/q^2$. In the context of these two sum rules it appears, from saturation by vector and axial vector mesons, that the second sum rule is not well satisfied (25).

Das Mathur and Okubo also generalized their results to $SU(3)$ by invoking asymptotic $SU(3)$ symmetry in the form

$$\lim_{q \to \infty} [\Delta_{\mu\nu}^{V,\pi}(q) - \Delta_{\mu\nu}^{V,K}(q)] \to 0 \tag{4.10}$$

where

$$\Delta_{\mu\nu}^{V;\pi,K}(q) = \int d^4x \, e^{-iq \cdot x} \langle 0| \, T\,(V_\nu^{\pi,K}(x), V_\nu^{\pi,K}(0)) \, |0\rangle. \tag{4.11}$$

Proceeding as before they obtain

$$\int d\mu^2 \, \frac{[\varrho_\pi^V(\mu^2) - \varrho_K^V(\mu^2)]}{\mu^2} = 0. \tag{4.12}$$

The application of asymptotic $SU(3)$ can easily be extended to the axial vector weak propagator as was discussed by Divakaran and Pandit (26). The relevant weak propagators are

$$\Delta_{\mu\nu}^{A;\pi,K}(q) = \int d^4x \, e^{-iqx} \langle 0| \, T\,(A_\mu^{\pi,K}(x), A_\nu^{\pi,K}(0)) \, |0\rangle. \tag{4.13}$$

Again, invoking asymptotic $SU(3)$ symmetry

$$\lim_{q \to \infty} [\Delta_{\mu\nu}^{A,\pi}(q) - \Delta_{\mu\nu}^{A,K}(q)] \to 0 \tag{4.14}$$

immediately leads to

$$\int d\mu^2 \, \frac{[\varrho_\pi^A(\mu^2) - \varrho_K^A(\mu^2)]}{\mu^2} = F_K^2 - F_\pi^2 \tag{4.15}$$

and

$$\int d\mu^- \, [\varrho_\pi^A(\mu^2) - \varrho_K^A(\mu^2)] = 0. \tag{4.16}$$

The second sum rule, Eq. (4.16) would require the stronger condition of convergence for Eq. (4.14).

Moffat and O'Donnell (27) have also derived further sum rules by appealing to the asymptotic $SU(3) \times SU(3)$ symmetry, following the method of DM0. Without making use of the stronger convergence, they derived the additional relation:

$$\int [\varrho_K^V(\mu^2) - \varrho_K^A(\mu^2)] \, d\mu^2 = 0. \tag{4.17}$$

It is worthwhile remarking here that while the pseudoscalar meson poles were explicitly separated in the spectral representation of the weak axial vector propagator, there is no corresponding pole for the vector propagator because of CVC. However, if one assumes the existence of scalar mesons defined in terms of the divergence of the vector current, one would include the scalar meson poles. This is what is done by Glashow, Schnitzer and Weinberg (28), who derived the spectral function sum rules for a general Lie algebra which is then applied to the vector and axial vector mesons coupled to the currents of $SU(3) \times SU(3)$. The sum rules so obtained seem to make sense in so far as the masses and coupling constants of the vector and axial vector mesons are concerned. Their derivation requires the stronger convergence conditions on the spectral functions following from asymptotic symmetry.

### 4.2  Wave function renormalization constants of $\pi$ and $K$ mesons

We shall now review a novel and interesting application (29) of the asymptotic symmetry idea to infer something about the wave function renormalization constants of pseudoscalar mesons.

It should be pointed out that there is an intimate connection between the wave function renormalization constants $Z_3$ and asymptotic symmetry for the two point functions defined in terms of currents. There are two simple reasons to substantiate this conjecture:

1) Renormalization constants can be defined as the asymptotic value of $q^2 \Delta(q)$ in the limit $q \to \infty$, where $\Delta(q)$ is the appropriate two-point function. This limit is also the domain where asymptotic symmetry is expected to be valid. The $Z$'s of different particles are thus related as a consequence of asymptotic symmetry.

2) Pseudoscalar mesons are described by means of interpolating fields via *PCAC* and vector mesons are equivalent to the vector currents in the sense of field-current identity (30). Any symmetry, therefore, which holds for a set of vector and axial vector currents should consequently imply some relationship among the different pseudoscalar meson fields and similarly for the vector meson fields.

First we shall consider a current algebra model which would allow us to relate the true wave function renormalization constants $Z_\pi$ and $Z_K$. We begin with Eq. (4.13). Multiplying by $q_\mu q_\nu$ and integrating by parts in the standard manner, we have

$$q_\mu q_\nu \left( \Delta_{\mu\nu}^{A,\pi}(q) - \Delta_{\mu\nu}^{A,K}(q) \right) = \int d^4x \, e^{-iq\cdot x} \{ \theta(x_0) \langle 0| \, [D^\pi(x), D^\pi(0)] \, |0\rangle$$

$$- \theta(x_0) \langle 0| \, [D^K(x), D^K(0)] \, |0\rangle - \delta(x_0) \langle 0| \, [D^\pi(x), A_0^\pi(0)] \, |0\rangle$$

$$+ \delta(x_0) \langle 0| \, [D^K(x), A_0^K(0)] \, |0\rangle - iq_\nu \delta(x_0) \langle 0| \, [A_0^\pi(x), A_\nu^\pi(0)] \, |0\rangle$$

$$+ iq_\nu \delta(x_0) \langle 0| \, [A_0^K(x), A_\mu^K(0)] \, |0\rangle \}. \tag{4.18}$$

We now consider the limit $q \to \infty$ of Eq. (4.18) and invoke the asymptotic $SU(3)$, Eq. (4.14). (We assume the strong convergence so that the left hand side of Eq. (4.18) vanishes.) The asymptotic symmetry automatically implies the equality of Schwinger terms as was seen earlier and together with CVC, we can pair off two of the equal time commutators in Eq. (4.18). We now make use of our second assumption, viz. the "$\sigma$"-terms of the $\pi$ and $K$ types are equal, in order to obtain some useful result. As a consequence, we arrive at

$$\int d^4x \, e^{-iq\cdot x} \theta(x_0) \{ \langle 0| \, [D^\pi(x), D^\pi(0)] \, |0\rangle - \langle 0| \, [D^K(x), D^K(0)] \, |0\rangle \} = 0. \tag{4.19}$$

Now utilising PCAC and the Källén–Lehman representation for the pseudo-scalar meson propagators

$$\Delta^{\pi,\,K}(q) = \int d\mu^2 \, \frac{\varrho_{\pi,K}^{(0)}(\mu^2)}{q^2 + \mu^2 - i\varepsilon} \tag{4.20}$$

and recalling the definition of $Z^{-1}$

$$Z_{\pi,K}^{-1} = \int \varrho_{\pi,K}^{(0)}(\mu^2) \, d\mu^2, \tag{4.21}$$

we immediately obtain

$$Z_K^{-1} m_K^4 F_K^2 = Z_\pi^{-1} m_\pi^4 F_\pi^2 \tag{4.22}$$

or, alternatively

$$\left( \frac{Z_\pi}{Z_K} \right) = \left( \frac{m_\pi}{m_K} \right)^4 \left( \frac{F_\pi}{F_K} \right)^2. \tag{4.23}$$

This result is certainly of considerable interest as it relates the renormalization constants to the masses of the particles themselves.

It may be mentioned in passing that Khuri had conjectured (31)

$$\frac{Z_\pi}{Z_K} \approx \left( \frac{m_\pi}{m_K} \right)^4 \tag{4.24}$$

from an entirely different set of assumptions. It should be pointed out that Eq. (4.24) is a statement of approximation and that the decay constants do not appear. Khuri's derivation is however very questionable, as has been pointed out (32) and it involves unjustifiable assumptions.

Acharya (33) has also obtained the result in Eq. (4.24).

It is easy to extend the above procedure to the wave function renormalization constant of the $\eta$ meson, $Z_\eta$. We would obtain a result analogous to Eq. (4.24) where $Z_\eta$ would be proportional to $(m_\eta)^4 F_\eta^2$, as a consequence of the asymptotic symmetry

$$\mathop{\text{Lim}}_{q \to \infty} [\Delta_{\mu\nu}^{A(3)}(q) - \Delta_{\mu\nu}^{A(8)}(q)] \to 0, \tag{4.25}$$

which is the analog of Eq. (3) of ref. 34. However, this derivation would depend crucially on the validity of *PCAC* for the $\eta$-type axial vector current.

## 4.3  Wave function renormalization constants of vector mesons

In order to apply the above method to a discussion of the vector meson renormalization constants, we shall appeal to a current algebra model where one introduces tensor currents.

Octet generalizations of the stress energy tensor are well known (35). One can also postulate commutation relations involving tensor currents and the usual vector and axial vector currents, which are supported by derivations in the framework of a quark model. To render this approach more useful one also appeals to the hypothesis of partial conservation of the tensor current (PCTC) (36) in exact analogy with the PCAC. In a recent application of this current algebra model, spectral function sum rules have been derived (37, 38) as a consequence of asymptotic $SU(3)$ symmetry for two point functions constructed from the tensor currents.

We shall work within the framework of the model used in ref. 37, where an octet of anti-symmetric tensor currents $\theta_{\mu\nu}^i(x)$ are introduced. Consider the two-point functions

$$\Delta_{\mu\nu\lambda\varrho}^i(q) = \int d^4x \, e^{-iq \cdot x} \langle 0| \, T\left(\theta_{\mu\nu}^i(x) \, \theta_{\lambda\varrho}^i(0)\right) |0\rangle \tag{4.26}$$

where $i = 1, \ldots 8$ are $SU(3)$ indices. We shall be concerned with the "$\pi$" and "$K$" type propagators. The tensor currents satisfy PCTC, which is expressed by

$$\partial_\mu \theta_{\mu\nu}^i(x) = D_\nu^i(x) = C^i \phi_\nu^i(x) \tag{4.27}$$

where $\phi^i_v(x)$ are octet vector meson fields and the constants $c^i$ are given by (39)

$$C^\pi = m^2_\varrho F_\pi, \quad C^K = m^2_{K*} F_\varkappa \qquad (4.28)$$

according to the universality argument of ref. 37.

Consider Eq. (4.26) for the $\pi$-type currents. Multiplying by $q_\mu q_\nu$ and integrating by parts, we get

$$q_\mu q_\nu \, \varDelta^\pi_{\mu\nu\lambda\varrho}(q) = -i \int d^4x \, e^{-iq \cdot x} \, [q_\lambda \delta \, (x_0) \, \langle 0 \, |[\theta^\pi_{0v}(x), \, \theta^\pi_{\lambda\varrho}(0)]| \, 0\rangle$$

$$+ \; \delta(x_0) \, \langle 0| \, [D^\pi_v(x), \, \theta^\pi_{0\varrho}(0)] \, |0\rangle$$

$$- \; \theta(x_0) \, \langle 0| \, [D^\pi_v(x), \, D^\pi_\varrho(0)] \, |0\rangle. \qquad (4.29)$$

We next write the corresponding reduction for the $K$-type propagator, subtract from Eq. (4.29) and take the limit $q \to \infty$. We now make the following assumptions: (a) The currents $\theta^i_{\mu\nu}$ and the spin-one fields satisfy the commutation relations

$$\delta(x_0) \, [\theta^i_{0\lambda}(x), \, \theta^i_{\mu\nu}(0)] = S^{ii}_{0\lambda, \, \mu\nu}(x) \qquad (4.30)$$

$$\delta(x_0) \, [\theta^i_{0\lambda}(x), \, D^i_\mu(0)] = \Sigma^{ii}_{0\lambda, \, \mu}(x), \qquad (4.31)$$

where the Schwinger terms $S$ and $\Sigma$ are symmetric in the $SU(3)$ indices. These commutation relations are those of ref. 37 when $D^i_\mu$ are replaced by $V^i_\mu$ and are justified since the invariant structure of $V^i_\mu$ and $D^i_\mu$ are the same. Alternatively, one can resort to the field current identity (30) to establish Eq. (4.31) from quark model commutation relations.

(b) The validity of asymptotic $SU(3)$:

$$\underset{q \to \infty}{\text{Lim}} \, [\varDelta^\pi_{\mu\nu\lambda\varrho}(q) - \varDelta^K_{\mu\nu\lambda\varrho}(q)] \to 0, \qquad (4.32)$$

where we shall invoke the stronger convergence requirement.

(c) A simplifying requirement that the $\Sigma$-terms of the $\pi$ and $K$ type are equal.

Under these assumptions and making use of Eq. (4.27), we immediately obtain

$$\underset{q \to \infty}{\text{Lim}} \, [(C^\pi)^2 \, \varDelta^\pi_{v\varrho}(q) - (C^K)^2 \, \varDelta^K_{v\varrho}(q)] \to 0 \qquad (4.33)$$

where $\varDelta^i_{v\varrho}(q)$ are the spin-one meson propagators. If we now employ the Källén–Lehman representation for these propagators and the definition of the wave-function renormalization constants $Z_\varrho$ and $Z_{K*}$ in terms of the corresponding spectral functions, we finally derive the result

$$\frac{Z_\varrho}{Z_{K*}} = \left(\frac{m_\varrho^2 F_\pi}{m_{K*}^2 F_K}\right)^2. \tag{4.34}$$

### 4.4   Consistency with field current identity

We now proceed to discuss the compatibility of the result in Eq. (4.34) with
the field-current identity (30) expressed by

$$V_\mu^i(x) = \frac{m_i^2}{g_i}\, \phi_\mu^i(x). \tag{4.35}$$

If we impose this identity on the two point functions $\Delta_{\mu\nu}^{V;\,i}(q)$ it is evident that
the true wave function renormalization constants are related to the corresponding current-renormalization constants $Z$ by

$$\begin{aligned}
Z_\varrho^{-1} &= (g_\varrho/m_\varrho^2)\, \bar{Z}_\varrho^{-1} \\
Z_{K*}^{-1} &= (g_{K*}/m_{K*}^2)\, \bar{Z}_{K*}^{-1}
\end{aligned} \tag{4.36}$$

where $\bar{Z}$ would be defined in terms of the spectral functions characterising
the weak-vector propagator.

Turning our attention to the renormalization constants $\bar{Z}$, we can of course
directly obtain the ratio $(\bar{Z}_\varrho/\bar{Z}_{K*})$ from the asymptotic $SU(3)$ imposed on the
two-point functions $\Delta_{\mu\nu}^V(q)$ introduced earlier. The current two-point function
has the representation

$$\Delta_{\mu\nu}^{V:\pi,K}(q) = \int \varrho_{\pi,K}^V(\mu^2)\, d\mu^2 \left[\delta_{\mu\nu} + \frac{q_\mu q_\nu}{\mu^2}\right] \frac{1}{q^2 + \mu^2 - i\varepsilon}. \tag{4.37}$$

The relation between the spectral functions and the renormalization constants
$\bar{Z}$ is expressed by the Johnson sum rule (40)

$$\bar{Z}_{\pi,K}^{V-1} = \int \sigma_{\pi,K}^V(\mu^2)\, d\mu^2 \tag{4.38}$$

where

$$\sigma_{\pi,K}^V(\mu^2) = \frac{\varrho_{\pi,K}^V(\mu^2)}{(\mu^2 - m_0^2)^2}. \tag{4.39}$$

The bare masses $m_0$ are assumed equal for the $\pi$ and $K$ type vector currents.
The constants $\bar{Z}$ are then determined by the asymptotic behaviour of $\Delta_{\mu\nu}$.
Accordingly we require the validity of

$$\lim_{q\to\infty} [\Delta_{\mu\nu}^{V;\,\pi}(q) - \Delta_{\mu\nu}^{V;\,K}(q)] \to 0. \tag{4.40}$$

Following Acharya's procedure (19) of lettering $m_0 \to \infty$, we thus obtain

$$Z_\varrho = Z_{K*}. \tag{4.41}$$

Let us now return to our discussion of the consistency of Eq. (4.34) with the field-current identity, Eq. (4.35). This reduces to the question of how far Eqs. (4.41) and (4.34) are compatible with Eq. (4.36).

Requiring the validity of all the three results gives the following consistency constraint:

$$\frac{m_\varrho^2 F_\pi}{m_{K*}^2 F_K} = \frac{m_\varrho^2}{g_\varrho} \cdot \frac{g_{K*}}{g_\varrho}. \tag{4.42}$$

Using Weinberg's relation (which follows from the spectral function sum rule in Eq. (4.12) when the integrals are saturated by vector mesons $\varrho$ and $K^*$)

$$\frac{g_{K*}}{m_{K*}} = \frac{g_\varrho}{m_\varrho}, \tag{4.43}$$

this reduces to

$$\frac{F_K}{F_\pi} = \left(\frac{m_\varrho}{m_{K*}}\right)^{3/2} \approx 0.8. \tag{4.44}$$

Considering the crude assumption of PCTC universality, the fact that $F_K/F_\pi$ required by Eq. (4.44) is close to unity is indeed an indication that Eqs. (4.34), (4.41) and field current identity are consistent. The error of approximation in Eq. (4.28) is reflected in the fact that the "determination" of $F_K/F_\pi$ expressed by Eq. (4.44) is not quite close to the accepted value of 1.28. Nevertheless even this reinforces the validity of the assumptions made in this section.


## 5   CONCLUDING REMARKS

In this review we have seen how the powerful techniques based on asymptotic symmetry can be employed to derive strong interaction sum rules in the context of $SU(2) \times SU(2)$ (chiral algebra). One of the interesting aspects that was discussed is the derivation of a new sum rule for $\pi\pi$ scattering, which is a slight modification of the sum rule originally derived by Adler. This calculation was based on the requirement that the proper amplitudes satisfy asymptotic chiral symmetry. Considering the improved agreement of the modified sum rule with experiment, it would appear that the use of asymptotic symmetry for the proper amplitudes is justified. On the other hand, it is believed that the asymptotic symmetry requirement for the weak amplitudes is equivalent to the validity of current commutation relations under certain conditions (7). It is thus important to know if the asymptotic symmetry for the proper amplitudes also involves some constraints. This question is particularly meaningful since this hypothesis leads to inconsistencies in the $\pi N$ case. We are currently investigating the physical constraints (in terms of coupling constant relations) which are required so that the Regge behaviour for the proper amplitudes and asymptotic symmetry would be physically compatible with each other. The understanding of the role played by Regge asymptotic behaviour will perhaps provide the raison d'être for the asymptotic symmetry hypothesis.

Another aspect which was discussed in detail was the applicability of asymptotic symmetry with regard to the renormalization constants. The derivation of the results seems to depend crucially on the use of stronger convergence requirements for the asymptotic behaviour of the two-point functions. One usually avoids the use of such stronger convergence on the ground that the resulting spectral function sum rules may not be satisfied when the integrals are saturated by vector and axial vector mesons (meson dominance). It must be stressed, however, that the meson dominance approximation may not be valid in every case. It may be valid only in a limited sector i.e., in the cases where $\varrho$ and $A_1$ mesons dominate the spectral functions which arise in the application of asymptotic $SU(2) \times SU(2)$. In these cases, we have avoided the use of strong convergence requirements and thus justify its use elsewhere.

### References and notes


1. S.L.Adler, *Phys. Rev. Letters*, **14**, 1051 (1965); *Phys. Rev.*, **140B**, 736 (1965); *Phys. Rev.*, **175**, 2224 (1968).
2. W.I.Weisberger, *Phys. Rev. Letters*, **14**, 1047 (1965); *Phys. Rev.*, **143**, 1302 (1966).
3. M.Gell–Mann, *Physics*, **1**, 63 (1964).
4. Fayyazuddin and Hussain, *Phys. Rev.*, **164**, 1864 (1967). This paper will be referred to as FH.
5. M.Gell–Mann and F.Zachariasen, *Phys. Rev.*, **123**, 1065 (1961).
6. See e.g., J.D.Bjorken, *Phys. Rev.*, **148**, 1467 (1966).
7. R.Acharya, *Phys. Rev.*, **164**, 1799 (1967).
8. For kinematical details, see e.g., ref. 4; also ref. 9 below.
9. The method of deriving superconvergent dispersion relations in this manner is discussed in S.Fubini, *Nuovo Cimento*, **43A**, 475 (1966).
10. See sec. 3 of ref. 4 for details.
11. I.S.Gerstein, *Phys. Rev.*, **161**, 1631 (1967).
12. S.L.Adler, *Phys. Rev.*, **143**, 1144 (1966).

13. See discussion in 'superconvergent dispersion relations and pion pion scattering sum rule'. R. Acharya, H. H. Aly and P. Narayanaswamy, preprint, March 1969.

14. G. W. Barry, G. J. Gounaris and J. J. Sakurai, *Phys. Rev. Letters*, **21**, 941 (1968).

15. H. J. Schnitzer and S. Weinberg, *Phys. Rev.*, **164**, 1828 (1968).

16. We have used the KSRF relation $g_\varrho^2 = 2m_\varrho^2 F_\pi^2$. See K. Kawarabayashi and H. Suzuki, *Phys. Rev. Letters*, **16**, 255 (1966); Riyazuddin and Fayyazuddin, *Phys. Rev.*, **147**, 1071 (1961).

17. S. Weinberg, *Phys. Rev. Letters*, **18**, 507 (1967).

18. T. D. Lee, S. Weinberg and B. Zumino, *Phys. Rev. Letters*, **18**, 1029 (1967).

19. R. Acharya, *Phys. Letters*, **24B**, 623 (1967).

20. T. Das, V. S. Mathur and S. Okubo, *Phys. Rev. Letters*, **18**, 761 (1967). This paper will be referred to as DMO.

21. This is the only equal time commutator we need, in addition to what is contained in Eq. (2.21).

22. R. Acharya, H. H. Aly and P. Narayanaswamy, "Inconsistency of asymptotic $SU(2) \times SU(2)$ for the proper amplitudes with local chiral algebra", preprint, 1969.

23. I. S. Gerstein, *Phys. Rev. Letters*, **21**, 1465 (1968).

24. P. Olesen, *Phys. Rev.*, **172**, 1461 (1968).

25. See the last section for a discussion on this point.

26. P. P. Divakaran and L. K. Pandit, *Phys. Rev.*, **166**, 1782 (1968).

27. J. W. Moffat and P. J. O'Donnell, *Canad. J. Phys.*, **45**, 3901 (1967).

28. S. L. Glashow, H. J. Schnitzer and S. Weinberg, *Phys. Rev. Letters*, **19**, 139 (1967).

29. H. H. Aly and P. Narayanaswamy, "Ratio of renormalization constants from asymptotic symmetry", preprint, 1969.

30. N. M. Kroll, T. D. Lee and B. Zumino, *Phys. Rev.*, **157**, 1376 (1967).

31. N. N. Khuri, *Phys. Rev. Letters*, **16**, 75 (1966).

32. N. N. Khuri, *Phys. Rev. Letters*, **16**, 601 (E) (1966).

33. R. Acharya, private communication.

34. T. Das, V. S. Mathur and S. Okubo, *Phys. Rev. Letters*, **19**, 470 (1967).

35. R. Delburgo, A. Salam and J. Strathdee, *Nuovo Cimento*, **49A**, 593 (1967).

36. W. Krolikowski, *Nuovo Cimento*, **42A**, 435 (1966); **44A**, 745 (1966); **46A**, 106 (1966).

37. M. Ademollo, G. Longhi, and G. Veneziano, *Nuovo Cimento*, **58A**, 540 (1968).

38. G. C. Joshi and L. K. Pande, *Phys. Rev. Letters*, **20**, 816 (1968).

39. It should be emphasized that PCTC in this form is an approximate statement and the error will be reflected in the final results as will be discussed later.

40. K. Johnson, *Nuc. Phys.*, **25**, 435 (1961).

# Some Applications
# of Spectral Function Sum Rules

R. ACHARYA

*University of Berne, Switzerland*

## Contents



## 1  INTRODUCTION

Spectral function sum rules for two-point functions of spin zero and spin one-half fields were first considered within the framework of a local, relativistic quantum field theory in the 1950's[1]. It was almost a decade later that the spin one field was considered by K. Johnson[2]. More recently, W. Zimmermann[3] has put forward a rigorous derivation of Johnson's sum rule which has been shown to be valid both for finite and infinite bare mass of the spin one field. These spectral function sum rules diverge in perturbation theory, with the possible exception of Johnson's sum rule for vector mesons. The

83

divergence of $Z_3^{-1}$ in perturbation theory is perhaps an indication of the limitation of the theory to yield a convergent answer for the wave function renormalization constant and is believed to have little bearing on the question of the composite condition $Z_3$ equals zero[4].

Spectral function sum rules came back into fashion in 1967. In an important publication, S. Weinberg[5] studied the spectral function sum rules for two-point functions of vector and axial-vector currents, following an earlier, far-reaching proposal of M. Gell–Mann[6] of a current algebra as an appropriate algebraic definition of broken symmetry (such as $SU(2) \otimes SU(2)$). Soon afterwards, T.D. Lee, S. Weinberg and B. Zumino[7] proposed the so-called algebra of fields which was a natural generalization to the non-abelian case of Johnson's original work on a single gauge field. These authors demonstrated that the algebra of fields could be successfully employed to rederive all results that follow from Gell-Mann's current algebra. In addition, the field-algebra provided useful information on the Schwinger terms[8] and on the space-space equal-time current commutators. Almost immediately after Weinberg's paper appeared in the "market", Das, Mathur and Okubo[9] introduced the concept of asymptotic symmetry and arrived at Weinberg's result without the explicit use of equal-time current commutation relations. The most important result to emerge from Weinberg's sum rules is the startlingly accurate prediction of the mass of the chiral partner of the rho meson. Weinberg's mass relation has since been rederived by entirely different methods such as superconvergence relations[10] and the Veneziano representation[11].

Weinberg's original proof of the $SU(2) \otimes SU(2)$ spectral function rules (there are two of them) assumed a conserved axial-vector current. This amounts to the assumption that in nature $SU(2) \otimes SU(2)$ symmetry is broken spontaneously. Goldstone's theorem then tells us that a zero mass pion must be present in the theory[12]. When we turn our attention to strangeness-changing $SU(2) \otimes SU(2)$ subgroup of $SU(3) \otimes SU(3)$, the situation is a little bit different: finite kaon and kappa meson corrections are expected to be significant. It turns out that although Weinberg's original proof of his first sum rule may be carried through with minor changes for a partially conserved current[13], the so-called second sum rule is almost certainly invalid for finite kappa and kaon masses[14]. Fortunately, however, there exists a possible alternative[15]. Presently, we shall consider this modified second sum rule which leads to a definite prediction of the mass of the strange, scalar meson in good agreement with the broken chiral symmetry prediction[16] and with preliminary experimental evidence[17] As a second example, we shall discuss a somewhat artificial model of spectral function sum rules for spin zero mesons, based on the assumed equality of the vacuum expectation value of the so-called sigma terms of "$\pi$" and "$K$" type. The motivation for presenting this particular example is to demonstrate the possibility of constructing a model in which the inclusion of the continuum contributions to the spin zero spectral functions completely alters the original prediction of the sum rule in the pole dominance approximation. As a final example, we shall briefly discuss the possibility of determining the width of the $A_1$ meson from Weinberg's sum rules[18] The procedure adopted here is a natural generalization of the delta function approximation, namely, the Brcit-Wigncr form for thc spcctral functions.

## 2   MASS OF KAPPA MESON FROM SPECTRAL FUNCTION SUM RULES

We start with a modified form of the second sum rule but retain the standard first sum rule for the strangeness-changing $SU(2) \otimes SU(2)$ subgroup of $SU(3) \otimes SU(3)$:

$$\frac{g_{K^*}^2}{m_{K^*}^2} - \frac{g_{K_A}^2}{m_{K_A}^2} = F_K^2 - F_\varkappa^2, \tag{2.1}$$

$$g_{K^*}^2 - g_{K_A}^2 = m_K^2 F_K^2 - m_\varkappa^2 F_\varkappa^2. \tag{2.2}$$

Although there exist no known models of field theory from which Eq. (2.2) may be rigorously derived, it is, nonetheless possible to advance plausibility arguments in its favour from asymptotic symmetry considerations of Ref. 9:
  Employing standard notation, we write:

$$\Delta_{\mu\nu}^{K^*}(q) - \Delta_{\mu\nu}^{K_A}(q) \equiv F(q^2)\, \delta_{\mu\nu} + G(q^2)\, q_\mu q_\nu \tag{2.3}$$

where

$$F(q^2) = \int d\mu^2\, \frac{\varrho_{K^*}(\mu^2) - \varrho_{K_A}(\mu^2)}{q^2 + \mu^2}, \tag{2.4}$$

$$G(q^2) = \int d\mu^2\, \frac{\varrho_{K^*}(\mu^2) - \varrho_{K_A}(\mu^2)}{\mu^2\,(q^2 + \mu^2)} + \frac{F_\varkappa^2}{q^2 + m_\varkappa^2} - \frac{F_K^2}{q^2 + m_K^2}. \tag{2.5}$$

The assumption that $G(q^2) \sim q^{-4-\varepsilon}$ gives us the desired modified second sum rule.

At this stage, we take into consideration the first sum rule for the $(\varrho, K^*)$ system and also the well-known KSRF relation:

$$g_\varrho^2 m_\varrho^{-2} - g_{K*}^2 m_{K*}^{-2} = F_\pi^2, \tag{2.6}$$

$$g_\varrho^2 = 2F_\pi^2 m_\varrho^2. \tag{2.7}$$

From Eqs. (2.1), (2.2), (2.6) and (2.7):

$$m_\varkappa^2 = m_{K*}^2 + 2\,\frac{F_\pi^2}{F_\varkappa^2}\,(m_{K_A}^2 - m_{K*}^2) - \frac{F_K^2}{F_\varkappa^2}\,(m_{K_A}^2 - m_K^2). \tag{2.8}$$

In view of the relation[19]

$$F_K^2 = F_\pi^2, \tag{2.9}$$

which is valid under pole dominance provided the KSRF relation is true, Eq. (8) reduces to the form

$$m_\varkappa^2 = m_{K*}^2 + \frac{F_\pi^2}{F_\varkappa^2}\,\{m_{K_A}^2 - 2m_{K*}^2 + m_K^2\}. \tag{2.10}$$

We observe that the kappa meson becomes essentially degenerate with the $K^*$ (890) in the limit of vanishing kaon mass (recall that $m_{K_A} = 2^{1/2} m_{K*}$). In order to estimate the finite kaon mass correction, we resort to the Glashow-Weinberg refinement[20] of the Ademollo-Gatto theorem:

$$f^+(0) = (F_K^2 + F_\pi^2 - F_\varkappa^2)\,(2F_K F_\pi)^{-1} \tag{2.11}$$

where $f^+(0)$ denotes the renormalized $K_{e3}$ form factor at $q^2 = 0$. Eq. (11) may be shown to be valid independent of the specific manner of chiral breakdown[21].

Eqs. (2.9) and (2.11) may be combined with the experimental result

$$\left|\frac{F_K}{F_\pi f^+(0)}\right| \approx 1.28 \tag{2.12}$$

to obtain $\dfrac{F_\pi^2}{F_\varkappa^2} \approx 2.2$.

Eq. (2.10) then gives $m_\varkappa \approx 1130$ MeV. This value is comparable with the broken chiral symmetry prediction[16] $m_\varkappa \approx 1080$ MeV and with the experimental result $m_\varkappa \approx 1100$–$1200$ MeV[17]. If future experiments do confirm the existence of the kappa meson around 1 GeV, then there is every reason to take the modified second sum rule Eq. (2.2) quite seriously.

Before concluding this discussion of the kappa meson, let us briefly consider the possibility of recovering the usual form of the second sum rule from

Eq. (2.2): this would happen provided

$$F_K^2 F_\varkappa^{-2} = m_\varkappa^2 m_K^{-2}.$$ (2.13)

Eq. (2.13) together with the Weinberg sum rules has the consequence that $m_\varkappa \approx 730$ MeV and $m_{K_A} \approx 1110$ MeV. Both these values are not favoured by experiment. Hence, it would appear that Eq. (2.13) which is a consequence of the "super-validity" of asymptotic $SU(2) \otimes SU(2)$ symmetry for the strangeness changing vector and axial-vector currents ($F(q^2) \sim q^{-2-\varepsilon}$ and $G(q^2) \sim q^{-4-\varepsilon}$) should be rejected on *experimental* grounds.

## 3  A MODEL OF SPIN ZERO SPECTRAL FUNCTIONS

The model is defined by

$$\int_0^\infty d\mu^2 \, [\varrho_\pi^{(0)}(\mu^2) - \varrho_K^{(0)}(\mu^2)] \, \mu^2 = 0$$ (3.1)

$$\int_0^\infty d\mu^2 \, [\varrho_\pi^{(0)}(\mu^2) - \varrho_K^{(0)}(\mu^2)] = 0$$ (3.2)

where $\varrho_{\pi,K}^{(0)}(\mu^2)$ stand for the spin zero spectral functions of "$\pi$" and "$K$" type axial-vector currents.

From the spectral representation for the commutator

$$\langle [A_\mu(x), A_\nu(y)] \rangle_0 = i \int_0^\infty d\mu^2 \left[ \left( \delta_{\mu\nu} + \frac{\partial_\mu \partial_\nu}{\mu^2} \right) \varrho^{(1)}(\mu^2) + \partial_\mu \partial_\nu \varrho^{(0)}(\mu^2) \right] \times$$
$$\times \Delta \, (x - y; \, \mu^2)$$ (3.3)

it is evident that the sum rule Eq. (3.1) is equivalent to the assumption that the "$\pi$" and "$K$" type sigma term equal-time vacuum expectation values are equal. The sum rule Eq. (3.2) states the equality of the pion and the kaon decay constants, *provided* one assumes the strong form of the PCAC hypothesis:

$$\varrho_{\pi,K}^{(0)}(\mu^2) = F_{\pi,K}^2 \, \delta \, (\mu^2 - m_{\pi,K}^2).$$ (3.4)

Clearly, Eqs. (3.1) and (3.2) are inconsistent with each other under pole dominance assumption Eq. (3.4), unless the pion and the kaon masses are degenerate. We now demonstrate that the inclusion of the continuum contributions to the spin zero spectral functions removes the discrepancy between Eqs. (3.1) and (3.2) provided a certain, strong technical assumption (to be specified) is made on the difference of the continuum contributions.

Separating out the pole terms, Eqs. (3.1) and (3.2) read

$$F_\pi^2 m_\pi^2 - F_K^2 m_K^2 + \int_{(3m_\pi)^2}^{(2m_\pi + m_K)^2} d\mu^2 \, \varrho_\pi^{(0)'}(\mu^2) \, \mu^2$$

$$+ \int_{(2m_\pi + m_K)^2}^{\infty} d\mu^2 \, [\varrho_\pi^{(0)}(\mu^2) - \varrho_K^{(0)}(\mu^2)]' \, \mu^2 = 0 \qquad (3.5)$$

$$F_\pi^2 - F_K^2 + \int_{(3m_\pi)^2}^{(2m_\pi + m_K)^2} d\mu^2 \, \varrho_\pi^{(0)'}(\mu^2)$$

$$+ \int_{(2m_\pi + m_K)^2}^{\infty} d\mu^2 \, [\varrho_\pi^{(0)}(\mu^2) - \varrho_K^{(0)}(\mu^2)]' = 0. \qquad (3.6)$$

We now assume that $[\varrho_\pi^{(0)}(\mu^2) - \varrho_K^{(0)}(\mu^2)]'$ is of one sign from $(2m_\pi + m_K)^2$ to $\infty$. This is certainly a very strong restriction but it enables us to employ the mean-value theorem to rewrite Eq. (3.5) as follows:

$$F_\pi^2 m_\pi^2 - F_K^2 m_K^2 + \int_{(3m_\pi)^2}^{(2m_\pi + m_K)^2} d\mu^2 \, \varrho_\pi^{(0)'}(\mu^2) \, \mu^2$$

$$+ \langle \mu^2 \rangle \int_{(2m_\pi + m_K)^2}^{\infty} d\mu^2 \, [\varrho_\pi^{(0)}(\mu^2) - \varrho_K^{(0)}(\mu^2)]' = 0 \qquad (3.7)$$

where $\langle \mu^2 \rangle_{\min.} = (2m_\pi + m_K)^2$.

From Eqs. (3.6) and (3.7), we obtain:

$$F_K^2 \, [\langle \mu^2 \rangle - m_K^2] - F_\pi^2 \, [\langle \mu^2 \rangle - m_\pi^2]$$

$$= \langle \mu^2 \rangle \int_{(3m_\pi)^2}^{(2m_\pi + m_K)^2} d\mu^2 \, \varrho_\pi^{(0)'}(\mu^2) - \int_{(3m_\pi)^2}^{(2m_\pi + m_K)^2} d\mu^2 \, \varrho_\pi^{(0)'}(\mu^2) \, \mu^2. \qquad (3.8)$$

The R.H.S. of Eq. (3.8) is positive for all $\langle \mu^2 \rangle$ including for $\langle \mu^2 \rangle_{\min.} = (2m_\pi + m_K)^2$.

Hence,

$$\frac{F_K^2}{F_\pi^2} > 1 + \frac{m_K^2 - m_\pi^2}{\langle \mu^2 \rangle_{\min} - m_K^2} = 1 + \frac{m_K - m_\pi}{4m_\pi} \approx (1.28)^2. \qquad (3.9)$$

We observe that the pole dominance prediction of Eq. (3.1) has been completely altered by the inclusion of the continuum contributions: Perhaps, this is not very surprising since the presence of the factor of $\mu^2$ in the integrand would tend to give weight to the large $\mu^2$ contributions.

## 4 THE $A_1$ WIDTH FROM WEINBERG'S SUM RULES

In this section, we present a calculation of the width of the $A_1$ meson within the framework of spectral function sum rules. The method adopted here is a straightforward generalization of the delta function approximation:

We take

$$\varrho_{V,A}(\mu^2) = \frac{1}{\pi} g_{\varrho,A}^2 \frac{\gamma_{\varrho,A} m_{\varrho,A_1}}{(\mu^2 - m_{\varrho,A_1}^2)^2 + \gamma_{\varrho,A}^2 m_{\varrho,A_1}^2} \tag{4.1}$$

where $\gamma_{\varrho,A}$ define the width of the $\varrho$ and the $A_1$ respectively and the rest of the notation is standard.

We take the lower limit of integration for the vector current two-point function to be $4m_\pi^2$ and for the axial-vector case at $9m_\pi^2$. These limits are suggested by the algebra of fields[7]. It is now straightforward to perform the elementary integrations to obtain the following expressions from the two Weinberg sum rules:

$$\frac{g_A^2}{g_\varrho^2} = \frac{\dfrac{\gamma_\varrho m_\varrho}{m_\varrho^4 + \gamma_\varrho^2 m_\varrho^2} \ln X + \dfrac{m_\varrho^2}{m_\varrho^4 + \gamma_\varrho^2 m_\varrho^2} A - \dfrac{\pi}{2} \dfrac{1}{m_\varrho^2}}{\dfrac{\gamma_A m_{A_1}}{m_{A_1}^4 + \gamma_A^2 m_{A_1}^2} \ln Y + \dfrac{m_{A_1}^2}{m_{A_1}^4 + \gamma_A^2 m_{A_1}^2} B}, \tag{4.2}$$

$$\frac{g_A^2}{g_\varrho^2} = \frac{A}{B}; \quad \left\{ \begin{aligned} &A = \frac{\pi}{2} + \tan^{-1} \frac{m_\varrho^2 - 4m_\pi^2}{\gamma_\varrho m_\varrho}, \quad B = \frac{\pi}{2} + \tan^{-1} \frac{m_{A_1}^2 - 9m_\pi^2}{\gamma_A m_{A_1}} \\ &X = \frac{1}{4m_\pi^2} \{(m_\varrho^2 - 4m_\pi^2)^2 + m_\varrho^2 \gamma_\varrho^2\}^{1/2} \\ &Y = \frac{1}{9m_\pi^2} \{(m_{A_1}^2 - 9m_\pi^2)^2 + m_{A_1}^2 \gamma_A^2\}^{1/2}. \end{aligned} \right.$$

$$\tag{4.3}$$

The transcendental Eqs. (4.2) and (4.3) are best handled on a computer after eliminating the ratio $g_A^2 g_\varrho^{-2}$. The details of this effort may be found in Ref. 18. Here, we summarize the result thus: it is possible to obtain a sensible value for the $A_1$ width ($\gamma_A < 200$ MeV) from Eqs. (4.2) and (4.3), by feeding in the experimental values for the masses of the $\varrho$ and the $A_1$ and the width of the $\varrho$ meson.

An interesting feature of this simple-minded calculation is the fairly successful determination of the $A_1$ width without resorting to the soft-pion approximation. The pion has been left on its mass shell, as evidenced by the finite lower limits of integration. If one were to take the limit of $m_\pi \to 0$, one is led to the enormous value $\gamma_A = \dfrac{9}{\sqrt{2}} \gamma_\varrho$ which clearly indicates the necessity of keeping the pion mass away from zero in the present calculation.

## 5 DISCUSSION AND CONCLUDING REMARKS

There are several interesting points which arise when one attempts to extend the $SU(2) \otimes SU(2)$ analysis of spectral function sum rules to $SU(3) \otimes SU(3)$. Okubo[22] has proposed a specific form of symmetry-breaking which corresponds to the presence of the extra term proportional to $d_{\alpha\beta\gamma}$ (where $d_{\alpha\beta 8}$ is the $SU(3)$ symmetric symbol) in addition to the term in $\delta_{\alpha\beta}$ where $\alpha$ and $\beta$ run from 0 to 8. In particular, Okubo has stressed the fact that the Weinberg sum rules are model-dependent. Certainly, the first sum rule is on a much sounder footing than the second. Furthermore, one does not expect the second sum rule to be valid in a model in which the first sum rule is violated. One arrives at this latter conclusion from asymptotic symmetry considerations provided one works with massless pions. Also, the pole dominance hypothesis is, perhaps, on a much safer footing in the case of the first sum rule than in the second, in view of the presence of the extra factor of $\mu^2$ in the denominator of the integrand of the first sum rule. It is quite straightforward to write down the continuum contributions to the spin zero spectral functions[23] which occur in the first sum rule but one faces practical difficulties in trying to estimate their importance relative to the pole contributions. The problem that one encounters here is one of estimating the multiparticle (i.e., three pion and higher states) contributions to the spin zero spectral functions and as is well-known this is a formidable, if not impossible task. The two-particle contributions to the spin one spectral functions have been investigated in special cases by Dietz and Pietschmann[24] and Lam and Raman[25].

## APPENDIX I

The chiral $SU(2) \otimes SU(2)$ equal-time commutations read

$$\delta(x_0 - y_0)\,[V_a^0(x),\, V_b^\mu(y)] = \delta(x_0 - y_0)\,[A_a^0(x),\, A_b^\mu(y)]$$
$$= i\varepsilon_{abc}\, V_c^\mu(x)\, \delta^4(x - y) + \text{S.T.}$$
$$\delta(x_0 - y_0)\,[V_a^0(x),\, A_b^\mu(y)] = \delta(x_0 - y_0)\,[A_a^0(x),\, V_b^\mu(y)]$$
$$= i\varepsilon_{abc}\, A_c^\mu(x)\, \delta^4(x - y) + \text{S.T.} \tag{1}$$

where S.T. are the non-covariant or Schwinger terms which are assumed to be $c$-numbers. This assumption that there are no operator Schwinger terms with $I = 1$ is an essential assumption which goes into the derivation of the first Weinberg sum rule. The presence of the S.T. in the time-space commutations may be shown to be a consequence of Lorentz invariance and positive metric together with the assumption of locality[26]. These latter assumptions imply that $J_\mu(x) \equiv 0$ unless the S.T. are present[27].

Eq. (1) is sometimes supplemented by

$$\delta\,(x_0 \,-\, y_0)\,[V_a^0(x),\, \partial_\mu A_b^\mu(y)] = i\varepsilon_{abc}\partial_\mu A_c^\mu\,(x)\,\delta^4\,(x \,-\, y)$$

$$\delta\,(x_0 \,-\, y_0)\,[A_a^0(x),\, \partial_\mu A_b^\mu(y)] = \sigma_{ab}(x)\,\delta^4\,(x \,-\, y).$$

$$(2)$$

Eq. (2a) may be verified in the free quark model and Eq. (2b) is true, e.g. in the $\sigma$ model in which $\sigma_{ab}(x) = \delta_{ab}\sigma\,(x)$.


## APPENDIX II

We present a proof of Weinberg's first sum rule. Details may be found in Ref. 13.

First of all, some definitions:

$$\varepsilon_{abc}M_{\mu\nu\lambda}\,(q,\,k) = i\int d^4x\,d^4y\,e^{i\,(qx+ky)}\,\langle T\,(V_\mu^a(x)\,A_\nu^b(y)\,A_\lambda^c(0))\rangle_0$$

$$A_{\nu\lambda}(q) = i\int d^4x\,e^{iqx}\,\langle T\,(A_\nu^a(x)\,A_\lambda^a(0))\rangle_0 \left. \right\}$$
$$V_{\nu\lambda}(q) = i\int d^4x\,e^{iqx}\,\langle T\,(V_\nu^a(x)\,V_\lambda^a(0))\rangle_0 \left. \right\} \text{no sum on } a.$$

$$W_\lambda(q) = \int d^4x\,e^{iqx}\,\langle T\,(\partial_\nu A_\nu^a\,(x)\,A_\lambda^a(0))\rangle_0$$

$$E_{0\lambda}^A(q) = \int d^4x\,e^{iqx}\,\delta(x_0)\,\langle 0|\,[A_0^a(x),\,A_\lambda^a(0)]\,|0\rangle$$

$$E_{0\lambda}^V(q) = \int d^4x\,e^{iqx}\,\delta(x_0)\,\langle 0|\,[V_0^a(x),\,V_\lambda^a(0)]\,|0\rangle.$$

$$(1)$$

Assuming CVC (and PCAC)

$$\partial_\mu V_\mu\,(x) = 0 \tag{2a}$$

$$\partial_\mu A_a^\mu\,(x) = F_\pi m_\pi^2 \phi_a\,(x) \tag{2b}$$

it is straightforward to arrive at

$$q_\mu k_\nu M_{\mu\nu\lambda} = W_\lambda\,(q + k) - W_\lambda(k) - q_\mu A_{\mu\lambda}\,(q + k) + q_\mu V_{\mu\lambda}\,(q) \tag{3}$$

where we have used Eqs. (1) and (2a) together with the $c$-no. nature of S.T., which have non-vanishing matrix elements only between vacuum states. For a more general discussion of the effect of $I = 1$ operator Schwinger terms, the reader may consult the work of R. Perrin[28].

We observe quite simply that

$$W_\lambda(k) = k_\nu A_{\nu\lambda}(k) - E^A_{0\lambda}(k) \tag{4}$$

where

$$E^A_{0\lambda}(k) = k_\lambda \left[ F^2_\pi + \int d\mu^2 \frac{\varrho_A(\mu^2)}{\mu^2} \right], \quad \lambda = 1, 2, 3 = 0, \quad \lambda = 0. \tag{5}$$

Since Eq. (3) is valid for all $q$ and $k$, we may let $k$ tend to zero to get

$$W_\lambda(q) - q_\mu A_{\mu\lambda}(q) + q_\mu V_{\mu\lambda}(q) = 0 \tag{6}$$

since the L.H.S. of Eq. (3) is smooth in this limit.

Eq. (6) is equivalent to

$$E^A_{0\lambda}(q) = E^V_{\lambda 0}(q).$$

## References


1. H. Lehmann, *Nuovo Cimento*, **11**, 342 (1954).
2. K. Johnson, *Nuclear Physics*, **25**, 435 (1961).
3. W. Zimmermann, *Commun. math. Phys.*, **8**, 66 (1968).
4. K. Hayashi, M. Girayama, T. Muta, N. Saito and T. Shirafuji, *Fortsch. d. Physik*, **15**, 626 (1967) and references therein.
5. S. Weinberg, *Phys. Rev. Letters*, **18**, 507 (1967).
6. M. Gell–Mann, *Physics*, **1**, 63 (1964).
7. T. D. Lee, S. Weinberg and B. Zumino, *Phys. Rev. Letters*, **18**, 1029 (1967); R. Acharya, *Phys. Letters*, **24B**, 623 (1967).
8. J. Schwinger, *Phys. Rev. Letters*, **3**, 296 (1969).
9. T. Das, V. Mathur and S. Okubo, *Phys. Rev. Letters*, **18**, 761 (1967).
10. F. Gilman and H. Harari, *Phys. Rev.*, **165**, 1803 (1968).
11. C. Lovelace, CERN preprint TH 950, (1968).
12. J. Goldstone, *Nuovo Cimento*, **19**, 154 (1961).
13. E. Lu, *Phys. Rev.*, **159**, 1427 (1967).
14. R. Jackiw, *Phys. Letters*, **27B**, 96 (1968).
15. R. Acharya, *Nuclear Physics*, in Press (1969).
16. L. Chang and Y. Leung, *Phys. Rev. Letters*, **21**, 122 (1968).
17. T. G. Trippe et al., *Phys. Letters*, **28B**, 203 (1968).
18. R. Acharya, H. H. Aly, H. Mavromatis and K. Schilcher, *Z. Physik*, **211**, 324 (1968).
19. R. Acharya and H. H. Aly, *Phys. Letters*, **27B**, 166 (1968).
20. S. Glashow and S. Weinberg, *Phys. Rev. Letters*, **20**, 225 (1968).
21. R. Arnowitt, M. Friedman and P. Nath, Northeastern University preprint, 1968.
22. S. Okubo, International Theoretical Conference on Particles and Fields, August 1967, at the University of Rochester, Rochester, N.Y., U.S.A.

23. R. Acharya; "Weinberg's First Sum Rule For Axial-vector Currents", American University of Beirut preprint, 1968.
24. K. Dietz and H. Pietschmann, University of Bonn preprint, 1967.
25. W. Lam and K. Raman, *Phys. Rev.*, **173**, 1497 (1968).
26. Y. Frishman, *Phys. Rev.*, **159**, 1428 (1967).
27. P. Federbush and K. Johnson, *Phys. Rev.*, **120**, 1926 (1960); R. Jost and B. Schroer (unpublished).
28. R. Perrin, *Phys. Rev. Letters*, **20**, 306, 781 E (1968).

## Acknowledgements

I take this opportunity to express my gratitude to Professor Hadi Aly for urging me to record the material presented here and for several illuminating discussions. It is a pleasure to express my sincere appreciation to Prof. Dr. A. Mercier of the Institute of Theoretical Physics, University of Berne, Berne, Switzerland for the kind hospitality extended to me during the academic year 1968–1969.

# The Veneziano Model for Meson Systems and its Connection with the Chiral Symmetry


FAYYAZUDDIN and RIAZUDDIN

*University of Islamabad, Rawalpindi, Pakistan*


## Contents



## 1  INTRODUCTION

In the last few years the finite-energy sum rules[1] (FESR) or super convergent dispersion relations have been used to determine the parameters for high-energy Regge-pole fits from low-energy data. These sum rules are obtained by exploiting the analytic properties and asymptotic behaviour of a scattering amplitude.

Let $A(v, t)$ be a scattering amplitude odd under crossing, and analytic in the energy variable $v$ for fixed momentum transfer $t$. Following Igi[2] we write $A(v, t) = A'(v, t) + R(v, t)$, so that $A(v, t) \to R(v, t)$ as $v \to \infty$.

$R(\nu, t)$ is determined by a sum of Regge poles:

$$R(\nu, t) = -\sum_i \frac{\beta_i(t)}{\sin \pi \alpha_i(t)} \left[P_{\alpha_i(t)}(-\nu) - P_{\alpha_i(t)}(\nu)\right] \sim \sum_i \beta_i(t) \frac{1 - e^{i\pi\alpha_i(t)}}{\sin \pi \alpha_i(t)} \nu^{\alpha_i(t)}$$

(1.1)

$$\text{Im } R(\nu, t) = \sum_i \frac{\beta_i(t)}{\sin \pi \alpha_i(t)} \sin \pi \alpha_i(t) P_{\alpha_i(t)}(\nu) \sim \sum_i \beta_i(t) \nu^{\alpha_i(t)}. \quad (1.2)$$

By including enough Regge poles in $R$, it is then possible to write a super-convergent dispersion relation for any scattering amplitude of the form

$$\int d\nu \text{ Im } [A(\nu, t) - R(\nu, t)] = 0 \qquad (1.3)$$

If one then divides the region of integration into the parts $\nu < N$ and $\nu > N$ such that the amplitude is well represented by its Regge expansion for $\nu > N$, one obtains the famous FESR:

$$\int_0^N d\nu \text{ Im } A(\nu, t) = \sum_i \beta_i(t) \frac{N^{\alpha_i(t)}}{\alpha_i(t) + 1} \qquad (1.4)$$

where we have used Eq. (1.2).

We note first the empirical fact that the superconvergent relation (1.4) is quite well satisfied even down to the low energies ($\sim 1$ GeV), assuming on the right-hand side only a few leading Regge poles. If, in addition, the integral on the left hand side is approximately saturated by resonances, one can consider the Regge poles as being "built up" by summing a series of direct channel resonances. Moreover, since the relation (1.4) is supposed to be valid for all $N$, the equivalence is "semi-local" meaning that the Regge poles approximate the contributions of resonances averaged over a small region in energy. This is the so-called Dolen-Horn-Schmid duality[3,4].

Clearly, one needs a model which can interpolate smoothly between direct channel resonances and Regge-poles exchange. Such a model was attempted by Veneziano[5]. He gave a simple formula which exhibits the Regge pole resonance duality. In his representation of a scattering amplitude, one has:

i) explicit crossing symmetry

ii) Regge asymptotic behaviour in all three channels,

iii) poles at points corresponding to resonances

iv) Validity of all super convergent dispersion relations at all values of $t$.

Recently Lovelace[6] has used the Veneziano model to obtain some remarkable results. He has shown the exciting prospect that it might be possible to unify the current-algebra approach and the Regge theory. He has applied the Veneziano model to $\pi$–$\pi$ scattering. Then by imposing the Adler[7] "self consistency condition", he has derived a number of relations among meson decay widths and their masses. These results are the same as implied by the current algebra or the chiral symmetry. The Veneziano representation may thus be connected with the chiral symmetry.

The purpose of this review is to explore this connection. The plan of this article is as follows. In Section 2, we review the Veneziano model for $\pi$–$\pi$ scattering, and derive the relations between $\varrho$ and $\sigma$ widths and their masses and also derive an expression for the ratio for $\pi$–$\pi$ scattering lengths in $I = 0$ and $I = 2$ states. In this section we also show how the constraints of the currunt algebra for $K \to 3\pi$ decay follow from the Veneziano representation for $\pi$–$\pi$ scattering. In Section 3, similar results are derived for $\pi$–$K$ scattering. In Section 4, we present an evidence that the current divergence belong to $(3, 3^*) + (3^*, 3)$ representation of the $SU(3) \times SU(3)$. We also give the various estimates of the parameter $c$ that describes the strength of the $SU(3)$-breaking term relative to the chiral symmetry-breaking term by applying the Veneziano representation for $\pi$–$\pi$ and $\pi$–$K$ scattering. The ratio $f_K/f_\pi$ is also estimated, where $f_K$ and $f_\pi$ are respectively the decay constants for $K \to l + \nu$ and $\pi \to l + \nu$. In Section 5 we further elucidate the connection between this model and the chiral symmetry. We show that the Veneziano model can accommodate both the versions of chiral symmetry—the gauge as well as the algebraic. In the last section, we briefly discuss the reaction $\pi K \to \pi K_A$ anf $K_{l4}$ form factors. It is shown that the constraints of the current algebra, namely, the zeros of the form factors $F_1$ and $F_2$ (defined in the text) in the soft pion limit are correctly given by the Veneziano model.

## 2  $\pi$–$\pi$ SCATTERING[6, 8, 4]

We consider the scattering process

$$\pi_l(q) + \pi_j(p) \to \pi_m(q') + \pi_k(p')$$

for which we can write the $T$-matrix as

$$T_{jk}^{lm}(s, u, t) = \frac{1}{(2\pi)^6}\, \frac{1}{(16 p_0 p_0' q_0 q_0')^{1/2}}\, A_{jk}^{lm}(s, u, t) \qquad (2.1)$$

where

$$s = -(p + q)^2 = -(p' + q')^2$$

$$u = -(p - q')^2 = -(p' - q)^2$$

$$t = -(p - p')^2 = -(q - q')^2$$

$$s + u + t = 4m_\pi^2 \tag{2.2}$$

The amplitude $A_{jk}^{lm}$ can be decomposed as follows:

$$A_{jk}^{lm}(s, u, t) = \delta_{lj}\delta_{mk}A_1(s, u, t) + \delta_{lm}\delta_{jk}A_2(s, u, t) + \delta_{lk}\delta_{mj}A_3(s, u, t). \tag{2.3}$$

The crossing symmetry implies the following relations:

$$\left. \begin{aligned} A_1(s, u, t) &= A_3(u, s, t) \\ A_2(s, u, t) &= A_2(u, s, t) \end{aligned} \right\} \tag{2.4a}$$

$$\left. \begin{aligned} A_1(s, u, t) &= A_2(t, u, s) \\ A_3(s, u, t) &= A_3(t, u, s) \end{aligned} \right\} \tag{2.4b}$$

$$\left. \begin{aligned} A_1(s, u, t) &= A_1(s, t, u) \\ A_2(s, u, t) &= A_3(s, t, u) \end{aligned} \right\} \tag{2.4c}$$

Also we note

$$A_1 = \tfrac{1}{3}[A^{(0)} - A^{(2)}]$$

$$A_2 = \tfrac{1}{2}[A^{(1)} + A^{(2)}] \tag{2.5}$$

$$A_3 = -\tfrac{1}{2}[A^{(1)} - A^{(2)}]$$

where $A^{(i)}$ ($i = 0, 1, 2$) represents the amplitude for a particular isospin state.

From Eqs. (2.4) and (2.5), we see that for $A_1$ amplitude in the $s$-channel, we have $I = 0$ and $I = 2$ exchanges, whereas in the $u$- and $t$-channels we have $I = 1$ and $I = 2$ exchanges. Since, there is no resonance with $I = 2$ in the $s$-channel, we have only the $f$ trajectory in the $s$-channel, whereas in the $u$- and $t$-channels we have the $\varrho$-trajectory. The amplitude $A_1$ should show a string of resonances (poles) in the $s$-, $u$- and $t$-channels located on the $f$ and $\varrho$ trajectories respectively. The amplitude $A_1$ should also have the asymptotic Regge behaviour in all the channels. We show that this is indeed the case for the following representation of the amplitude $A_1$

$$A_1(s, u, t) = -\gamma \left[ \frac{\Gamma(1 - \alpha_f(s))\,\Gamma(1 - \alpha_\varrho(t))}{\Gamma(1 - \alpha_f(s) - \alpha_\varrho(t))} - \frac{\Gamma(1 - \alpha_\varrho(u))\,\Gamma(1 - \alpha_\varrho(t))}{\Gamma(1 - \alpha_\varrho(u) - \alpha_\varrho(t))} \right.$$

$$\left. + \frac{\Gamma(1 - \alpha_f(s))\,\Gamma(1 - \alpha_\varrho(u))}{\Gamma(1 - \alpha_f(s) - \alpha_\varrho(u))} \right] \tag{2.6}$$

where

$$\alpha(s) = \alpha(0) + \alpha' \cdot s, \quad \alpha' > 0. \tag{2.7}$$

The Veneziano representation for the amplitudes $A_2$ and $A_3$ are obtained from $A_1$ by using the crossing relations (2.4). We also note that for the reaction $\pi^+\pi^- \to \pi^+\pi^-$, the relevant amplitude is $A_1 + A_2$, which is symmetric under $s \leftrightarrow t$, and has $I = 2$ in $u$-channel. Hence it follows from Eq. (2 6) that the $\varrho$ and $f$ trajectories must be exchange degenerate so that $\alpha_f = \alpha_\varrho = \alpha$. Thus Eq. (2.6) becomes

$$A_1(s, u, t) = -\gamma \left[ \frac{\Gamma(1 - \alpha(s))\,\Gamma(1 - \alpha(t))}{\Gamma(1 - \alpha(s) - \alpha(t))} - \frac{\Gamma(1 - \alpha(u))\,\Gamma(1 - \alpha(t))}{\Gamma(1 - \alpha(u) - \alpha(t))} \right.$$

$$\left. + \frac{\Gamma(1 - \alpha(s))\,\Gamma(1 - \alpha(u))}{\Gamma(1 - \alpha(s) - \alpha(u))} \right] \tag{2.8}$$

and

$$A_1(s, u, t) + A_2(s, u, t) = -2\gamma \left[ \frac{\Gamma(1 - \alpha(s))\,\Gamma(1 - \alpha(t))}{\Gamma(1 - \alpha(s) - \alpha(t))} \right]. \tag{2.8a}$$

First we note that $\Gamma(z)$ has poles at $z = -n, n = 0, 1, 2, \ldots$ with residues $(-1)^n/n!$. Thus the $s$-channel resonances, which have zero widths, are given by the poles of $\Gamma(1 - \alpha(s))$. The $\Gamma$-function in the denominator is just right to make the residues of these poles polynomials in $t$. The polynomials are not Legendre polynomials, so the daughters appear, but with unit spacing $\Delta J = 1$. Thus for $\alpha(s) = J \geq 1$, we have together with the leading particle exchanged with spin $J$ a series of daughters with spins $J-1, J-2, \ldots, 0$.

In the neighbourhood of an $s$-channel pole $\alpha(s) \sim I$, one may write $\Gamma(1 - \alpha(s))$ as:

$$\Gamma(1 - \alpha(s)) \sim \frac{(-1)^{J-1}}{(J - 1)!\,(M_J^2 - s)\,\alpha'} \tag{2.9}$$

with $\alpha(M_J^2) = J$. As a result, the pole residues in Eq. (2.8) at $\alpha(s) = 1$ and $\alpha(t) = 1$ are respectively given by

$$R_{A_1} = \frac{\gamma}{\alpha'} \left[\alpha(t) + \alpha(u)\right] = \frac{\gamma}{\alpha'} \left[2\alpha(0) + \alpha'\, (4m_\pi^2 - s)\right]_{s=m_\varrho^2}, \quad (2.10a)$$

$$R_{A_1} = \frac{\gamma}{\alpha'} \left[\alpha(s) - \alpha(u)\right] = \gamma\, [s - u]_{t=m_\varrho^2}. \quad (2.10b)$$

We now show that the reprerentation (2.8) gives the correct asymptotic behaviour. From the relation

$$\frac{\Gamma(z + a)}{\Gamma(z + b)} \xrightarrow[z \to \infty]{} z^{a-b}$$

for $-\pi < \arg z < \pi$, the asymptotic behaviour of the first term of Eq. (2.8) is given by

$$-\gamma\, (-\alpha(s))^{\alpha(t)}\, \Gamma(1 - \alpha(t)) = -\frac{\pi\gamma}{\Gamma(\alpha(t))\, \sin \pi\alpha\,(t)}\, (-\alpha' s)^{\alpha(t)}. \quad (2.11)$$

This is the Regge behaviour. Moreover, the residue contains the required factor $\dfrac{1}{\Gamma(\alpha(t))}$ and no further $t$-dependence. To see that one gets the correct signature, we consider the first two terms of Eq. (2.8), namely

$$-\gamma \left[\frac{\Gamma(1 - \alpha(s))\, \Gamma(1 - \alpha(t))}{\Gamma(1 - \alpha(s) - \alpha(t))} - \frac{\Gamma(1 - \alpha(u))\, \Gamma(1 - \alpha(t))}{\Gamma(1 - \alpha(u) - \alpha(t))}\right]. \quad (2.12)$$

The second term is purely real (for positive $s$) and goes like

$$\frac{\pi}{\Gamma(\alpha(t)\, \sin \pi\alpha\,(t)}\, (\alpha' s)^{\alpha(t)}.$$

Now using the relation

$$\Gamma(z)\, \Gamma(1 - z) = \frac{\pi}{\sin \pi z}$$

the first term of Eq. (2.12) can be written as

$$\frac{\pi}{\Gamma(\alpha(t))\, \sin \pi\alpha\,(t)} \left[\frac{\pi}{\sin \pi\alpha\,(s)}\, \frac{\sin \pi\,(\alpha(s) + \alpha(t))}{\pi}\, \frac{\Gamma(\alpha(s) + \alpha(t))}{\Gamma(\alpha(s))}\right]$$

$$= \frac{\pi}{\Gamma(\alpha(t))} \left[\frac{\cos \pi\alpha\,(t)}{\sin \pi\alpha\,(t)} + \cot \pi\alpha\,(s)\right] (\alpha' s)^{\alpha(t)}$$

so that Eq. (2.12) has the following behaviour

$$-\frac{\gamma\pi}{\Gamma(\alpha(t))}\left[\frac{\cos \pi\alpha(t) - 1}{\sin \pi\alpha(t)}(\alpha's)^{\alpha(t)} + \cot \pi\alpha(s)(\alpha's)^{\alpha(t)}\right] \qquad (2.13)$$

for $|s| \to \infty$, $\arg s \to 0^+$. The first term corresponds to the properly behaved real part with a correct signature factor. The second term corresponds to the absorptive part. It has a succession of poles on the real axis. At this stage one may smooth them out, letting the imaginary part of $\alpha$ to increase with $s$ as it normally should. One then has $\cot \pi\alpha(s) \to -i$ as $s \to \infty$ and obtains from Eq. (2.13), the standard Regge behaviour:

$$\frac{\gamma\pi}{\Gamma(\alpha(t))}\frac{1 - e^{-i\pi\alpha(t)}}{\sin \pi\alpha(t)}(\alpha's)^{\alpha(t)}. \qquad (2.14)$$

It can be shown that the third term in Eq. (2.8) also does not violate the Regge behaviour.

The Veneziano representation is only strictly valid when the trajectories are exactly linear, so that the resonances are approximated by poles. To give the resonances finite widths, we must add an imaginary part to $\alpha(s)$ for $s > 4m_\pi^2$. Unfortunately there is no consistent way of doing this. The resonances at the same mass then acquire the same total width. But formula (2.8) gives different elastic widths for resonances at the same mass. So we get contradiction with elastic unitarity. We shall not worry too much about this, as we shall confine ourselves to the low energy because we are interested in the relationship between this model and the chiral symmetry.

Now the Adler's self consistency condition requires (2.8) to vanish when $s = t = u = m_\pi^2$ and one of the pion has zero mass. This can happen if, at $s = u = t = m_\pi^2$, the denominator in each of the three terms of (2.8) vanishes, which it does provided that[6]

$$\alpha(m_\pi^2) = \tfrac{1}{2}. \qquad (2.15)$$

This gives

$$\alpha(0) = \tfrac{1}{2} - \alpha'm_\pi^2 \approx \tfrac{1}{2}$$

$$\qquad (2.16)$$

$$\alpha' = \frac{1}{2(m_\varrho^2 - m_\pi^2)} \approx \frac{1}{2m_\varrho^2}.$$

Now by the Feynman rules, the contributions of $\varrho$ and $\sigma$ poles to the amplitude $A$ in the $s$- and $t$-channels are given by

$$\gamma_{\varrho\pi\pi}^2 \left(\delta_{ml}\delta_{kj} - \delta_{mj}\delta_{kl}\right)(t - u)\,\frac{1}{m_\varrho^2 - s}$$

$$\gamma_{\varrho\pi\pi}^2 \left(\delta_{lj}\delta_{mk} - \delta_{mj}\delta_{kl}\right)(s - u)\,\frac{1}{m_\varrho^2 - t} \tag{2.17a}$$

$$\gamma_{\sigma\pi\pi}^2 \delta_{lj}\delta_{mk}\,\frac{1}{m_\sigma^2 - s}$$

$$\gamma_{\sigma\pi\pi}^2 \delta_{lm}\delta_{jk}\,\frac{1}{m_\sigma^2 - t} \tag{2.17b}$$

First we note that there cannot be any $\varrho$-pole in the $s$-channel for the amplitude $A_1$. As is clear from Eq. (2.10a) the Veneziano model gives a pole at $s = m_\varrho^2$ which contributes to the $s$-wave. Therefore the Veneziano model predicts $\sigma$ degenerate with $\varrho$ (daughter of $\varrho$). Now comparing the residues of the poles from Eq. (2.17) with those given in Eq. (2.10) and using Eq. (2.16), we obtain

$$\gamma = \gamma_{\varrho\pi\pi}^2, \tag{2.18}$$

$$\gamma_{\sigma\pi\pi}^2 = m_\varrho^2 \gamma_{\varrho\pi\pi}^2. \tag{2.19}$$

This gives

$$\Gamma(\sigma \to \pi\pi) = \tfrac{9}{2}\,\Gamma(\varrho \to \pi\pi) = 608\ \text{MeV} \tag{2.19a}$$

for $\varrho$ width of 135 MeV.

Also from Eqs. (2.8) and (2.4), we find the ratio of $I = 0$ and $I = 2$ $s$-wave scattering lengths to be $a_0/a_2 \approx -\tfrac{7}{2}$ in agreement with the current algebra result[9]. To proceed further, we note that for the linear approximation around $s = t = u = m_\pi^2$

$$\frac{\Gamma\left(1 - \alpha(s)\right)\Gamma\left(1 - \alpha(t)\right)}{\Gamma\left(1 - \alpha(s) - \alpha(t)\right)} \approx \frac{\Gamma(\tfrac{1}{2})\,\Gamma(\tfrac{1}{2})}{\Gamma\left(2\alpha' m_\pi^2 - \alpha'(s + t)\right)} = -\pi\alpha'\left(s + t - 2m_\pi^2\right)$$

so that

$$A_1 = 2\pi\gamma_{\varrho\pi\pi}^2\alpha'\left(s - m_\pi^2\right).$$

Hence using Eq. (2.4), we get in the linear approximation

$$A_{jk}^{lm} = 2\pi\alpha'\gamma_{\varrho\pi\pi}^2 \left[(s - m_\pi^2)\,\delta_{lj}\delta_{mk} + (u - m_\pi^2)\,\delta_{lk}\delta_{mj} + (t - m_\pi^2)\,\delta_{lm}\delta_{jk}\right]. \tag{2.20}$$

Now the Weinberg[9] amplitude for $\pi$–$\pi$ scattering is given by

$$A_{jk}^{lm} = \frac{2}{f_\pi^2} \left[(s - m_\pi^2)\,\delta_{lj}\delta_{mk} + (u - m_\pi^2)\,\delta_{lk}\delta_{mj} + (t - m_\pi^2)\,\delta_{lm}\delta_{jk}\right]. \quad (2.21)$$

The structure of (2.21) follows from the Adler relation, the $SU(3) \times SU(3)$ current algebra and the absence of $I = 2\sigma$ term. This last assumption is translated in our approach by the absence of force in the $I = 2$ state which is due to the absence of any resonance in this channel. We can, therefore, compare Eq. (2.20) with Eq. (2.21) to get the relation

$$2\pi\alpha'\gamma_{\varrho\pi\pi}^2 = \frac{2}{f_\pi'^2}$$

or

$$\gamma_{\varrho\pi\pi}^2 = \frac{2}{2\pi\alpha'}\frac{1}{f_\pi^2} \approx \frac{2}{\pi}\frac{m_\varrho^2}{f_\pi^2}. \quad (2.22)$$

This differs from the KSRF value[10] by a factor $\dfrac{2}{\pi}$ ; the $\varrho$ width obtained from the Veneziano formula being smaller than the one obtained from the KSRF relation.

As emphasized by Lovelace[6], the Veneziano amplitude can be used to describe the three pion final state interaction in a pseudoscalar meson decay process (e.g. $K \to 3\pi$). We can continue one of the external pions to a large positive mass, i.e. we consider the $3\pi$ decay of a particle with the same quantum members as the pion (such as $K \to 3\pi$ decay with $\Delta I = \frac{1}{2}$ rule). This will have the same Regge trajectories as $\pi$–$\pi$ scattering so that the corresponding Veneziano formulas can only differ in the values of $\gamma$ in (2.8). We can then predict every thing about this decay except the total decay rate.

We can visualize that $K \to 3\pi$ decay proceeds as shown in Fig. 1. The matrix elements for $K \to 3\pi$ is defined by

$$-i\,\langle\pi_1\pi_2\pi_3|\,H_W\,|K\rangle = \frac{1}{(2\pi)^6}\frac{1}{(16K_0 q_{10}q_{20}q_{30})^{1/2}}\,A\,(s_1, s_2, s_3) \quad (2.23)$$

where

$$s_1 = -(K - q_1)^2 = -(q_2 + q_3)^2$$

$$s_2 = -(K - q_2)^2 = -(q_1 + q_3)^2$$

$$s_3 = -(K - q_3)^2 = -(q_1 + q_2)^2. \quad (2.24)$$

From Fig. 1 it is clear that $A\,(s_1,s_2,s_3)$ is determined apart from a constant by

$$\langle \pi_2\pi_3| J_\pi |\pi_1\rangle = \gamma^2_{\varrho\pi\pi}\,\{[C\,((1-\alpha(s_1)),(1-\alpha(s_2)))$$
$$-\,C\,((1-\alpha(s_3)),(1-\alpha(s_2)))+C\,((1-\alpha(s_1)),(1-\alpha(s_3)))]\times$$
$$\times\,\delta_{lj}\delta_{mk}+(s_1\leftrightarrow s_2)\,\delta_{lm}\delta_{jk}+(s_1\leftrightarrow s_3)\,\delta_{lk}\delta_{mk}\}\qquad(2.25)$$

where we have used Eqs. (2.8) and (2.4) and

$$C\,((1-\alpha(s_1)),(1-\alpha(s_2)))=\frac{\Gamma\,(1-\alpha(s_1))\,\Gamma\,(1-\alpha(s_2))}{\Gamma\,(1-\alpha(s_1)-\alpha(s_2))}.\qquad(2.26)$$

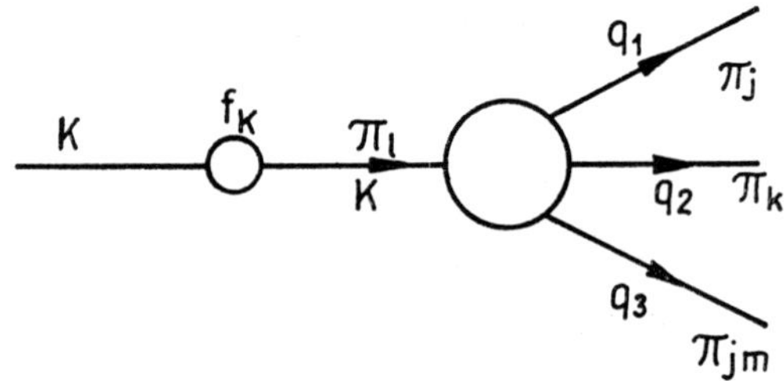


**Figure 1**   $K\to 3\pi$ Decay

We now consider the particular processes:

$$K^+ \to \pi^+(q_1) + \pi^+(q_2) + \pi^-(q_3)\qquad(2.27\,a)$$
$$K^+ \to \pi^0(q_1) + \pi^0(q_2) + \pi^+(q_3)\qquad(2.27\,b)$$
$$K^0 \to \pi^+(q_1) + \pi^-(q_2) + \pi^0(q_3)\qquad(2.27\,c)$$
$$K^0 \to \pi^0(q_1) + \pi^0(q_2) + \pi^0(q_3).\qquad(2.27\,d)$$

For these processes we have respectively the following Veneziano amplitudes

$$A^{++-}\,(s_1,s_2,s_3)=2N\,[C\,((1-\alpha(s_1)),(1-\alpha(s_2)))]\qquad(2.28\,a)$$
$$A^{00+}\,(s_1,s_2,s_3)=N\,[C\,(1-\alpha(s_3)),(1-\alpha(s_2)))$$
$$-\,C\,((1-\alpha(s_1)),(1-\alpha(s_2)))$$
$$+\,C\,((1-\alpha(s_1)),(1-\alpha(s_3)))]$$
$$=A^{+-0}\,(s_1,s_2,s_3)\qquad(2.28\,b)$$
$$A^{000}\,(s_1,s_2,s_3)=N\,[C\,((1-\alpha(s_1)),(1-\alpha(s_2)))$$
$$+\,(s_1\to s_3)+(s_2\to s_3)]\qquad(2.28\,c)$$

where $N$ is a constant involving $\gamma^2_{\varrho\pi\pi}$ and an effective weak coupling constant. From Eqs. (2.28 a) and (2.28 b) we see that the amplitude $A^{++-}$ vanishes when $q_3 \to 0$ ($s_1 = m^2_\pi$, $s_2 = m^2_\pi$) whereas the amplitude $A^{00+}$ vanishes when either $q_1 \to 0$ ($s_2 = m^2_\pi$, $s_3 = m^2_\pi$) or $q_2 \to 0$ ($s_1 = m^2_\pi$, $s_3 = m^2_\pi$). From Eq. (2.28 c), it is clear that the amplitude $A^{000}$ does not vanish when any of the three pions is soft. These results are in agreement with the current algebra results[11,12]. Further, if we write the amplitude $A^{++-}$ in the linear approximation as $[s_0 = \tfrac{1}{3}m^2_K + m^2_\pi]$

$$A^{++-}(s_1, s_2, s_3) = A^{++-}(0)\left[1 + \frac{\sigma}{2m^2_\pi}(s_3 - s_0)\right] \qquad (2.29)$$

it follows from Eq. (2.28 a) that

$$\sigma = -\frac{3m^2_\pi}{m^2_K - m^2_\pi} \qquad (2.30)$$

in agreement with the current algebra result[12].

## 3  $\Pi$–$K$ SCATTERING[6, 8]

We now examine the Veneziano model for pion-kaon scattering. The scattering amplitude for his case is decomposed as follows:

$$T_{ji}(s, u, t) = \delta_{ji}A^{(+)}(s, u, t) + \tfrac{1}{2}[\tau_j, \tau_i]A^{(-)}(s, u, t). \qquad (3.1)$$

We write the following Veneziano representation for the amplitude $A^{(\pm)}$

$$A^{(\pm)} = -\frac{1}{2}\gamma'\left[\frac{\Gamma(1 - \alpha_{K*}(s))\,\Gamma(1 - \alpha(t))}{\Gamma(1 - \alpha_{K*}(s) - \alpha(t))} \pm \frac{\Gamma(1 - \alpha_{K*}(u))\,\Gamma(1 - \alpha(t))}{\Gamma(1 - \alpha_{K*}(u) - \alpha(t))}\right].$$
$$(3.2)$$

In order to get the right signature in the Regge behaviour of the amplitudes $A^{(\pm)}$, the slope of the $K^*$-trajectory must be the same as for the $\varrho$-trajectory. Further, the Adler's consistency condition $A^{(+)}(m^2_K, m^2_K, m^2_K) = 0$ implies that $\alpha_{K*}(m^2_K) = \tfrac{1}{2}$, which gives

$$\frac{1}{\alpha'} = 2(m^2_{K*} - m^2_K) = 2(m^2_\varrho - m^2_\pi) \simeq 2m^2_\varrho,$$
$$(3.3)$$
$$\alpha_{K*}(0) = \frac{1}{2} - \frac{m^2_K}{2(m^2_{K*} - m^2_K)} \simeq \frac{1}{2} - \frac{m^2_K}{2m^2_\varrho}.$$

The contributions of $K^*$ and $\kappa$ mesons to the amplitudes $A^{(\pm)}$ in the $s$-channel are given by

$$A_{K*}^{(\pm)} = \gamma_{K*K\pi}^2 \frac{4q^{*2}\cos\theta}{m_{K*}^2 - s} \tag{3.4}$$

$$A_\kappa^{(\pm)} = \gamma_{\kappa K\pi}^2/(m_\kappa^2 - s) \tag{3.5}$$

where $q^*$ is the c.m. momentum at $s = m_{K*}^2$. On the other hand, the contribution of $\varrho$ and $\sigma$ mesons to these amplitudes in the $t$-channel are given by

$$A_\varrho^{(-)} = \frac{\gamma_{\varrho\pi\pi}\gamma_{\varrho K\bar{K}}}{m_\varrho^2 - t}(u - s), \quad A_\varrho^{(+)} = 0 \tag{3.6}$$

$$A_\sigma^{(+)} = \frac{\gamma_{\sigma\pi\pi}\gamma_{\sigma K\bar{K}}}{m_\sigma^2 - t}, \quad A_\sigma^{(-)} = 0. \tag{3.7}$$

From Eq. (3.2), we see that residues $R$ at $s = m_{K*}^2$ and $t = m_\varrho^2$ are given by

$$R_{K*}^{(\pm)} = \frac{1}{4}\gamma'\left[4q^{*2}\cos\theta + m_{K*}^2 - \frac{(m_K^2 - m_\pi^2)^2}{m_{K*}^2}\right] \tag{3.8}$$

$$R_\varrho^{(+)} = \tfrac{1}{2}\gamma' m_\varrho^2 \tag{3.9}$$

$$R_\varrho^{(-)} = \tfrac{1}{2}\gamma'(s - u). \tag{3.10}$$

Comparing Eq. (3.8) with Eqs. (3.4) and (3.5), we see that the $s$-wave contribution in Eq. (3.8) must be interpreted as arising due to $\kappa$ meson with its mass equal to $m_{K*}$ (daughter of $K^*$), so that we get by this comparison

$$\gamma_{K*K\pi}^2 = \tfrac{1}{4}\gamma' \tag{3.11}$$

$$\gamma_{\kappa K\pi}^2 = \frac{1}{4}\gamma'\left[m_{K*}^2 - \frac{(m_K^2 - m_\pi^2)^2}{m_{K*}^2}\right]. \tag{3.12}$$

Comparing Eqs. (3.9) and (3.10) with Eqs. (3.6) and (3.7) we see that $\sigma$ meson must be degenerate with $\varrho$ meson and we obtain

$$\gamma_{\varrho\pi\pi}\gamma_{\varrho K\bar{K}} = -\tfrac{1}{2}\gamma'$$

$$\gamma_{\sigma\pi\pi}\gamma_{\sigma K\bar{K}} = \tfrac{1}{2}\gamma' m_\varrho^2. \tag{3.13}$$

From Eq. (3.12), we get[13]

$$\frac{\Gamma(\kappa \to K\pi)}{\Gamma(K^* \to K\pi)} = 3\frac{[m_{K*}^4 - (m_K^2 - m_\pi^2)^2]}{[(m_{K*}^2 - m_K^2 - m_\pi^2)^2 - 4m_\pi^2 m_K^2]} \tag{3.14}$$

which gives

$$\Gamma(\kappa \to K\pi) = 345 \text{ MeV} \quad \text{for} \quad \Gamma(K^* \to K\pi) = 50 \text{ MeV}.$$

## 4  SYMMETRY BREAKING[14]

In this section we apply the Veneziano model for $\pi$–$\pi$ and $\pi$–$K$ scattering to get information about the symmetry-breaking parameter $c$. We present some evidence in support of the commutation relations[15] of scalar and pseudo-scalar densities $S_i$, $P_i$ with vector and axial vector charges, namely

$$[F_i(t), S_j(\varkappa, t)] = if_{ijk}S_k(\varkappa) \tag{4.1a}$$

$$[F_i(t), P_j(\varkappa, t)] = if_{ijk}P_k(\varkappa) \tag{4.1b}$$

$$[F_i^5(t), S_j(\varkappa, t)] = id_{ijk}P_k(\varkappa) \tag{4.1c}$$

$$[F_i^5(t), P_j(\varkappa, t)] = -id_{ijk}S_k(\varkappa) \tag{4.1d}$$

which specify the transformation properties of the scalar and pseudosclar nonets $S_i$ and $P_i$ as belonging to the representation $(3, 3^*) + (3^*, 3)$ of $SU(3) \times SU(3)$. Assumption is also made that $S_0$ and $S_8$ components of the nonet $S_i$ specify the symmetry-breaking terms in the hadron energy density except for scale factors i.e. one writes the hadron energy density as

$$H = H_0 + \delta m_0 S_0 + \delta m S_8 \tag{4.2}$$

where $H_0$ is invariant under $SU(3) \times SU(3)$ and $\delta m_0$ and $\delta m$ express respectively the scales of $SU(3) \times SU(3)$ and $SU(3)$ symmetry breaking. The current divergences can be simply obtained in terms of $S_i$ and $P_i$ from Eq. (4.2) and the commutation relations (4.1a,c) and are given by

$$\partial_\mu V_{i\mu} = -i[F_i, H] = -\delta m f_{8ik}S_k \tag{4.3}$$

$$\partial_\mu A_{i\mu} = -i[F_i^5, H] = \delta m_0 \left(\sqrt{\tfrac{2}{3}}\, P_i + cd_{8ik}P_k\right). \tag{4.4}$$

Recently attempts have been made to estimate the scale parameter[16,17] $c$ $\left(c = \dfrac{\delta m}{\delta m_0}\right.$ measures the strength of the $SU(3)$ symmetry-breaking term relative to the $SU(3) \times SU(3)$ symmetry breaking term$\left.\right)$. It is easy to see from Eq. (4.3) that[18]

$$(m_K^2 - m_\pi^2) = -\sqrt{\tfrac{3}{2}}\, \delta m s_d' \tag{4.5}$$

where $s'_d$ is defined by ($P_i$ denotes the octet of pseudoscalar mesons)

$$\langle P_k(p')| \, S_i \, |P_j(p)\rangle = \frac{1}{(2\pi)^3} \, \frac{1}{\sqrt{4p'_0 p_0}} \left[ \frac{1}{\sqrt{2}} \, s''_d \delta_{i0}\delta_{jk} + d_{ijk}s'_d \right] \quad (4.6)$$

and $j, k = 1, \ldots, 8$, and $i = 0, \ldots, 8$. By using the soft meson technique partial conservation of axial vector current (PCAC) and the commutation relation (4.1c), $s''_d$ and the parameter $c$ have been estimated by Gell-Mann, Oakes and Renner[16] to be

$$s''_d \approx 0, \quad C = -2\sqrt{2} \, \frac{m_K^2 - m_\pi^2}{2m_K^2 + m_\pi^2} \approx -1.256. \quad (4.7)$$

It is important to note that in this estimate the $\eta$–$\eta'$ mixing has been neglected.

It is desirable to confirm the universality of $c$. Moreover, one should point out that in the above estimate of $c$, the commutation relation (4.1d) has not been used. It is, therefore, important that one should consider such processes where the commutation relation (4.1d) is explicitly used and obtain an independent estimate of the parameter $c$. It is well known[19] that, by using the commutation relation (4.1d), PCAC and the soft meson technique, one can express a scattering amplitude involving at least two pseudoscalar mesons (where these two pseudoscalar mesons are off the mass shell) in terms of $S_0$ and $S_8$ and the parameter $c$. An attempt to estimate $c$ in this way has been able to establish only a lower bound on $c$, $|c| \geq 0.7$. The Veneziano model for $\pi$–$\pi$ and $\pi$–$K$ scattering enables us to extrapolate the amplitude concerned to the unphysical values of the scattering variables $s, u, t$ at which values the amplitude is expressible in terms of $c$ and thus provides a measure of $c$. This value of $c$ is in an excellent agreement with that given in Eq. (4.7). The ratio $f_K/f_\pi$ can also be estimated, where $f_K$ and $f_\pi$ are the decay constants for $K \to \mu\nu$ and $\pi \to \mu\nu$ respectively.

We consider the scattering amplitude for the process

$$P_i(q) + P_j(p) \to P_m(q') + P_k(p') \quad (4.8)$$

for which we can write the $T$-matrix as

$$T_{jk}^{lm}(s, u, t) = \frac{1}{(2\pi)^6} \, \frac{1}{(16p_0 p'_0 q_0 q'_0)^{1/2}} \, A_{jk}^{lm}(s, u, t). \quad (4.9)$$

The soft meson technique and the use of PCAC

$$\partial_\lambda A_\lambda^0 = \partial_\lambda A_{1+i2,\,\lambda} = f_\pi m_\pi^2 \pi^-$$

$$\partial_\lambda A_\lambda^1 = \partial_\lambda A_{4+i5,\,\lambda} = f_K m_K^2 K^- \quad (4.10)$$

give

$$A_{jk}^{\pi^-\pi^-}(s = m_j^2, u = m_j^2, t = 0)$$

$$= \frac{1}{f_\pi^2}(2\pi)^3\, 2p_0 i\, \langle P_k(p)|\,[\partial_\lambda A_\lambda^0, F_{4-i5}^5]\,|P_j(p)\rangle \qquad (4.11)$$

$$A_{jk}^{K^-K^-}(s = m_j^2, u = m_j^2, t = 0)$$

$$= \frac{1}{f_K^2}(2\pi)^3\, 2p_0 i\, \langle P_k(p)|\,[\partial_\lambda A_\lambda', F_{4-i5}^5]\,|P_j(p)\rangle. \qquad (4.12)$$

Taking in (4.11), $j, k = \dfrac{1 \pm i2}{\sqrt{2}}$ and $\dfrac{4 \pm i5}{\sqrt{2}}$ in turn and in (4.12) $j, k = \dfrac{1 \pm i2}{\sqrt{2}}$,

we obtain, on using (4.4) and the commutation relation (4.1d)

$$A_{\pi^+\pi^+}^{\pi^-\pi^-}(s = u = m_\pi^2, t = 0)$$

$$= -\frac{1}{f_\pi^2}\frac{2}{3}\left(1 + \frac{\sqrt{2}}{c}\right)\delta m\,\langle \pi^+(p)|\,S_8 + \sqrt{2}\,S_0|\,\pi^+(p)\rangle$$

$$= \frac{1}{f_\pi^2}\frac{4}{3}\left(1 + \frac{\sqrt{2}}{c}\right)(m_K^2 - m_\pi^2)\left[1 + \frac{2}{\sqrt{3}}\frac{s_d''}{s_d'}\right] \qquad (4.13\text{a})$$

$$A_{K^+K^+}^{\pi^-\pi^-}(s = u - m_K^2, t = 0) = \frac{1}{f_\pi^2}\frac{2}{3}\left(1 + \frac{\sqrt{2}}{c}\right)(m_K^2 - m_\pi^2)\left[1 + \frac{2}{\sqrt{3}}\frac{s_d''}{s_d'}\right]$$

$$(4.13\text{b})$$

$$A_{\pi^+\pi^+}^{K^-K^-}(s = u = m_\pi^2, t = 0) = \frac{1}{f_K^2}\left(\frac{2\sqrt{2}}{c} - 1\right)\frac{1}{3}(m_K^2 - m_\pi^2)\left[1 + \frac{2}{\sqrt{3}}\frac{s_d''}{s_d'}\right]$$

$$(4.13\text{c})$$

where we have made use of Eqs. (4.5) and (4.6).

We now use the Veneziano representations (2.8 a) and (3.2) to estimate $A_{\pi^+\pi^+}^{\pi^-\pi^-}(s = u = m_\pi^2, t = 0)$, $A_{K^+K^+}^{\pi^-\pi^-}(s = u = m_K^2, t = 0)$ and $A_{\pi^+\pi^+}^{K^-K^-}(s = u = m_\pi^2, t = 0)$ and obtain respectively

$$A_{\pi^+\pi^+}^{\pi^-\pi^-}(s = u = m_\pi^2, t = 0) = -2\gamma_{\varrho\pi\pi}^2\,\frac{\Gamma\left(\dfrac{1}{2}\right)\Gamma\left(\dfrac{1}{2} + \dfrac{m_\pi^2}{2m_\varrho^2 - m_\pi^2}\right)}{\Gamma\left(m_\pi^2/2\,(m_\varrho^2 - m_\pi^2)\right)}$$

$$= -2\gamma_{\varrho\pi\pi}^2 \times 0.0505 \qquad (4.14\text{a})$$

$$A_{K^+K^+}^{\pi^-\pi^-}\ (s = u = m_K^2, t = 0) = -4\gamma_{K^*K\pi}^2 \times 0.0505 \tag{4.14b}$$

$$A_{\pi^+\pi^+}^{K^-K^-}\ (s = u = m_\pi^2, t = 0) = -4\gamma_{K^*K\pi}^2 \times 0.5266. \tag{4.14c}$$

From Eqs. (4.13 a, b) and (4.14 a, b), we obtain

$$\gamma_{\varrho\pi\pi}^2/4\gamma_{K^*K\pi}^2 = \frac{1 + \dfrac{1}{\sqrt{3}} \dfrac{s_d''}{s_d'}}{1 + \dfrac{2}{\sqrt{3}} \dfrac{s_d''}{s_d'}}. \tag{4.15}$$

This gives, in the $SU(3)$ limit in which case $4\gamma_{K^*K\pi}^2 = \gamma_{\varrho\pi\pi}^2$

$$s_d'' = 0 \tag{4.16}$$

in agreement with Eq. (4.7). Experimentally, however,

$$4\gamma_{K^*K\pi}^2/\gamma_{\varrho\pi\pi}^2 \approx 1.5 \tag{4.17}$$

which gives

$$s_d''/s_d' = \sqrt{3}. \tag{4.18}$$

Now we can write $\gamma_{\varrho\pi\pi}^2$ as

$$\gamma_{\varrho\pi\pi}^2 = a\,\frac{m_\varrho^2}{f_\pi^2} \tag{4.19}$$

where as we have seen in the Veneziano model $a = 2/\pi$ while the current algebra gives $a = 1$.

Table 1 Estimated values of the parameter $c$

|  | $a$ | |
|---|---|---|
| | $2/\pi$ | $1$ |
| $s_d''/s_d'$ | | |
| $a$ | $-1.256$ | $-1.20$ |
| $\sqrt{3}$ | $-1.33$ | $-1.29$ |

From Eqs. (4.13 a) and (4.14 a), we obtain

$$\left(1 + \frac{\sqrt{2}}{c}\right) = -2\gamma_{\varrho\pi\pi}^2\,(0.05)\,\frac{3}{4}\,\frac{f_\pi^2}{m_K^2 - m_\pi^2}\left[1 + \frac{1}{\sqrt{3}}\,\frac{s_d''}{s_d'}\right]^{-1}. \tag{4.20}$$

In order to evaluate $c$ from Eq. (4.20), we make use of Eqs. (4.16), (4.18) and (4.19). The various values of $c$ obtained in this way are given in Table 1.

These values are compatible with that given in Eq. (4.7). Note also that the value of $c$ is not very sensitive to the value of $s''_a$ nor to that of $a$. In particular the value of $c$ obtained for $a = 2/\pi$ (which probably, one should use for consistency since it is the one to which the Veneziano model leads) and $s''_a = 0$ is in excellent agreement with (4.7) obtained in an entirely different way. This supports not only the universality character of $c$ but also the commutation relations (4.1).

From (4.13 b,c) and (4.15 b,c), we obtain

$$\frac{(0.0505)}{(0.5266)} = \frac{2\left(c + \sqrt{2}\right) f_K^2}{\left(2\sqrt{2} - c\right) f_\pi^2}.$$

This gives

$$\frac{f_K}{f_\pi} = 1.12 \quad \text{for} \quad c = -1.256$$

$$= 1.25 \quad \text{for} \quad c = -1.29.$$

If one uses (4.13 a,c) and (4.15 a,c), one again obtains the same results showing that these equations are mutually consistent. For $c = -1.33$, $f_K/f_\pi$ comes out to be quite large. It should be noted that $f_K/f_\pi$ is sensitive to the values of $c$ and in particular when $c$ approaches $-\sqrt{2}$.


## 5   $A_1\varrho\pi$ SYSTEM[13, 20]

In this section we apply the Veneziano representation to $A_1\varrho\pi$ system. The study of this system gives more insight into the relationship between the Veneziano model and chiral symmetry.

It is sufficient to consider the reaction

$$\pi^-(q) + \pi^+(p) \to \pi^-(q') + A_1^+(k). \tag{5.1}$$

For this process there are two amplitudes which we write (with $k \cdot \eta = 0$)

$$T(s, t) = i\left[A(s, t)(q - q')_\mu + B(s, t)(q + q')_\mu\right]\eta_\mu \tag{5.2}$$

where $\eta$ is the polarization vector for $A_1$. The partial wave expansion for the invariant amplitudes $A$ and $B$ in the $t$- and $s$-channels are given by

$$\sqrt{2}\, q_t B = -\sum_J \frac{2J + 1}{[J(J + 1)]^{1/2}} \langle 1, 0|\, T_J\, |0, 0\rangle P'_J(\cos\theta_t)$$

$$\frac{2}{m_{A_1}}\left[\omega_t k_t A + q_t \varepsilon_t \cos\theta_t B\right] = \sum_J (2J + 1)\langle 0, 0|\, T_J\, |0, 0\rangle P_J(\cos\theta_t] \tag{5.3}$$

$$\frac{1}{\sqrt{2}} \, q_s \, (A + B) = -\sum_J \frac{(2J + 1)}{[J(J + 1)]^{1/2}} \langle 1, 0 | T_J | 0, 0 \rangle P'_J (\cos \theta_s)$$

$$\times \frac{1}{m_{A_1}} \, [(3B - A) \, k_s \omega_s + q_s \varepsilon_s \, (A + B) \cos \theta_s]$$

$$= \sum_J (2J + 1) \langle 0, 0 | T_J | 0, 0 \rangle P_J (\cos \theta_s) \tag{5.4}$$

where $q$, $k$ are the centre of mass momenta of pion and kaon and $\omega$ and $\varepsilon$ are the centre of mass energies of initial pion and kaon. We shall confine ourselves to the amplitude $B$ only. It is clear from Eqs. (5.3) and (5.4) that the amplitude $B$ has the following Regge behaviour

$$B(s, t) \sim s^{\alpha(t) - 1}, \quad s \to \infty, \; t \text{ fixed}$$

$$B(s, t) \sim t^{\alpha(s)}, \quad t \to \infty, \; s \text{ fixed} \tag{5.5}$$

The contributions from $\varrho$ and $\sigma$ mesons to the amplitude $B$ in the $s$-channel are given by

$$B_\varrho (s, t) = -\frac{1}{2} \, \gamma_{\varrho\pi\pi} \, [G_S + G_D \, (m_{A_1}^2 + 3m_\pi^2 - m_\varrho^2 - 2t)] \frac{1}{m_\varrho^2 - s} \tag{5.6}$$

$$B_\sigma (s, t) = -\gamma_{A_1\sigma\pi} \gamma_{\sigma\pi\pi} / (m_\sigma^2 - s). \tag{5.7}$$

The $\sigma$ meson does not contribute to the amplitude $B$ in the $t$-channel whereas the $\varrho$ meson contribution to this amplitude is simply given by

$$B_\varrho (s, t) = -\gamma_{\varrho\pi\pi} G_S / (m_\varrho^2 - t). \tag{5.8}$$

The coupling constants used in the Eqs. (5.6), (5.7) and (5.8) are defined as follows:

$$\langle \varrho^0(K) | j_\pi^- | A_1^+(k) \rangle = -i \, [G_S \eta \cdot \varepsilon + 2G_D \eta \cdot K k \cdot \varepsilon]$$

$$\langle \pi^+(p) | j_\pi^+ | \varrho^0(K) \rangle = -\gamma_{\varrho\pi\pi} \, (K - 2p) \cdot \varepsilon$$

$$\langle \pi^+(p) | j_\pi^+ | \sigma^0(K) \rangle = -i \gamma_{\sigma\pi\pi}$$

$$\langle \sigma^0(K) | j_\pi^- | A_1^+(k) \rangle = \gamma_{A_1\sigma\pi} \, (k - 2K) \cdot \eta. \tag{5.9}$$

We note that $\varrho$ and $f$ trajectories are degenerate and there are no $u$-channel resonances, since these would have isospin two. It follows from the above properties that the Veneziano formula for the amplitude $B$ involving only one term can be written as:

$$B(s, t) = -(\gamma_1 + \gamma_2 t) \frac{\Gamma(1 - \alpha(s)) \, \Gamma(1 - \alpha(t))}{\Gamma(2 - \alpha(s) - \alpha(t))}. \tag{5.10}$$

We now make use of the Adler's consistency condition which requires that $B(s, t) = 0$, when one pion is off the mass shell ($p = 0$; $s = u = m_\pi^2$, $t = m_{A_1}^2$). This can happen if, at $s = u = m_\pi^2$, $t = m_{A_1}^2$, $\Gamma(2 - \alpha(s) - \alpha(t))$ becomes infinite, which it does when

$$2 - \alpha(s) - \alpha(t) = 0. \tag{5.11}$$

Using Eq. (2.16), we see that at $s = m_\pi^2$, $t = m_{A_1}^2$, Eq. (5.11) gives:

$$m_{A_1}^2 + m_\pi^2 = 2m_\varrho^2 \tag{5.12}$$

which is the Weinberg mass formula for $A_1$ meson.

From Eq. (5.10), we see that the residue at the pole $s = m_\varrho^2$ is $-\dfrac{1}{b}(\gamma_1 + \gamma_2 t)$.

In the Veneziano model $\sigma$ is the daughter of $\varrho$ with $m_\sigma = m_\varrho$. Therefore, the residue $-\dfrac{1}{b}(\gamma_1 + \gamma_2 t)$ also contains the contribution of $\sigma$ meson. The residue at $t = m_\varrho^2$ is $-\dfrac{1}{b}(\gamma_1 + m_\varrho^2 \gamma_2)$. Comparing these residues with those from Eqs. (5.6), (5.7) and (5.8), we get:

$$-\frac{1}{b}(\gamma_1 + m_\varrho^2 \gamma_2) = -\gamma_{\varrho\pi\pi} G_S$$

$$-\frac{1}{b}\gamma_2 = \gamma_{\varrho\pi\pi} G_D$$

$$-\frac{1}{b}\gamma_1 - -\frac{1}{2}\gamma_{\varrho\pi\pi}[G_S + G_D(m_{A_1}^2 - m_\varrho^2 + 3m_\pi^2)] - \gamma_{\sigma\pi\pi}\gamma_{A_1\sigma\pi}. \tag{5.13}$$

Using Eq. (5.12) and taking $m_\pi = 0$, we obtain from Eq. (5.13):

$$\tfrac{1}{2}\gamma_{\varrho\pi\pi}[G_S + m_\varrho^2 G_D] \approx \gamma_{\sigma\pi\pi}\gamma_{A_1\sigma\pi}. \tag{5.14}$$

From Eq. (2.19), it follows that $\gamma_{\sigma\pi\pi}^2 = m_\varrho^2 \gamma_{\varrho\pi\pi}^2$ so that Eq. (5.14) becomes:

$$\tfrac{1}{2}[G_S + m_\varrho^2 G_D] \approx m_\varrho \gamma_{A_1\sigma\pi}. \tag{5.15}$$

The decay widths for $A_1 \to \varrho\pi$ and $A_1 \to \sigma\pi$ in terms of the coupling constants $G_S$, $G_D$ and $\gamma_{A_1\sigma\pi}$ are given by

$$\Gamma_{A_1 \to \varrho\pi}(\lambda = 0) \approx \frac{G_D^2}{4\pi} \frac{(3r - m_\varrho^2)^2}{96\sqrt{2}\, m_\varrho}$$

$$\Gamma_{A_1 \to \varrho\pi}(\lambda = +1) \approx \frac{G_D^2}{4\pi} \frac{r^2}{6\sqrt{2}\, m_\varrho}$$

$$\Gamma(A_1 \to \sigma\pi) \approx \frac{\gamma_{A_1\sigma\pi}^2}{4\pi} \frac{1}{3} \frac{m_\varrho}{16\sqrt{2}} \tag{5.16}$$

where $r = G_S/G_D$ and $\lambda$ denotes the helicity. We now discuss three possible cases (i) $\gamma_{A_1\sigma\pi} = 0$ so that from Eq. (5.15) $r = -m_\varrho^2$. This corresponds to the value of $\delta = -1$ in the gauge[21] theory of chiral symmetry and one obtains $\Gamma(A_1 \to \sigma\pi) = 0$, $\Gamma_{A_1\to\varrho\pi}(\lambda = 0)/\Gamma_{A_1\to\varrho\pi}(\lambda = +1) \approx 1$. In this case $\gamma_1$ is just proportional to $m_\pi^2$ so that the amplitude for the reaction $\pi^-\pi^+ \to \pi^- A_1^+$ in the forward direction will be negligibly small as compared with the other two cases. But unfortunately this amplitude cannot be experimentally measured. (ii) $G_D = 0$ $(\delta = 0)$ so that from Eq. (5.15), we get $\gamma_{A_1\sigma\pi} \approx \dfrac{1}{2m_\varrho} G_S$. The case $G_D = 0$ corresponds to $\delta = 0$ in the gauge theory of chiral symmetry. Here we get $\Gamma_{A_1\to\varrho\pi}(\lambda = 0) = 0$,

$$\Gamma_{A_1\to\varrho\pi}(\lambda = +1)/\Gamma(A_1 \to \sigma\pi) \approx 32$$

(iii) $G_S = 0$ $(\delta = -2)$, so that from Eq. (5.15) we get $\gamma_{A_1\sigma\pi} = \tfrac{1}{2}m_\varrho G_D$. This corresponds to algebraic realization[22] of the chiral symmetry. In this case

$$\Gamma_{A_1\to\varrho\pi}(\lambda = +1) = 0$$

$$\Gamma_{A_1\to\varrho\pi}(\lambda = 0)/\Gamma_{(A_1\to\varrho\pi)} \approx 2.$$

It follows that the Veneziano model can accomodate all the versions of the chiral symmetry. Only experimental value for the parameter $\delta$ can distinguish between all these cases.


## 6   $K_{l4}$ FORM FACTORS[13]

In this section we first briefly discuss the Veneziano amplitude for the reaction

$$\pi_i(q) + K(p) \to \pi_j(q') + K_A(k). \tag{6.1}$$

The interest in this problem is twofold, namely to get information about the mass and the anomalous magnetic moment of $K_A$ and to extract information about the $K_{l4}$ form factors. As before the amplitude $T$ for the reaction (6.1) can be written as

$$T(s, u, t) = i\,[A(s, u, t)(q - q')_\mu + B(s, u, t)(q + q')_\mu]\,\eta_\mu. \tag{6.2}$$

The isospin decomposition of the amplitudes $A$ and $B$ is given by

$$\begin{aligned} A &= \delta_{ji}A^{(+)} + \tfrac{1}{2}[\tau_j, \tau_i]\,A^{(-)} \\ B &= \delta_{ji}B^{(+)} + \tfrac{1}{2}[\tau_j, \tau_i]\,B^{(-)}. \end{aligned} \tag{6.3}$$

The crossing symmetry ($s \leftrightarrow u$) gives the relations

$$A^{(\pm)}(s, u, t) = \pm A^{(\pm)}(u, s, t)$$

$$B^{(\pm)}(s, u, t) = \mp B^{(\pm)}(u, s, t). \tag{6.4}$$

From Eqs. (5.3) and (5.4) and (6.4) it follows that the Veneziano representation for the amplitudes $A$ and $B$ can be written as:

$$A^{(\pm)}(s, u, t) = [(\gamma_1^{\pm} s + \gamma_2^{\pm} t + \gamma_3^{\pm}) F(\alpha_{K^*}(s), \alpha(t)) \pm (s \to u)] \tag{6.5a}$$

$$B^{(\pm)}(s, u, t) = [(\beta_1^{\pm} + \beta_2^{\pm} t) F(\alpha_{K^*}(s), \alpha(t)) \mp (s \to u)] \tag{6.5b}$$

where

$$F(\alpha_{K^*}(s), \alpha(t)) = \frac{\Gamma(1 - \alpha_{K^*}(s)) \, \Gamma(1 - \alpha(t))}{\Gamma(2 - \alpha_{K^*}(s) - \alpha(t))}. \tag{6.6}$$

Now the Adler's consistency condition requires that $B(s, t) = 0$, when the kaon is off the mass shell, which as before (see Section 5) gives the mass relation

$$m_{K_A}^2 + m_K^2 = 2m_{K^*}^2. \tag{6.7}$$

This gives $K_A$ mass to be about 1160 MeV. Further the Adler's consistency condition, when one of the pion is off the mass shell, gives the constraint

$$\gamma_1 m_K^2 + (\gamma_2 - \beta_2) m_\pi^2 + (\gamma_3 - \beta_1) = 0. \tag{6.8}$$

Using Eq. (6.8) and comparing the residues of the poles as obtained from Eq. (6.5) with those by means of Feynman rules, we get the following relations:

$$\gamma_1 = \frac{1}{2} \gamma_{K^*K\pi} \frac{1}{m_{K^*}^2} \left[ G_D' - \frac{1}{m_\varrho^2} G_S' \right] + \gamma_{\kappa K\pi} \gamma_{K_A \kappa\pi} \frac{1}{m_\varrho^4} \tag{6.9}$$

$$\gamma_2 = -\frac{1}{2m_\varrho^2} \gamma_{K^*K\pi} G_D' = \beta_2 \tag{6.10}$$

$$\gamma_3 = -\frac{1}{2} \gamma_{K^*K\pi} \frac{1}{2m_{K^*}^2} [(m_{K^*}^2 + m_K^2) G_D' + G_S'] - \frac{1}{2} \frac{m_{K^*}^2 + m_K^2}{m_\varrho^4} \gamma_{\kappa K\pi} \gamma_{K_A \kappa\pi} \tag{6.11}$$

$$\beta_1 = -\frac{1}{2} \gamma_{K^*K\pi} \frac{1}{2m_{K^*}^2} \left[ m_\varrho^2 G_D' + \frac{m_{K^*}^2 + m_K^2}{m_\varrho^2} G_S' \right] - \frac{1}{2} \frac{1}{m_\varrho^2} \gamma_{\kappa K\pi} \gamma_{K_A \kappa\pi} \tag{6.12}$$

$$[G_S' + m_\varrho^2 G_D'] = \gamma_{\sigma\pi\pi} \gamma_{K_A K\sigma} \tag{6.13}$$

$$\frac{1}{2} \gamma_{K^*K\pi} \left[ -2G'_D + \frac{2}{m_\varrho^2} G'_S \right] = -\frac{1}{2} \gamma_{\varrho\pi\pi} \left[ H_D - \frac{1}{m_\varrho^2} H_S \right] \qquad (6.14)$$

$$\frac{1}{2} \gamma_{K^*K\pi} \frac{2m_K^2}{m_{K^*}^2} \left[ -G'_D + \frac{1}{m_\varrho^2} G'_S \right] + \frac{2}{m_\varrho^2} \gamma_{\kappa K\pi} \gamma_{K_A\kappa\pi} = \frac{1}{2} \gamma_{\varrho\pi\pi} \left[ H_D + \frac{1}{m_\varrho^2} H_S \right]$$

$$\qquad (6.15)$$

where we have used the mass relations (6.7) and (3.3) and have neglected the pion mass. The coupling constants $G'_S$ and $G'_D$ and $H_S$ and $H_D$ refer to $K_A \to K^*\pi$ and $K_A \to \varrho K$ decays respectively, whereas $\gamma_{K_A\kappa\pi}$ and $\gamma_{K_A K\sigma}$ refer to $K_A \to \kappa + \pi$ and $K_A \to \sigma + K$ decays respectively. If, in particular, $\gamma_{K_A K\sigma} = 0 = \gamma_{K_A \kappa\pi}$ then from Eqs. (6.13), (6.14) and (6.15), we get

$$G'_S/G'_D \approx -m_\varrho^2$$

$$H_S/H_D \approx -(m_{K^*}^2 + m_K^2)$$

$$2m_\varrho^2 \gamma_{K^*K\pi} G'_D = m_{K^*}^2 \gamma_{\varrho\pi\pi} H_D. \qquad (6.16)$$

Finally the Veneziano representation for the amplitudes $A$ and $B$ becomes

$$A^{(\pm)}(s, u, t) = [(\gamma_1 s + \gamma_2 t + \gamma_3) F(\alpha_{K^*}(s), \alpha(t)) \pm (s \to u)]$$

$$B^{\pm}(s, u, t) = [(\gamma_1 m_K^2 + \gamma_3 - \gamma_2 t) F(\alpha_{K^*}(s), \alpha(t)) \mp (s \to u)] \qquad (6.17)$$

where $\gamma_1, \gamma_2$ and $\gamma_3$ are given in Eqs. (6.9), (6.10) and (6.11). We shall make use of these results in the discussion of the $K_{l4}$ form factors.

We now discuss the $K_{l4}$ form factors within this model. The axial-vector $K_{l4}$ form factors $F_1$, $F_2$ and $F_3$ are defined by

$$\langle \pi_1(q_1) \pi_2(q_2) | A_\mu(0) | K(p) \rangle = \frac{1}{(2\pi)^{9/2}} i \frac{1}{(8q_{10}q_{20}p_0)^{1/2}} \frac{1}{m_K} \times$$

$$\times [F_1 (q_1 + q_2)_\mu + F_2 (q_1 - q_2)_\mu + F_3 (p - q_1 - q_2)_\mu] \qquad (6.18)$$

where $A_\mu$ is the hypercharge-changing axial vector current. We now make use of the field current identity:

$$A_\mu = \frac{1}{\sqrt{2}} g_{K_A} a_\mu \qquad (6.19)$$

where $a_\mu$ refers to $(K, K_A)$ field and $g_{K_A}$ is a constant of proportionality. Now the pure spin 1 part of $A_\mu$ contributes to $F_1$ and $F_2$ whereas the spin zero component contributes to $F_3$ which may be well approximated by the $K$-

meson pole. Thus $F_3$ must be proportional to the amplitude for $K\pi$ scattering[23] and, within the framework of the Veneziano model has been, discussed in reference (23). Here we concentrate on $F_1$ and $F_2$ and write the field current identity for the pure spin 1 part of $A_\mu$ as:

$$A_\mu^{(1)} = \frac{1}{\sqrt{2}} g_{K_A} a_\mu^{(1)} \tag{6.20}$$

where $a_\mu^{(1)}$ is now the field of $K_A$ particle. Hence we can write

$$\frac{1}{(2\pi)^{9/2}} \frac{i}{(8q_{10}q_{20}p_0)^{1/2}} \frac{1}{m_K} [F_1 (q_1 + q_2)_\mu + F_2 (q_1 - q_2)_\mu]$$

$$= \langle \pi_1(q_1) \pi_2(q_2)| A_\mu^{(1)} |K(p)\rangle = \frac{1}{\sqrt{2}} g_{K_A} \langle \pi_1(q_1) \pi_2(q_2)| a_\mu^{(1)} |K(p)\rangle$$

$$= \frac{1}{\sqrt{2}} g_{K_A} \frac{1}{k^2 + m_{K_A}^2} \langle \pi_1(q_1) \pi_2(q_2)| J_\mu^{(1)} |K(p)\rangle \tag{6.21}$$

where $J_\mu^{(1)}$ is the source current of $K_A$. Now $\langle \pi_1(q_1) \pi_2(q_2)| J_\mu^{(1)} |K(p)\rangle$ is nothing but the scattering matrix for the process

$$\bar{\pi}_1(-q_1) + K(p) \to \pi_2(q_2) + K_A(k) \tag{6.22}$$

in the $s$-channel, where $k = p - q_1 - q_2$. Thus using (6.2) we have

$$\frac{1}{m_K} [F_1 (q_1 + q_2)_\mu + F_2 (q_1 - q_2)_\mu] = \frac{1}{\sqrt{2}} g_{K_A} \frac{1}{k^2 + m_{K_A}^2} \times$$

$$\times [A(s,u,t)(-q_1 - q_2)_\mu + B(s,u,t)(-q_1 + q_2)_\mu] \tag{6.23}$$

where now

$$s = (p - q_1)^2 = -(k + q_2)^2$$

$$u = -(p - q_2)^2 = -(k + q_1)^2$$

$$t = -(q_1 + q_2)^2 = -(k - p)^2. \tag{6.24}$$

Thus

$$F_1 = -\frac{1}{\sqrt{2}} g_{K_A} \frac{m_K}{k^2 + m_{K_A}^2} A$$

$$F_2 = -\frac{1}{\sqrt{2}} g_{K_A} \frac{m_K}{k^2 + m_{K_A}^2} B \tag{6.25}$$

where $A$ and $B$ are to be evaluated at $k^2 = m_{K_A}^2$.

118                                                              Lectures on Particles and Fields

We now consider the particular processes:

$$K^+(p) \to \pi^+(q_1) + \pi^-(q_2) + l^+ + \nu \tag{6.26a}$$

$$K^+(p) \to \pi^0(q_1) + \pi^0(q_2) + l^+ + \nu \tag{6.26b}$$

$$K^0(p) \to \pi^-(q_1) + \pi^0(q_2) + l^+ + \nu. \tag{6.26c}$$

The reaction (6.26c) is related to (6.26a, b) by $\Delta I = \frac{1}{2}$ rule and therefore we need to consider only (6.26a) and (6.26b). Making use of (6.25) it is easy to see from (6.3) and (6.17) that for (6.26a)

$$F_1^{+-} = -\frac{1}{\sqrt{2}} g_{K_A} \frac{m_K}{k^2 + m_{K_A}^2} (A^{(+)} + A^{(-)})$$

$$= -\sqrt{2} g_{K_A} \frac{m_K}{k^2 + m_{K_A}^2} [(\gamma_1 s + \gamma_2 t + \gamma_3) F(\alpha_{K*}(s), \alpha(t))]_{k^2 = -m_{K_A}^2}$$

$$\tag{6.27a}$$

$$F_2^{+-} = -\frac{1}{\sqrt{2}} g_{K_A} \frac{m_K}{k^2 + m_{K_A}^2} (B^{(+)} + B^{(-)})$$

$$= -\sqrt{2} g_{K_A} \frac{m_K}{k^2 + m_{K_A}^2} [(\gamma_1 m_K^2 - \gamma_2 t + \gamma_3) F(\alpha_{K*}(s), \alpha(t))]_{k^2 = -m_{K_A}^2}$$

$$\tag{6.27b}$$

while for (6.26b):

$$F_1^{00} + F_2^{00} = -\frac{1}{\sqrt{2}} g_{K_A} \frac{m_K}{k^2 + m_{K_A}^2} (A^{(+)} + B^{(+)}) = -\frac{1}{\sqrt{2}} g_{K_A} \frac{m_K}{k^2 + m_{K_A}^2} \times$$

$$\times [(\gamma_1 (s + m_K^2) + 2\gamma_3) F(\alpha_{K*}(s), \alpha(t))$$

$$+ (\gamma_1 (u - m_K^2) + 2\gamma_2 t) F(\alpha_{K*}(u), \alpha(t))]_{k^2 = -m_{K_A}^2} \tag{6.28a}$$

$$F_1^{00} - F_2^{00} = -\frac{1}{\sqrt{2}} g_{K_A} \frac{m_K}{k^2 + m_{K_A}^2} [(\gamma_1 (s - m_K^2) + 2\gamma_2 t) F(\alpha_{K*}(s), \alpha(t))$$

$$+ (\gamma_1 (u + m_K^2) + 2\gamma_3) F(\alpha_{K*}(u), \alpha(t))]_{k^2 = -m_{K_A}^2}. \tag{6.28b}$$

In the limit $q_1 \to 0$, $s \to m_K^2$, $u \to -k^2 = m_{K_A}^2$, $t \to m_\pi^2 \approx 0$ while for $q_2 \to 0$, $s \to -k^2 = m_{K_A}^2$, $u \to m_K^2$, $t \to m_\pi^2 \approx 0$. Further we note that since $\varrho$ and $K^*$

trajectories are parallel, it follows from (2.15) and (6.7) that $[2 - \alpha_{K*}(u) - \alpha(t)]$ and hence $F(\alpha_{K*}(u), \alpha(t))$ vanishes for $u = m_{K_A}^2$, $t = 0$ while $[2 - \alpha_{K*}(s) - \alpha(t)]$ and hence $F(\alpha_{K*}(s), \alpha(t))$ vanishes for $s = m_{K_A}^2$, $t = 0$. Therefore denoting

$$\tilde{F} = \frac{1}{2}\left(\lim_{q_1 \to 0} F + \lim_{q_2 \to 0} F\right) \tag{6.29}$$

it follows from (6.27) and (6.28) that

$$\tilde{F}_1^{+-} = \tilde{F}_2^{+-} = C \tag{6.30a}$$

$$\tilde{F}_1^{00} + \tilde{F}_2^{00} = C \tag{6.30b}$$

$$\tilde{F}_1^{00} - \tilde{F}_2^{00} = C \tag{6.30c}$$

where

$$C = -\frac{1}{\sqrt{2}} g_{K_A} \frac{m_K}{k^2 + m_{K_A}^2} (\gamma_1 m_K^2 + \gamma_3) F(\alpha_{K*}(m_K^2), \alpha(0)). \tag{6.31}$$

From (6.30), it follows that

$$\tilde{F}_1^{+-} = \tilde{F}_2^{+-} = \tilde{F}_1^{00} = C$$
$$\tilde{F}_2^{00} = 0. \tag{6.32}$$

These are the same constraints as implied by the current algebra[24] except that $C$ in the current algebra is fixed in terms of $f_\pi$, the pion decay constant, and $f_+(-m_K^2)$ where $f_+$ is one of the $K_{l3}$ form factors. These results further support the conclusion that the Veneziano model is connected with the chiral symmetry.

Since $k^2 \approx 0$, Eq. (6.3) becomes:

$$C = -\frac{1}{\sqrt{2}} g_{K_A} \frac{m_K}{m_{K_A}^2} (\gamma_1 m_K^2 + \gamma_3) F(\alpha_{K*}(m_K^2), \alpha(0))$$

which takes a particular simple form, when $\kappa$ and $\sigma$ couplings to $K_A$ are neglected. In this case:

$$C = -\frac{1}{2\sqrt{2}} \frac{m_K^3}{m_{K_A}^2 m_{K*}^2} g_{K_A} \gamma_{K*K\pi} G_D' [F(\alpha_{K*}(m_K^2), \alpha(0))]. \tag{6.33}$$

## 7  SUMMARY AND CONCLUSIONS

We may summarize the situation regarding the connection between the Veneziano model and current algebra or chiral symmetry as follows:

I.   There are three kinds of predictions of current algebra:

i) Those which follow from the soft pion theorems and depend only on current commutators. No assumption in saturating the sum rules by few particles is made.

ii) The sum rules, although they follow from the current commutators but depend on the assumption of saturation with few particles.

iii) The predictions which are sensitive to chiral symmetry breaking, that is, depend upon the commutator $[X_i, [H, X_j]]$, where $X_i$, $T_i$ are the generators of $SU(2) \times SU(2)$ algebra.

There does not appear to be any inconsistency between the predictions in category (i) and the Veneziano model, although detailed comparison is not possible except for the constraints of current algebra in $K \to 3\pi$ and $K_{l4}$ decays which are just the consequence of Adler's condition in the Veneziano model. In the case of scattering lengths, current algebra value is used to fix the scale in the Veneziano model. This comparison gives $\gamma^2_{\varrho\pi\pi} = \dfrac{2}{\pi} \dfrac{m^2_\varrho}{f^2_\pi}$.

In the category (ii) let us consider the Adler sum rule for $\pi$–$\pi$ scattering. Saturation of this sum rule with $\varrho$ and $\sigma$ can be achieved provided $m_\varrho = m_\sigma$, $\gamma^2_{\varrho\pi\pi} = m^2_\varrho/f^2_\pi$ and $\Gamma_\sigma/\Gamma_\varrho = \frac{9}{2}$. Except for the value of $\gamma^2_{\varrho\pi\pi}$, the other values are the same as in the Veneziano model. It may be that the contribution from the higher mass states in saturating this sum rule affect the value of $\gamma^2_{\varrho\pi\pi}$ so as to bring it to the value $\dfrac{2}{\pi} \dfrac{m^2_\varrho}{f^2_\pi}$.

There is a version of chiral symmetry[22] (algebraic realization of chiral symmetry) whose predictions fall into the categories (ii) and (iii). In this model one assigns the mesons $\pi$, $\sigma$, $\varrho$ and $A_1$ (with helicity zero) in the representation which is the direct sum of chiral four-vector and chiral antisymmetric tensor. One has the following predictions:

$$m^2_{A_1} = m^2_\varrho + m^2_\sigma$$

$$m^2_\varrho/m^2_\sigma = \tan^2 \theta$$

$$\Gamma_\sigma / \Gamma_\varrho = \frac{9}{2} \frac{m_\sigma^3}{m_\varrho^3} \tan^2 \theta$$

$$\gamma_{\varrho\pi\pi}^2 = \frac{2m_\varrho^2}{f_\pi^2} \cos^2 \theta$$

$$\gamma_{A_1\sigma\pi}^2 = 2m_{A_1}^2 / f_\pi^2 \cos^2 \theta$$

where $\theta$ is the mixing angle between $\pi$ and $A_1$. These predictions with $\theta = 45°$ agree with those of the Veneziano model for $\pi$–$\pi$ scattering and $\pi\pi \to \pi A_1$ except for $\gamma_{\varrho\pi\pi}^2$ and $\gamma_{A_1\sigma\pi}^2$. Here $\gamma_{A_1\sigma\pi}^2 = m_{A_1}^2 / f_\pi^2$, whereas in the Veneziano model it is undetermined.

However, since $\theta$ is not determined in this theory, the agreement may be superficial. It has been shown explicitly[26] that this theory with $\theta = 45°$ is not compatible with the Veneziano model. The reason is that for this version of chiral symmetry, the chiral symmetry-breaking is large. Probably one has to enlarge the multiplet[26,27] $\pi$, $\sigma$, $\varrho$ and $A_1$. This is understandable, since in the Veneziano model one takes into account a large number of resonances.

As has been discussed in this review, the Veneziano model is consistent with the chiral symmetry-breaking mechanism proposed by Gell-Mann, Oakes and Renner. In this approach one does not assign the particles to any representation of chiral algebra.

II.   The $A_1\varrho\pi$ system has provided some useful information:

$$m_{A_1}^2 = 2m_\varrho^2$$

$$\tfrac{1}{2} [G_S + m_\varrho^2 G_D] \approx m_\varrho \gamma_{A_1\sigma\pi}.$$

This can accommodate the gauge version of chiral symmetry, although in the latter case one does not know how to incorporate $\sigma$. The case $\gamma_{A_1\sigma\pi} = 0$ corresponds to the gauge theory of chiral symmetry with $\delta = -1$.

III.   The Adler consistency condition, imposed on the Veneziano representation of those invariant amplitudes for which it is non-trivial, leads to the mass formulae[28]. In particular, we have derived the mass relations

$$m_{A_1}^2 + m_\pi^2 = 2m_\varrho^2$$

$$m_{K^*}^2 - m_K^2 = m_\varrho^2 - m_\pi^2, \quad m_{K_A}^2 + m_K^2 = 2m_{K^*}^2$$

$$m_f^2 = 3m_\varrho^2 - 2m_\pi^2$$

(this follows from the fact that $\varrho$ and $f$ trajectories are exchange degenerate). Similarly, by considering other reactions, one can derive the mass relations

$$m_\omega^2 = m_\varrho^2, \quad m_{A_2}^2 = 3m_\varrho^2 + m_\eta^2 - 3m_\pi^2.$$

The last formula is not unambiguous because some reactions also give $m_{A_2}^2 = m_f^2$. This difficulty seems to be due to the fact that the Veneziano model implies nonet symmetry for the mesons but, as is well known, the nonet mass formula for the pseudoscalar mesons is not good.

One is impressed by the similarity between the predictions of the Veneziano model and current algebra or chiral symmetry. Whether the connection between these two approaches is deep or just superficial one does not know.

## References


1. C.H.Mo, *Proceedings of 14th International Conference on High Energy Physics, Vienna (1968)*. W.R.Frazer *ibid*. The references to the original papers on this subject can be found in these papers.
2. K.Igi, *Phys. Rev. Letters*, **9**, 76 (1962).
3. R.Dolen, D.Horn and C.Schmid, *Phys. Rev. Letters*, **19**, 402 (1967); G.F.Chew, *Comments on Nuclear and Particle Physics*, **2**, 74 (1968).
4. For a general review of this subject, see M.Jacob, Duality in Strong Interaction Physics, Lecture Notes CERN, TH. 1010 (unpublished).
5. G.Veneziano, *Nuovo Cimento*, **57A**, 190 (1960).
6. C.Lovelace, *Phys. Letters*, **28B**, 265 (1968).
7. S.Adler, *Phys. Rev.*, **137**, B1022 (1965).
8. K.Kawarabayashi, S.Kitakado and H.Yabuki, *Phys. Letters*, **28B**, 432 (1969).
9. S.Weinberg, *Phys. Rev. Letters*, **17**, 616 (1966).
10. K.Kawarabayashi and M.Suzuki, *Phys. Rev. Letters*, **16**, 255 (1966); Riazuddin and Fayyazuddin, *Phys. Rev.*, **147**, 1071 (1966).
11. C.G.Callan and S.B.Treiman, *Phys. Rev. Letters*, **16**, 153 (1966).
12. Y.Hara and Y.Nambu, *Phys. Rev. Letters*, **16**,875 (1966); D.K.Elios and J.C.Taylor, *Nuovo Cimento*, **44A**, 518 (1966) and **48A**, 814 (E) (1967).
13. Fayyazuddin and Riazuddin, Veneziano Model for Meson Systems, University of Islamabad preprint to be published.
14. Riazuddin and Fayyazuddin, University of Islamabad preprint to be published.
15. M.Gell–Mann, *Physics*, **1**, 63 (1964).
16. M.Gell–Mann, R.J.Oakes and B.Renner, *Phys. Rev.*, **175**, 2195 (1968).
17. C.H.Chan and F.T.Meiere, *Phys. Rev. Letters*, **22**, 737 (1969); F. von Hippel and J.K.Kim, *Phys. Rev. Letters*, **22**, 740 (1969).
18. Riazuddin and K.T.Mahanthappa, *Phys. Rev.*, **147**, 972 (1966).
19. K.T.Manhanthappa and Riazuddin, *Nuovo Cimento*, **45A**, 252 (1966); K.Kawarabayashi and W.W.Wada, *Phys. Rev.*, **146**, 1209 (1966).
20. Fayyazuddin and Riazuddin, *Phys. Letters*, **28B**, 561 (1969).

21. H. Schnitzer and S. Weinberg, *Phys. Rev.*, **164**, 1828 (1967); J. W. Wess and B. Zumino, *Phys. Rev.*, **163**, 1727 (1967); J. Schwinger, *Phys. Letters*, **24B**, 473 (1967). B. W. Lee and H. T. Nieh, *Phys. Rev.*, **166**, 1507 (1968).

22. F. Gillman and H. Harari, *Phys. Rev.*, **165**, 1803 (1968); S. Weinberg, *Phys. Rev.*, Jan. 25 (1969); Fayyazuddin and Riazuddin, Chiral Dynamics and Spectral Function Sum Rules, to be published in Phy. Rev.

23. R. G. Roberts and F. Wagner, CERN preprint.

24. S. Weinberg, *Phys. Rev. Letters*, **17**, 336 (1966).

25. M. Ademallo, G. Veneziano and S. Weinberg, *Phys. Rev. Letters*, **22**, 83 (1969).

26. Fayyazuddin, Riazuddin and Masud Ahmad, *Phys. Rev. Letters*, **23**, 103 (1969).

27. S. Weinberg, *Phys. Rev. Letters*, **22**, 1023 (1969).

28. M. Ademollo, S. Weinberg and G. Veneziano, *Phys. Rev. Letters*, **22**, 83 (1969).

# Phenomenological Hadron Couplings

A. N. MITRA

*University of Delhi, India*

## Contents

## 1   INTRODUCTION AND BACKGROUND

When in 1935 Yukawa[1] first proposed the existence of the "heavy electron" as the basic medium for nuclear interactions, he had to wait for 13 years before a suitable candidate was found in the $\pi$-meson. However, the development during the next two decades must have surpassed even his wildest expectations at that time. We now have an ever-increasing number of new particles or resonant states, strange and non-strange, in the multi-GeV range. The discovery of these resonant states of baryons ($B$) and mesons ($M$) has taken us considerably far from the simple Yukawa picture of protons and neutrons interacting through the exchange of $\pi$-mesons, for this concept must now be extended to include the couplings of the resonant states with various mesons and baryons in some systematic manner. During the present decade, a great deal of effort, theoretical and experimental, has been made in this direction. However, the basis of construction of these couplings has still remained essentially the same as originally proposed by Yukawa, viz., three-particle vertices of the form $\bar{B}BM$ for meson-baryon couplings, and those of the form $MMM$ for pure meson couplings.

The theoretical tools for the construction of these couplings, based largely on group theory, have been closely linked with the question of assignments of the particles involved, to the appropriate multiplet or supermultiplet representations. Conversely the strengths of these couplings, as determined from the *experimental* two-body decay modes of the various resonances, have helped greatly in checking on the authenticity of the assignments for specific cases. For the internal degrees of freedom, characterized by charge and hypercharge, the natural group structure is provided by $SU(3)$, first suggested by Sakata[2], and subsequently adapted to elementary particle classifications by Gell-Mann[3] and Neeman[4], through the ingenuous device of replacing the triplet by the octet as the basic representation for both baryons and mesons. The fact that $SU(3)$ is the appropriate tool for providing the charge-hypercharge multiplicities for all types of hadrons, stable ones and resonances, is now generally taken as granted. However, the question of whether or not $SU(3)$ is a good *symmetry* for the construction of hadron vertices is still largely open. For purposes of nomenclature, it may be convenient to use the phrase "$SU(3)$-symmetry" in two different senses, depending on the context in which it is being employed. Thus "$SU(3)$-symmetry" will be considered *valid*, for the (limited) purpose of assignement of various resonances to appropriate multiplet-structures. On the other hand, "$SU(3)$-symmetry" will be considered *broken* in the (move dynamical) context of Yukawa couplings involving the same hadrons.

## 1.1  Supermultiplet classifications

The way to the investigation of still higher group structures for hadrons was opened up with the discovery of $SU(6)$ symmetry, by Gursey and Radicati[5], as a direct extension of Wigner's idea of $SU(4)$ symmetry[6] for nucleons which had led to supermultiplet classifications for nuclear states through the combination of isospin $-SU(2)$ and spin $-SU(2)$ degrees of freedom under the bigger symmetry group $SU(4)$. While the important concept of Wigner has had only limited usefulness in nuclear physics, the corresponding extension of isospin to the $SU(3)$ degrees of freedom[5], has proved particularly exciting in the domain of hadron physics, the familiar examples being the **56** states of the octet and demplet of baryons, and the **36** states of pseudo-scalar and vector nonets of mesons. These states do not need the concept of internal orbital excitations which may be taken as zero for practical purposes.* However, the extension of $SU(6)$-type groups structures to higher resonances must take account of this extra degree of freedom, for a successful supermultiplet classification. One way of doing this could be through the introduction of higher representations of $SU(6)$, (e.g., 1134 instead of 70,

---

* This statement is not really in conflict with the requirement of $p$-wave resonances, if one is willing to think of pseudoscalar mesons as equivalent to axial vector particles in the PCAC spirit. See Ref. (5).

etc.) but such a procedure would not be particularly illuminating, as this would bring in a large number of "unwanted" states (from the experimental point of view), without shedding much light on the basic mechanism of orbital excitations. A more rational modification, first suggested by Mahan-thappa and Sudarshan[7], would be to simulate the orbital excitations through the rotation group $O(3)$, and thus obtain the extended group $SU(6) \times O(3)$, instead of mere $SU(6)$  The great advantage of this group lies in the comparative economy of its representations and its remarkable capacity to account for the multiplicities of various hadronic resonances up to excitations as high as $L = 2$, without bringing in more than a marginal number of additional states. As this group will play a central role in the development of this article, it will be discussed at greater length in the subsequent sections.

## 1.2  Relativistic extensions

With the continued increase in the energies of the observed resonances, the problem of relativistic extensions of these purely non-relativistic groups has received a good deal of attention in this decade. These extensions have necessarily been non-compact in nature, because of the expected role of the Lorentz and Poincaré groups. The earliest extension of $SU(6)$ to a complete relativistic theory was attempted by Salam *et al.*[8] who arrived at the group structure $\tilde{U}(12)$ through a marriage of the $SU(3)$ group with the special Lorentz group $\tilde{U}(4)$ generated by all the sixteen $4 \times 4$ Dirac matrices. This group was remarkably successful in respect of three-point functions, especially in predicting a constant value for the ratio $G_E/G_M$ of the electric and magnetic form factors of the proton, and the value $-\frac{3}{2}$ for the ratio of the magnetic form factors of the proton and the neutron, at *all values* of the momentum transfer. However this theory got plagued by the apparent impossibility of reconciling the unitary requirement for scattering problems (involving four-point functions) with the group theoretical requirement of staying within a particular representation. A similar form of relativistic extension, suggested by Budini and Fronsdal[9], was the group $SL(6, C)$, obtained by imbedding $SU(3)$ into the (more conventional) homogeneous Lorentz group $SL(2, C)$.

A particularly simple relativistic extension of $SU(6)$ with the additional feature of compactness, is the group $SU(6)_W$, proposed by Lipkin and collaborators[10]. This group which provides a valid description for collinear processes, has generators which commute with Lorentz transformations in

a particular $(Z)$ direction. This fact ensures a momentum-independent classification of all particle states with moments only in the $Z$-direction. As will be
discussed in more details later, this group has considerable points of similarity with $SU(6) \times O(3)$ in respect of their predictions on three-point functions. A different form of relativistic extension of $SU(6)$, which is also com-
pact, is the non-chiral group $U(6) \times U(6)$ proposed by various authors[11],
based on the algebra of positive parity operators. The two states of operators
associated with this group are of the forms $\lambda_\alpha \sigma_j (1 \pm \gamma_4)$, where $\lambda_\alpha$
$(\alpha = 1, \ldots, 8)$ are the Gell-Mann matrices and $\sigma_j (j = 0, 1, 2, 3)$ are the Pauli
spin matrices. The two $U(6)$ spins associated with the two sets of (positive
parity) operators together form the total $U(6)$-spin for the system under
consideration. This group, which is a sub-group of $\tilde{U}(12)$, remains a good
symmetry for mesons and baryons *at rest*, and is thus useful for their classification in a broken $\tilde{U}(12)$ scheme.

The essential difficulties associated with relativistic generalizations of
$SU(3)$ are theoretical in character. First, one is up against the Michel-O'Raifertaigh theorem[12] according to which, any such generalization that contains
$SU(3) \times P$ as a subgroup (where $P$ is the Poincaré group), must pay the price
of more than four components for the energy-momentum vector associated
with it. Even if one can formally avoid this difficulty by combining the relativistic spinor indices with the $SU(3)$ indices and considering only those
transformations which are independent of the momenta, e.g., in the $\tilde{U}(12)$
theory of Salam *et al.*[8], there is the further problem of incompatibility with
the unitarity condition for multiplets with *finite* dimensionality. Finally, the
causality requirement, under conditions of exact group symmetry, seems to
rule out Fermi statistics for any spin (including half-integral values)[13]. For
most of these reasons, the applications of most relativistic groups have usually
been confined to the study of vertex functions, such as electromagnetic form
factors[8] and strong decay rates[14].

Unfortunately, the predictions of the different groups are not widely
different for the decay rates so that little discriminating criteria are available
from such comparisons, without going into greater experimental details. As
a result, the formal relativistic generalizations of $SU(6)$ have largely receded
into the background, making way for less ambitions symmetry principles,
such as current algebra[15] whose successes are too numerous to be elaborated
in this article. Yet, the predictive powers of current algebra in respect of
higher decays of hadron resonances are so much limited than those of more
formal groups, that even restricted symmetry groups enjoy a certain amount

of advantage over the methods of current algebra in this regard. Of course a restricted symmetry group requires some extra parametrizations over and above those required for more formal groups (like $\tilde{U}(12)$), but the extra parameters may will be put to advantage by employing them to extract more physical information from experiment. Two natural candidates for such phenomenology are $SU(6) \times O(3)$ and $SU(6)_W$ which have in a sense, the significance of "least common denominators" among the more formal groups, and are also largely free from the theoretical objections mentioned above) against the latter. While the phenomenology of resonance decays based on the use of smaller groups can never be a substitute for a more formal theory, it may at least serve as a guide to the experimental features which a future theory should incorporate. One of the most important of these features is the very nature and extent of symmetry-breaking in the strong couplings arising out of mass splittings among hadrons, whether at the $SU(6)$ or at the $SU(3)$ levels.

### 1.3  The quark model and $SU(6) \times O(3)$

As is well known, a rather concrete realization of $SU(6)$ is obtained through the quark model, first proposed by Gell-Mann[16] and Zweig[17] in order to provide a natural explanation for the occurrence of only the **1, 8** and **10** representations of baryons. In a sense the quarks have helped restore the conditions necessary for the usefulness of Sakata's original triplet representation of $SU(3)$ which had later to be rejected as the basis of $SU(3)$ representations for baryons, in favour of the octet version. This resurrection of the triplet representation was however achieved at a rather heavy price, viz., by endowing the quarks with fractional values of charge, hypercharge and baryon numbers. It is not the purpose or scope of this article to review anew the philosophy and applications of the quark model for which excellent reviews[18-20] exist. In particular, the question of whether (or not) the quarks are heavy, point objects moving non-relativistically in deep potential wells, is largely academic for our present purposes. On the other hand, the "index" aspects of this objects, which label their spin and $SU(3)$ states, are much more relevant for the construction of various hadron states and their couplings. Thus, without the introduction of the orbital degrees of freedom, each quark has six possible states in $SU(6)$ space, so that the possible baryon representations are contained in the reduction:

$$6 \times 6 \times 6 = 56 + 70' + 70'' + 20. \tag{1.1}$$

In contrast to the symmetric ($s$) **56** and antisymmetric ($a$) **20** representations, which appear once each, the **70** representations of mixed symmetry ($m$) seem to appear twice. This however is misleading when it is remembered, that unlike the **56** and **20** states, which correspond to $1 \times 1$ representations of the permutation operators for three objects, the two **70** states together are needed to form basis of a $2 \times 2$ representation of these operators. Indeed, each of these **70** states must be an *integral* part of any overall three-quark state which should also include the spatial degrees of freedom. Thus if one demands an overall antisymmetry in the $SU(6)$ and spatial degrees of freedom, in accordance with the assumption of Fermi statistics for the quarks, then the above $SU(6)$ functions (1.1) must be multiplied by spatial functions of $a$ ($m'$, $m''$) and $s$ symmetries respectively, to obtain the resultant functions:

$$\mathbf{56} \times a; \ 2^{-1/2} \ (\mathbf{70'} \times m'' - \mathbf{70''} \times m'); \ \mathbf{20} \times s. \tag{1.2}$$

For mesons the corresponding $SU(6)$ representations of $Q\bar{Q}$ states are given by

$$\mathbf{6} \times \mathbf{6^*} = \mathbf{35} + \mathbf{1}, \tag{1.3}$$

but since in this case, the individual $Q$ and $\bar{Q}$ states stand for *different* particles, there is no further requirement of any overall symmetry or antisymmetry, even after the spatial degrees of freedom are included. Thus each of the **35** and **1** states can be associated with $s$ and/or $a$ functions of $Q$ and $\bar{Q}$ in the spatial degrees of freedom.

The quark model provides a particularly transparent interpretation for the groups structure of $SU(6) \times O(3)$, in which the subgroup $O(3)$ refers to the internal orbital momenta ($L$) of the $QQQ$ or $\bar{Q}Q$ states, according as one considers baryons or mesons respectively. For baryons, the total $L$-value is made up of two independent quantities $l_{ij}$ and $l_k$, according to

$$\mathbf{L} = \mathbf{l}_{ij} + \mathbf{l}_k, \tag{1.4}$$

where $\mathbf{l}_{ij}$ is the orbital momentum of the ($ij$) pair of quarks, and $\mathbf{l}_k$ that of quark $\#\, k$ with respect to the c.m. of quarks $i$ and $j$. The parity of this baryon state is given by

$$P_B = (-1)^{l_{ii} + l_k}, \tag{1.5}$$

so that the $L$-value, *by itself*, does *not* determine the parity. For meson states as $Q\bar{Q}$ pairs, on the other hand, the $L$-value is merely the orbital momentum of the $Q\bar{Q}$ composite itself, so that the parity is directly reflected in the $L$-

value:

$$P_M = (-1)^{L+1}, \qquad (1.6)$$

the extra factor $(-1)$ signifying opposite parity assignments to $Q$ and $\bar{Q}$.

## 1.4   Quark statistics

The above statements are largely independent of dynamical considerations since the same ideas could as well be formulated without postulating the existence of quarks. On the other hand, if such a model is taken literally, it must be supplemented by more precise statements about the statistics of quarks as well as the forces governing their interactions. As for statistics, the easiest alternative, prima facie, would be Fermi, in accordance with the traditional association of spin-$\frac{1}{2}$ with antisymmetric functions. This means that the **56** baryons must have spatially antisymmetric ($a$) wave functions of $L^P = 0^+$. Since this implies *odd* partial waves in all $QQ$ pairs, one would expect little attraction in such states, in accordance with the traditional concept of two-body forces as the driving mechanism. On the other hand, a totally symmetric ($s$) function of $L^P = 0^+$ would have *even* partial waves in all $QQ$ pairs, and in this distribution, $s$-waves would have the dominant contribution.[21-25] It is only for such states that strong binding can be achieved in a natural manner with attractive $QQ$ forces. This preference for $s$-function of $L^P = 0^+$, over corresponding $a$-functions to be associated with **56** states is thus a dynamical requirement bearing on the nature of quark forces, but is nonetheless supported by our intuition as to the main sources of attraction which can be generated by two-body forces.

A less dynamical argument for preferring $s$- to $a$-functions for the **56** states, comes from the momentum dependence of the electromagnetic form factors of the nucleon. It is possible to show that an $a$-function, by its very structure, would give rise to *nodes* in these form factors even at moderate values of the momentum transfer ($q^2 \approx (20\text{–}50)\ \mathrm{F}^{-2}$),[26] at complete variance with the experimental facts even up to $q^2 \sim 600\ \mathrm{F}^{-2}$.[27] Physically, the "nodal" behavior of the form factor arises from the effect of a high centrifugal barrier associated with the $a$-function of $L^P = 0^+$. Since, according to (1.4), this zero-value of $L$ is made up of $l_{12}$ and $l_3$, and the $a$-function must have at least $l_{12} = 1$, it is clear that $l_3$ is also equal to unity. These non-zero values of $l_{12}$ and $l_3$ (or any other cyclic permutations of these sub-angular momenta) are the direct manifestations of centrifugal barriers, and hence of nodes in the wave function as well as in the form factors. This argument is largely kinematical since it depends only on considerations of angular momentum in relation to a given three-particle symmetry. Clearly no such problem arises with an $s$-function of $L^P = 0^+$, since in this case, *both* $l_{12}$ and $l_3$ can afford to be zero. Indeed, the fact that the electromagnetic form factors of the nucleon do not show any nodes up to $q^2 \sim 600$ F$^{-2}$, is strongly suggestive of $l_{12} = l_3 = 0$ in a quark model.

From this discussion it would appear reasonable to conclude that an $s$-function of $L^P = 0^+$ for **56** states is preferable to an $a$-function, both on the dynamical grounds of possible sources of attraction, as well as the more kinematical considerations governing the occurrence of nodes in the form factors due to the centrifugal effects of non-zero values of internal angular momenta. This preference of $s$ over $a$ for **56** states has a direct effect on the choice of quark statistics within the Gell-Mann-Zweig (GMZ) version of the quark model, since Fermi statistics is no longer compatible with such a preference. To retain Fermi statistics, with $s$ functions for **56** states, one must either extend the GMZ model to include additional quantum numbers such as triality,[28] or introduce additional quarks through, e.g., the three-triplet model (nine quarks).[29-32] Such extensions could make almost any symmetry ($s$, $a$ or $m$) compatible with Fermi statistics, but only at the cost of losing the basic simplicity of the original GMZ model.

A different suggestion, due to Greenberg[33], is to invoke Green's[34] para-statistics idea for quarks, as this allows enough flexibility in the overall symmetry scheme for the $QQQ$ state, so that the latter can have any one of $s$, $m$ or $a$ symmetry, depending on the type of parastatistics chosen. The simplest possibility is to *assume $s$-symmetry* for the $QQQ$ state which allows the **56** states to have the much desired $s$-function of $L^P = 0^+$. Formally this effect is the same as that of Bose statistics, as far as $QQQ$ states are concerned, but this analogy does not extend beyond the three-particle level, since it can be shown[35] that parastatistics by itself gives rise to saturation at the three-quark level in a particular quantum state, even without invoking any specific form of dynamics. In the language of Young diagrams, parastatistics, unlike Bose statistics, does not allow more than three boxes in any row. Thus for a two-baryon state (of six quarks), parastatistics would require a [3, 3] symmetry in the joint space of $SU(6)$ and the orbital degrees of freedom, while Bose statistics would predict a [6] symmetry for the same system[36].

### 1.5  Phenomenology with quarks

The question now arises as to the possible *phenomenological* manifestations of the "Fermi" versus "para" assumption on the choice of quark statistics, for a $QQQ$ system. Certainly, any dynamical calculation, such as that of the energy levels of successive resonances, would depend vitally on the choice of quark statistics as well as quark forces. Indeed, any physical quantity which depends on the *detailed* structure of the $QQQ$ wave function, such as the momentum dependence of a proton's electromagnetic form factors, must be sensitive to the type of symmetry assumed for that function. On the other hand, there are many processes whose amplitudes, while being formally expressible in terms of quark wave functions, nevertheless involve these functions only in the form of certain overlap integrals. It is frequently useful to give a direct parametrization to such integrals especially if the experimental data are not of such a type as to warrant a more detailed study of these quantities. Processes of these types involve transitions between single baryon states looked upon as $QQQ$ composites, in which only one quark at a time is assumed to participate in the relevant transition; the total amplitude is then obtained by folding the sum of the amplitudes for these individual quark transitions, into the $QQQ$ wave functions of the initial and final baryon states. Some examples are provided by mason-baryon and photoproduction processes of the respective types

$$PB \rightarrow PB, \quad PB \rightarrow PB^*; \quad PB \rightarrow VB, \quad PB \rightarrow VB^*; \qquad (1.7)$$

$$\gamma B \rightarrow PB, \quad \gamma B \rightarrow VB, \quad \gamma B \rightarrow PB^*, \quad \gamma B \rightarrow VB^*. \qquad (1.8)$$

Here $\gamma$, $P$ and $V$ are photons, pseudoscalar mesons and vector mesons respectively; $B$, $B^*$ represent the **8** baryons and their **10** resonances respectively. The "elementary" mechanism for any one of the processes (1.7) and (1.8) is obtained through the replacement of the "baryon" on each side by the constituent quark, assuming $P$, $V$ and $\gamma$ as "elementary" objects.[37,38] For such processes, the predictions of Fermi and parastatistics are formally equivalent in the following sense. The algebraic structures of the various amplitudes obtained from Fermi statistics bear a one-to-one correspondence to those derived from parastatistics, provided that the orbital functions of different symmetry types occurring in the various overlap integrals are replaced according to the following prescriptions:

$$s \rightarrow a, \quad a \rightarrow s, \quad m' \rightarrow m'', \quad m'' \rightarrow -m', \qquad (1.9)$$

or vice versa. The essential point of this argument is that the evaluation of the detailed structures of the overlap functions is probably not worth while at the present (imperfect) state of knowledge of the quark philosophy, and that it makes much better sense to parametrize these integrals as such. The case of hadron couplings, which will form the central theme of this article, also belongs to this category, when it is remembered that the basic vertices are provided by the (virtual) processes

$$Q \to Q + P \quad \text{or} \quad Q \to Q + V.$$

Taking such a pragmatic view of the quark model, it is possible to confine attention to a certain (outer) level of hadronic investigations which depend primarily on certain symmetries in the initial and final (multiquark) states, as well as the "elementary" amplitudes governing the quark transitions, without having to make a prior commitment to a more specific form of quark dynamics. This point of view allows, in principle, a *gradual* unfolding of the quark philosophy in relation to the experimental situation, starting first with its "index" aspects (dealing mainly with symmetries), and successively extending its tentacles to incorporate more specific dynamical assumptions, only if the experimental situation warrants such optimism. At the moment the nature and extent of the experimental data are too superficial to facilitate any meaningful comparison on the basis of a full-fledged dynamical quark model, since these are still largely confined to the level of sum rules, density matrices and at most hopeful conjectures on many "expected" resonances. Under such conditions not much concrete criteria are available for distinguishing between "mathematical" quarks of the Gell-Mann school (which mainly makes use of the "index" aspect) and the "real" quarks of the Dalitz school (which relies more heavily on quantitative dynamics). For the problem of hadron couplings in particular, the quark model, whether with mathematical or real ones, plays only a limited role, viz., as a means for correlating and parametrizing the data on hadron vertices through the use of (i) certain symmetries in the wave functions and (ii) certain rules of construction of vertices based on a single-quark model.

Within these restricted premises, the effect of the quark assumption on the construction of hadron couplings is to predict results which are largely equivalent to those obtainable from the group $SU(6)$ or its appropriate extensions. Even so, there are some important differences from the predictions of pure group theory. Thus while $SU(6)$ or its relativistic extensions (such as $SU(6)$) do not predict any correlations between baryon ($\bar{B}BP$) and meson ($MMP$) couplings with a particular type of mesons (such as the pseudoscalar octet $P$), the quark model relates their couplings[39] through the assumption of a *common* $Q\bar{Q}P$ vertex for *both* the baryon ($QQQ$) and meson ($Q\bar{Q}$) cases. Another notable example of the stronger predictive power of the quark model than that of group theory, is provided by the prediction of a $\frac{3}{2}$-ratio[40] between $p-p$ and $\pi-p$ total cross sections, obtained from the simple assumption of additivity of the different quark-scattering amplitudes.

Now the question of whether or not, the extra predictive power of the quark model over pure group is a manifestation of some sort of dynamical effect, could by itself be an interesting academic exercise. It is interesting to note that this extra predictive power arises from the *additivity assumption* on certain *single-particle* amplitudes, whether in hadron scattering or in hadron coupling problems. This is somewhat reminiscent of an allied situation in the nuclear shell model where the empirical but impressive successes of the independent particle model had remained quite a puzzle for a long time, until the more formal Brueckner–Bethe theory provided a qualitative dynamical understanding of the phenomena. It might be interesting to speculate as to whether or not this analogy will extend to the quark model at the dynamical level. However at the present stage such questions are not so relevant as the facilities which the model provides as a pedagogical device for the correlation of a large number of phenomena, covering a much wider domain than can be achieved within pure group theory.

### 1.6  Scope of the article

In this article we shall try to present a unified review of baryon and meson couplings at the phenomenological levels, using the quark model mainly as a tool for the evaluation of the "index" structures of the couplings. The most important manifestation of these couplings comes from the decay rates of the hadron resonances. While the theoretical methods for the construction of the couplings require input assumptions on the supermultiplet assignments for the hadrons, the latter in turn must be checked through their predictions on the strong decay rates themselves. This mutual feedback between the coupling schemes and supermultiplet assignments is an essential requirement for any successful investigation of hadron couplings. Unlike the mass levels, the decay rates depend very sensitively, not only on the supermultiplet assignments, but even on small admixtures of states caused by $SU(6)$ and $SU(3)$ breaking effects. Thus the decay rates offer a much more sensitive probe on the multiplet or supermultiplet assignments than do investigation of mass levels, spin-orbit splittings and similar phenomena which moreover depend on many more dynamical assumptions.

The symmetry breaking effects on the couplings which are caused mainly by the mass splittings in the relevant hadron constituents, are most conveniently parametrized by certain form factors which are taken to depend on the masses and c.m. momenta of the decay products. In some more fortunate cases, where angular distributions of the products of a sequential decay[41] are available, one obtains further information on the supermultiplet assignments via the coupling schemes. Study of coupling schemes has the further advantage that since these depend on rather little dynamical assumptions, the information they provide on the supermultiplet assignments is generally more valuable than the corresponding information from quantities which depend on more quantitative dynamical assumptions, such as the splittings of mass levels due to spin-orbit, $SU(3)$—breaking, etc. forces.

Since the quark model will mainly be used as a guide to evaluation of Clebsch–Gordan structures, the more esoteric questions like the nature of quarks (real versus mathematical), statistics of these objects, the theoretical problem of reconciling the non-relativistic quark motion with the relativistic $Q$-values of the decays, and so on, will not play any important role in the development of the article. The aim of this review will be mainly to develop a pedagogical device for a systematic evaluation of coupling schemes for successively higher resonances to their more familiar hadron counterparts, and to confront their predictions with experiment in some detail. It will be shown that even a very general form of parametrization for the form factors already yields valuable information on (i) the potential roles of representation mixings among appropriate members of different spin and unitary spin multiplets, and (ii) the effects of the "direct" and "recoil" terms in the quark-meson couplings on the relative magnitudes of the light and heavy meson modes of decay in relation to experiment. With more specialized forms of parametrizations for these quantities through suitable assumptions on their mass dependence, it is even possible to give a quantitative and unified description of decay widths for both types of hadrons in a relativistic manner using as inputs only single overall constants governing entire supermultiplet transitions. It is hoped that such descriptions would go a long way towards bridging the gap between quark phenomenology and more formal theories.

In Section 2 we review the evaluation of couplings among the **56** baryon and **36** meson states in terms of the quark-meson vertex as the basic mechanism. The various phase conventions to be used for the "direct" and "recoil" terms associated with the quark ($Q$) and antiquark ($\bar{Q}$) transitions are clarified, and a systematic prescription is outlined for relating the matrix elements for hadronic transitions in the (composite) quark space to the corresponding elements in the space of hadrons looked upon as elementary particles. The role of the form factors in simulating $SU(3)$—breaking due to various mass differences among hadrons is discussed with special reference to the van Royen–Weisskopf prescription for the normalization of the meson $(Q\bar{Q})$ wave functions. Finally a method of relativistic extension is outlined to bring out some extra mass dependence in the couplings based on the idea of normalizations of hadron fields and the conservation of energy in the PCAC limit of zero meson four-momentum. The predictions of these mass structures in the couplings are shown to be well substantiated by experiment.

Section 3 gives a short account of the classification for the meson and baryon resonances according to an $SU(6) \times O(3)$ scheme, with special reference to the various quantum numbers needed for the purpose. It also summarises the current status of the better known (i) meson resonances of $L^P = 1^+$ in terms of nonet structures and (ii) baryon resonances of $L^P = 1^-$ and $2^+$ in terms of certain "harmonic oscillator" representations of $SU(6) \times O(3)$. In particular, the occurrence of only certain natural parity baryon states of the types $(\mathbf{56}, 2l^+)$ and $(\mathbf{70}, (2l + 1)^-)$ is emphasized in relation to a more specific dynamical mechanism based on $s$-wave interactions. Section 4 outlines a general scheme of couplings between different hadron supermultiplets $B_L$ and $M_L$. The various ingredients for the evaluation of $M_L PM$ and $\bar{B}_L PB$ couplings, including (i) the wave functions for the $M_L$ and $B_L$ states and (ii) the roles of the direct and recoil terms on the structures of the $(L \pm 1)$ wave amplitudes are described in some detail. It is shown that all transitions can be described in terms of certain overlap integrals over the quark wave functions for the initial and final states. Section 5 is devoted to a phenomenological analysis of hadron decays from $L = 1$ and 2 supermultiplet states, in terms of certain general forms of parametrizations of the overlap integrals representing the direct and recoil transitions. The role of the recoil term in bringing about enhanced heavy meson modes and the sensitiveness of several decay modes to small effects of configuration mixings among suitable $SU(3)$ and internal spin states are brought out in some detail. The recoil term is also found to be helpful for understanding the observed angular distributions in certain sequential decays, notably $B \to \pi\omega$, $\omega \to 3\pi$. Finally, the points of similarity and contrast between $SU(6) \times O(3)$ and $SU(6)_W$ are discussed.

Section 6 is devoted to an improved form of parametrization of the $SU(6) \times O(3)$ coupling scheme so as to incorporate (i) relativistic requirements,

(ii) enhanced heavy meson modes in $s$-wave decays, and (iii) a more explicit role of the meson mass in bringing about $SU(3)$ breaking. Such a model which allows a very unified treatment of both baryon and meson couplings with a single adjustable parameter for each supermultiplet transition, is shown in Section 7 to work extremely well for a large number and variety of decay modes. The model is also shown to bring out the feature of universal couplings of baryon Regge trajectories. Section 8 summarises the principal results and conclusions.

## 2   COUPLINGS AMONG $L = 0$ HADRONS

The hadrons of zero orbital excitation are the **56** baryons of $L^P = 0^+$ and the two meson nonets, pseudoscalar $(P)$ and vector $(V)$, of $L^P = 0^-$. We use the notation $B$, $B^*$ for the **8** and **10** baryons respectively. The main couplings under this head are*

$$\bar{B}BP, \quad \bar{B}B^*P; \quad VPP, \quad VVP; \tag{2.1}$$

which are manifestations of strong interactions. Each of these which involves at least one $P$-meson, can be pictured as the result of folding a $\bar{Q}QP$ vertex into the appropriate hadron structures in the initial and final states. In momentum space, this vertex for a particular quark $i$ is of the form[39,42,43]

$$H_I^{(i)}(Q_q) = f_q \mu^{-1} \bar{Q}_q^{(i)} \sigma^{(i)} \cdot (\mathbf{K} - M_Q^{-1} \omega_k \mathbf{P}_i)\, \lambda_\alpha^{(i)} \pi_\alpha Q_q, \tag{2.2}$$

for the emission of a $P$-meson $\pi_\alpha$ $(\alpha = 1, 2, \ldots, 8)$ of 4-momentum $(\mathbf{k}, \omega_k)$ and mass $\mu$ by the quark field $Q_q^{(i)}$, $\lambda_\alpha^{(i)}$ being the Gell–Mann matrices[44] associated with quark $i$. The term involving $\mathbf{P}_i$ (the momentum of the quark) represents the recoil effect on the quark (mass $M_Q$) in accordance with the principle of Galilean invariance.[42] In a similar way, the vertex for the emission of $\pi_\alpha$ by an antiquark field $Q_a^{(i)}$ (momentum $-\mathbf{P}_i$) is[43]

$$H_I^{(i)}(Q_a) = -f_q \mu^{-1} \bar{Q}_a^{(i)} \sigma^{(i)} \cdot (\mathbf{K} + M_Q^{-1} \omega_k \mathbf{P}_i)\, \bar{\lambda}_\alpha^{(i)} \pi_\alpha Q_a^{(i)}, \tag{2.3}$$

where $\bar{\lambda}_\alpha^{(i)}$ are the Gell–Mann matrices associated with the antiquark field $Q_a^{(i)}$. Note that while the direct term enters with opposite signs in (2.2) and (2.3), the recoil term has the same sign in both.

---

* We are not considering $\bar{B}BV$ or $\bar{B}B^*V$ couplings, since the physical interest in these lies mostly in the electromagnetic processes like $\Delta \to N + \gamma$, etc., generated through the linear couplings of photons to neutral $V$-particles.

In writing the above vertices the tacit assumption is made that the $P$-meson must be regarded as an elementary radiation quantum, rather than a $Q\bar{Q}$ composite. This amounts to giving the $P$-meson a special status over the other hadrons. Such an assignment can be defended on the ground that the vertex $Q\bar{Q}P$ is basically a manifestation of a weak effect, if the $P$-meson is regarded as a "soft" pion (having a small momentum). Such a concept is consistent with the "PCAC" equivalence of the pion with the axial vector field which couples weakly to the baryon field.[43] This provides the essential justification for the *additive* structures the total $MMP$ or $\bar{B}BP$ vertices in terms of the individual $\bar{Q}QP$ vertices. However, these arguments are probably over simplified. For, while a pion may be regarded as soft (due to its very small mass), the corresponding assumption for the (more massive) $K$- and $\eta$-mesons is much less defensible. Again the emission of rather energetic pions from, say $\varrho$- and $K^*$-mesons, seems to invalidate further the assumption of "softness" in many actual cases. These difficulties indicate that for actual applications to decay processes, this simple picture of $P$-meson emission from hadrons looked upon as quark composites must be supplemented by certain "realistic" requirements on the structures of the hadron vertices, these hadrons now being looked upon as elementary, relativistic objects. The explicit prescriptions for incorporating these modifications which are discussed later in this section, must of course be judged by their results, rather than on purely theoretical grounds. For $MMP$ couplings there is the further problem of deciding which meson in the final state must be regarded as a $Q\bar{Q}$ composite and which one as elementary, especially if both happen to be the same type of meson. For baryons, this formal problem is not so vexing since one can at least pretend that $B$ and $P$ are entirely different kinds of objects. Further, since in a non-relativistic theory particles and antiparticles have nothing to do with each other, one can argue that it is largely a matter of taste as to whether $Q$ and $\bar{Q}$, *or* $Q$ and $P$ are regarded as the two independent *entities**.

For meson couplings on the other hand, the corresponding ambiguity is of a more overlapping nature. Fortunately, even in the meson case, it is immaterial for the purpose of calculating the Clebsch–Gordan coefficients, to know which meson is which. However it is important that (i) a label on each be

---

attached for definiteness and that (ii) the amplitudes obtained from the two independent possibilities be *not* added.[43] Certain additional conventions which are necessary for the case of two *unequal* *P*-mesons in the final state will be described in a later section.

## 2.1  56 baryon couplings

For the sake of definiteness, we shall use the symmetrical quark model as the basis of calculations though, as pointed out in Section 1, the algebraic form of the results for the coupling constants is independent of the choice of statistics. The wave functions of the **8** and **10** states of the **56** baryons are

$$\Psi_8 = \psi^S \left( \chi' \phi' + \chi'' \phi'' \right)/\sqrt{2}, \tag{2.4}$$

$$\Psi_{10} = \psi^S \chi^S \phi^S, \tag{2.5}$$

where $\chi$, $\phi$, $\psi$ represent the normalized spin, $SU(3)$ and spatial functions respectively[23], and the superscripts on these functions indicate their symmetry types with respect to the permutation operators for three objects. The $\eta$-functions $(\chi', \chi'')$ correspond to spin-$\frac{1}{2}$, while the *s*-function $\chi^S$ has spin-$\frac{3}{2}$; likewise the $SU(3)$ functions $(\phi', \phi'')$ and $\phi^S$ give **8** and **10** states respectively.[46] The spatial function $\psi^S$ of *s*-symmetry and $L^P = 0^+$ depends externally on the normalized c.m. momentum **P** of *s*-symmetry given in terms of the individual quark momenta $\mathbf{P}_i$ by

$$\mathbf{P} = 3^{-1/2} \left( \mathbf{P}_1 + \mathbf{P}_2 + \mathbf{P}_3 \right). \tag{2.6}$$

Its dependence on the internal momenta is expressed in terms of two normalized relative momenta $\mathbf{P}'$, $\mathbf{P}''$ of *m*-symmetry, according to the definitions[46]

$$\mathbf{P}' = (\mathbf{P}_3 - \mathbf{P}_2)/\sqrt{2}, \tag{2.7}$$

$$\mathbf{P}'' = (-2\mathbf{P}_1 + \mathbf{P}_2 + \mathbf{P}_3)/\sqrt{6}. \tag{2.8}$$

These two functions, like any pair of *m*-functions, form a basis for the following $2 \times 2$ representations of the permutation operators $(ij)$ of three objects:

$$(23) = \begin{pmatrix} -1 & 0 \\ 0 & 1 \end{pmatrix}; \quad (12), (13) = \begin{pmatrix} \frac{1}{2} & \pm\sqrt{3}/2 \\ \pm\sqrt{3}/2 & -\frac{1}{2} \end{pmatrix}. \tag{2.9}$$

The normalization of the internal part of $\psi^S$ is given by

$$\iint d\mathbf{P}' d\mathbf{P}'' |\psi^S (\mathbf{P}', \mathbf{P}'')|^2 = 1. \tag{2.10}$$

The vertex functions for $\bar{B}BP$ and $\bar{B}B^*P$ couplings are now obtained by evaluating the matrix elements, between appropriate $B$ of $B^*$ states, of the operator

$$H_I(Q) = \sum_{i=1}^{3} H_I^{(i)}(Q), \tag{2.11}$$

where the various terms on the right are given by (2.2). To illustrate the procedure, consider, e.g., $\bar{N}N\pi$ coupling, which is most conivenently obtained by evaluating the matrix elements of (2.11) for the transition $p \to \pi^0 + p$, with no spin flip, viz.,*

$$\tfrac{3}{2} f_q \mu^{-1} \langle \chi'\phi' + \chi''\phi'' | \sigma_z^{(1)} k_z \lambda_3^{(1)} \pi_3 | \chi'\phi' + \chi''\phi'' \rangle \times$$

$$\times \iint d\mathbf{P}' \, d\mathbf{P}'' \, \psi^s (\mathbf{P}', \mathbf{P}'') \, \psi^s (\mathbf{P}', \mathbf{P}'' + \sqrt{\tfrac{2}{3}}\, \mathbf{k}). \tag{2.12}$$

To evaluate (2.12), one observes that for moderate values of $\mathbf{K}$, the second line is nearly unity, according to (2.10). Secondly by virtue of the results

$$\langle \chi'' | \sigma_z^{(1)} k_z | \chi'' \rangle = -\tfrac{1}{3} \langle \chi' | \sigma_z^{(1)} k_z | \chi' \rangle \tag{2.13}$$

and

$$\langle \phi'' | \lambda_3^{(1)} \pi_3 | \phi'' \rangle = -\tfrac{1}{3} \langle \phi' | \lambda_3^{(1)} \pi_3 | \phi' \rangle, \tag{2.14}$$

it is possible to express these spin and $SU(3)$ matrix elements defined in the composite $QQQ$ space, directly in terms of the spin and $SU(3)$ variables of the baryon *regarded as an elementary particle*. Thus we have the correspondence

$$\langle \chi' | \boldsymbol{\sigma}^{(1)} \cdot \mathbf{k} | \chi' \rangle \Leftrightarrow \langle \chi | \boldsymbol{\sigma} \cdot \mathbf{k} | \chi \rangle, \tag{2.15}$$

$$\langle \phi' | \tau_a^{(1)} \pi_a | \phi' \rangle \Leftrightarrow \langle \phi | \tau_a \pi_a | \phi \rangle, \tag{2.16}$$

where now $\chi, \phi$ are the spin and $SU(3)$ functions of the baryon as a *whole*, $\boldsymbol{\sigma}$ is its spin operator and $\tau_a$ the isospin matrices associated with it (summation over $a = 1, 2, 3$ is implied). These identifications thus obviate the need for keeping the quark labels any longer. Using the above results one is led immediately to the following operator for $\bar{N}N\pi$ coupling

$$\tfrac{5}{3} f_q \mu^{-1} \bar{N} \tau_a \pi_a \sigma_i k_i N F_B (k^2), \tag{2.17}$$

---

* Note that parity considerations do not allow any contribution from the recoil term in this case where the spatial functions of the initial and final states are the same.

where the form factor $F_B(k^2)$ is normalized by $F_B(0) = 1$, and $\sigma_i k_i$ is the tensor form of $\boldsymbol{\sigma} \cdot \mathbf{k}$. Taking $F_B(k^2) \approx 1$, one obtains the result

$$f_\pi = \tfrac{5}{3} f_q, \tag{2.18}$$

where $f_\pi$ is the effective $\bar{N}N\pi$ coupling constant in $ps-pv$ coupling. In a similar way one recovers all the $SU(3)$ coefficients $(F/D = \tfrac{2}{3})$ for the other $\bar{B}_8 B_8 P_8$ coupling constants.[47]

For the couplings of **8** with **10** baryons, it is clearly convenient to consider a process like $\varDelta^{++} \to p + \pi^+$, which involves a flip in each of the spin and isospin variables. Proceeding in exactly the same way as above for the elimination of the quark indices, one obtains the $\varDelta\pi N$ coupling (again with $F_B(k^2) \approx 1$)

$$2\sqrt{2} f_q \mu^{-1} \varDelta_i^{+a} k_i \pi_a N. \tag{2.19}$$

Here $\varDelta_i^a$ is a non-relativistic Rarita–Schwinger field[48] in both the spin ($i$) and isospin ($a$) indices, through the identification that $a = 1$, $i = 1$ correspond to the normalized $\varDelta^{++}$ state of $S_z = \tfrac{3}{2}$. The couplings for the other **10** states, which are obtained in the same way, lead to the well known $SU(6)$ Clebsch–Gordan coefficients for $\bar{B}BP$ and $\bar{B}B^*P$ coupling constants, all expressed in terms of the $\bar{Q}QP$ constant $f_q$. The only approximation involved in this quark model derivation of the $SU(6)$ results lies in the neglect of the variation of the form factor $F_B(K^2)$ with the momentum transfer $K$ for the different cases. This variation in the form factor represents an important refinement over $SU(6)$ which the quark model is in principle capable of producing. However in the absence of a more fundamental (dynamical?) theory, this refinement must be expressed in a parametric form. The most natural candidates for such a parametrization are clearly the masses of the particles involved. This question is considered later in this section for both the baryon and meson cases.

The derivations of (2.17) and (2.19) suffer from the obvious disadvantage, characteristic of the quark model, of being non-relativistic in character. For actual applications to decay processes, this defect must first be remedied through suitable relativistic modifications. Because of the formidable problems (discussed in Section 1) associated with generalizations to bigger relativistic symmetries, it is prudent to consider a more modest, but nevertheless numerically successful symmetry, first suggested by Beg and Pais.[47] This symmetry is merely the direct product of $SU(3)$ and the Lorentz group $0(3, 1)$, in which the coupling constants are so adjusted as to reproduce

$SU(6)$ symmetry in the limit of zero momentum transfer. In such a spirit, all that is necessary is to generalize the 3-vector indices like $i$ in (2.17) and (2.19) to 4-vector indices like $\mu$, and introduce obvious Dirac covariance which is facilitated through the successive replacements

$$\sigma_i \rightarrow i\gamma_i\gamma_4\gamma_5, \quad \text{and} \quad i \rightarrow \mu. \tag{2.20}$$

These modifications result in the following structures of (2.17) and (2.19) respectively

$$2mf_q\mu^{-1}\left(\tfrac{5}{3}\right)\overline{\psi}\gamma_5\tau_a\pi_a\psi, \tag{2.17$'$}$$

$$2\sqrt{2}f_q\mu^{-1}\overline{\psi}_\mu^a k_\mu\pi_a\psi, \tag{2.19$'$}$$

where $\psi$ is a Dirac field, $\psi_\mu$ a *relativistic* Rarita–Schwinger field[48], and the appearance of the nucleon mass $m$ in (2.17)$'$ is a result of application of the Dirac equation to the $ps-pv$ coupling term $\overline{\psi}\gamma_\mu\gamma_5 k_\mu\psi$. Such modifications which are beyond the strict scope of the quark model, are nevertheless necessary for any meaningful encounter of the model with experiment.

A numerical estimate of $f_q$ can be made from (2.17)$'$ on the basis of its relation to the nucleon mass $m$ and the effective pseudoscalar coupling constant $G_{NN\pi}$, given by

$$G_{NN\pi} = \tfrac{5}{3}f_q\,(2m/\mu), \tag{2.21}$$

so that

$$f_q^2/4\pi \approx 0.03 \quad (\text{from} \quad G^2/4\pi \approx 15.5). \tag{2.22}$$

Unfortunately this value leads to a rather low estimate of the width calculated from (2.19)$'$, viz

$$\Gamma_{\Delta \rightarrow N\pi} \approx 77 \text{ Mev}, \tag{2.22$'$}$$

compared with its experimental value of $(120 \pm 1)$ MeV. Note that while this estimate takes account of the symmetry breaking in the phase space, it does not as yet incorporate this breaking in the coupling constants caused by the variation of the form factor $F_B(k^2)$ due to the actual masses of the particles involved. Some simple parametrizations towards this end, which help in improving the results for the decay widths, will be discussed later in this section.

### 2.2  Couplings of $V$ and $P$ meson monets

We now consider the cases of $VPP$ and $VVP$ couplings among the familiar **36** meson states ($P$ and $V$) in which one of the $P$-mesons is regarded as an elementary (radiation) quantum. This specifies the convention uniquely for

$VVP$ couplings, where the two $V$-mesons must be regarded as $Q\bar{Q}$ composites in $3_{S_1}$ states. For the $VPP$ case, both the $V$-meson and the second $P$-meson must also be regarded as $Q\bar{Q}$ composites in $3_{S_1}$ and $1_{S_0}$ states respectively. Since in this case, it is necessary to specify *which* $P$-meson is elementary, one must use a further convention which can be justified only by its experimental support. We specify this convention by regarding the *heavier* of the two $P$-mesons as the "quantum". This rather unusual choice has the advantage that it happens to obviate the need for further conventions on the dimensional factor (mass) required to compensate for the momentum dependence of the ($p$-wave) $\bar{Q}QP$ interaction (2.2) or (2.3), (in which the factor $\mu$ represents the mass of the emitted quantum).

For the $VVP$ case, the transition is characterised by the initial $V$-meson ($3_{S_1}Q\bar{Q}$) emitting a $P$-meson (quantum) in a non-spin-flip fashion so that it remains a $V$-meson. For the $VPP$ case on the other hand, the initial $V$-meson must emit the quantum with a spin-flip transition to become a $P$-meson ($1_{S_0}Q\bar{Q}$). The Clebsch–Gordan coefficients for both types of transitions must be calculated in exactly the same way as described for baryons (Section 2.1), viz., by adding the matrix elements for the individual $Q$ and $\bar{Q}$ transitions, except that the $\bar{Q}$-transitions are governed by the operator (2.3) rather than (2.2). Further, since the radial function for both types of couplings ($VVP$ and $VPP$) are the same in the initial and final states, the recoil terms in (2.2) or (2.3) do not have any effect on these couplings, exactly as in the **56** baryon cases (Section 2.1). However, unlike the **56** baryon ($QQQ$) cases, the $Q\bar{Q}$ structures of the initial and final meson states are no longer inhibited by any symmetry requirements, since quarks and antiquarks are two entirely different types of particles in a non-relativistic framework.

For the evaluation of these couplings, the cases of special interest are the following:

$$\varrho\pi\pi, \quad K^*K\pi, \quad \phi K\bar{K}; \quad \omega\varrho\pi, \quad \phi\varrho\pi.$$

As in the baryon case, the coupling scheme must be supplemented by certain relativistic prescriptions before the resulting expressions can be applicable to decay processes. Indeed, these relativistic prescriptions are even more necessary in the meson case not only because the energies involved are more relativistic, but (in addition) because these Yukawa type meson-couplings involve *boson* operators whose relativistic normalizations ($\sim (2\omega)^{-1/2}$) are *dimensional*, unlike the fermion operators in baryon couplings whose normalizations ($\sim \sqrt{m/E}$) are dimensionless.

To illustrate the procedure, consider the case of $\varrho\pi\pi$ coupling. The matrix element for the transition $\varrho^+ \to \pi^+ + \pi^0$ (quantum) is obtained by adding the matrix elements of the sum of (2.2) and (2.3) between the initial $3_{S_1}$ state of $\varrho^+$, viz.,

$$(\bar{n}p)\, 2^{-1/2}\, (\alpha\bar{\beta} + \bar{\alpha}\beta)\, \psi(\mathbf{p})\, \delta(\mathbf{P}), \tag{2.23}$$

and the final $1_{S_0}$ state of $\pi^+$, viz.,

$$(\bar{n}p)\, 2^{-1/2}\, (\alpha\bar{\beta} - \bar{\alpha}\beta)\, \psi(\mathbf{p})\, \delta\,(\mathbf{P} - \mathbf{k}); \tag{2.24}$$

here $\mathbf{k}$ is the momentum of the emitted quantum; $\mathbf{P} = \mathbf{P}_1 + \mathbf{P}_2$ is the c.m. momentum of the $Q\bar{Q}$ state in terms of the individual quark momenta ($\mathbf{P}_1$ and $\mathbf{P}_2$); and $\pm\mathbf{p} = \pm(\mathbf{P}_1 - \mathbf{P}_2)$ are the momenta of $Q$ and $\bar{Q}$ in their c.m. frame. The symbols $(p, n, \lambda)$ and $(\bar{p}, \bar{n}, \bar{\lambda})$ are the $SU(3)$ states of $Q$ and $\bar{Q}$ respectively; similarly $(\alpha, \beta)$ and $(\bar{\alpha}, \bar{\beta})$ are the spin-states (up and down) of the same particles. The factors $\psi(\mathbf{p})$ and $\delta\,(\mathbf{P} - \mathbf{k})$ are respectively the internal and external parts of the $Q\bar{Q}$ orbital wave function ($L^P = 0^-$). The matrix element of the transition $\varrho^+ \to \pi^+ + \pi^0$ is given by[43]

$$\langle \varrho^+ | H(Q) + H(\bar{Q}) | \pi^+ \rangle = f_q \mu^{-1}\, (k_Z^{(+)} - k_Z^{(0)})\, F_M(k^2), \tag{2.25}$$

where $\mathbf{k}^{(+,0)}$ are the momenta of $\pi^+$ and $\pi^0$ respectively, and $F_M(k^2)$ is the form factor defined by

$$F_M(k^2) = \int d\mathbf{p}\, \psi(p)\, \psi\,(\mathbf{p} - \tfrac{1}{2}\mathbf{k}). \tag{2.26}$$

The quantity (2.25) must now be equated to the corresponding matrix element of the *relativistic* $\varrho\pi\pi$ interaction defined by[43]

$$if_{\varrho\pi\pi}\varrho_\mu^a \pi^b \partial_\mu \pi^c \varepsilon_{abc}, \tag{2.27}$$

in which the operators are normalized by

$$\langle 0 |\, \pi(x)\, | \pi \rangle = (2\omega_k)^{-1/2} e^{ik \cdot x}, \quad \text{etc.} \tag{2.28}$$

Equating the two results for the case of the $\varrho$-meson decaying at rest, one obtains

$$f_{\varrho\pi\pi}\, (2m_\varrho)^{-1/2}\, (2\omega_{\pi+})^{-1/2} = \mu^{-1} f_q\, F_M(k^2). \tag{2.29}$$

The interpretation of the body form factor $F_M(k^2)$ in (2.29) is similar to the baryon case, viz., that it represents the effect of $SU(6)$ symmetry breaking due to the masses. In the absence of any reliable "theory" of the quark wave functions, this effect can at most be parametrized in a phenomenlogical manner, with the normalization

$$F_M(0) = 1,  \tag{2.30}$$

following from the condition

$$\int |\psi(p)|^2 \, d\mathbf{p} = 1. \tag{2.31}$$

While this problem is discussed in the next sub-section, it may be noted in the meantime that the prescription (2.29) is still not free from ambiguities, arising from the question of what numerical value should be assigned to the factor $(2\omega_{\pi+})^{-1/2}$. Note that this factor represents the relativistic normalization arising out of the final pion which has been taken as the $Q\bar{Q}$ composite. One prescription, due to Becchi and Morpurgo (BM) is to take the *actual* energy associated with this pion, in which case $\omega_{\pi+} - \frac{1}{2}m_\varrho$. This prescription happens to give a reasonably good value for the $\varrho \to \pi\pi$ decay width, viz.,

$$\Gamma_{\varrho\pi\pi} \approx 93 \text{ Mev}; \quad (BM). \tag{2.32}$$

A different (and perhaps more rational) prescription was given by Van Royen and Weisskopf (VW)[43] who suggested that this normalization factor must be evaluated in the exact PCAC limit, in which case the pion represents merely the spin-flip state of rho, without any change of energy. In this case one obtains

$$\omega_{\pi+} = m_\varrho \tag{2.33}$$

and

$$f_{\varrho\pi\pi} = 2m_\varrho\mu^{-1}f_q. \tag{2.34}$$

This however gives for the $\varrho\pi\pi$ width, double the value of (2.32), viz.,

$$\Gamma_{\varrho\pi\pi} \approx 185 \text{ Mev}; \quad (VW). \tag{2.35}$$

Inspite of this "bad" result for $\varrho\pi\pi$, this alternative prescription seems to work much better for most other cases.*

---

* Actually, Becchi and Morpurgo[39] obtained four times the value (2.32) due to their (erroneous) prescription that the two pions should be regarded as radiation quanta by turn, and the resulting amplitudes *added*. This was refuted by Van Royen and Weisskopf[43] who based their suggestion against this addition, on the PCAC hypothesis. A third prescription was given by Cook[49] who took the value $\omega_\pi = m_\pi$, instead of (2.32) or (2.33), using the argument that while the normalization should be determined at the zero-momentum limit, it should still correspond to the actual pion mass. He obtained a surprisingly good value for the $\varrho$-width, but only at the cost of repeating the Becchi–Morpurgo error of adding the above two amplitudes.

For the couplings of $\phi$ and $\omega$ mesons, it is first necessary to consider their $SU(3)$ structures as[19]

$$\phi = \bar{\lambda}\lambda; \quad \omega = (\bar{n}n + \bar{p}p)/\sqrt{2}, \tag{2.36}$$

based on the ideal mixing angle of artan $\sqrt{2}$, rather than represent them by the pure $SU(3)$ states

$$\omega_8 = (\bar{p}p + \bar{n}n - 2\bar{\lambda}\lambda)/\sqrt{6}; \quad \omega_0 = (\bar{p}p + \bar{n}n + \bar{\lambda}\lambda)/\sqrt{3}. \tag{2.37}$$

As is well known, this automatically leads to

$$f_{\phi\varrho\pi} = 0, \tag{2.38}$$

just from Clebsch–Gordan arguments. For the $\omega\varrho\pi$ coupling, the non-relativistic matrix elements for $\omega \to \varrho + \pi$ in terms of $Q\bar{Q}$ transitions must first be calculated as in the $\varrho \to \pi\pi$ case, and the result equated to the corresponding quantity obtained from the relativistic $\omega\varrho\pi$ coupling[43]

$$im_\omega^{-1}f_{\omega\varrho\pi}\varepsilon_{\mu\nu\lambda\sigma}\partial_\mu \omega_\nu(x)\, \partial_\lambda \varrho_\sigma^a(x)\, \pi^a(x), \tag{2.39}$$

where the fields are represented by the particle symbols. Using the V–W prescription[43] one would then obtain

$$f_{\omega\varrho\pi} = 4f_q m_\omega \mu^{-1}. \tag{2.40}$$

The cases of $K^*K\pi$ and $\phi K\bar{K}$ couplings which require additional symmetry-breaking effects due to masses are discussed in the next sub-section.

### 2.3 Symmetry-breaking effects (due to masses)

As we have seen in the last two subsections, the quark model provides a natural outlet for the incorporation of symmetry-breaking effects due to mass differences in the particles involved, even before the inclusion of relativistic effects in the interactions. This feature is provided, in principle, by the body form factors $F_B(k^2)$ and $F_M(k^2)$ in both the baryon and meson cases. Since a literal evaluation of these quantities in terms of quark wave functions would not make much sense without a quantitative commitment to the detailed dynamical aspects of the model, it is perhaps more convenient to consider direct parametrizations for these quantities in terms of the obviously available parameters (the masses). In this respect certain guidelines are available on the basis of an important observation, due to VW[43], that the amplitudes for processes where a meson disappears depend, among other quantities, on the value of $\psi(0)$ which is the $Q\bar{Q}$ wave function at zero distance. For weak and electromagnetic processes, this factor was found by VW to be almost exactly proportional to the square root of the meson mass. It is tempting to extend this idea to the case of strong couplings where the meson as the radiation quantum may be assumed to play the VW role. It is therefore natural to suppose that the form factors $F_B(k^2)$ and $F_M(k^2)$ provide the necessary carriers for this parametrization, so that we *assume*, à la VW,

$$F(k^2) = \sqrt{\mu/m_\pi}, \qquad (2.41)$$

where $\mu$ is the emitted meson mass, and $m_\pi$ the pion mass. The normalization in (2.41) becomes consistent with our earlier remarks on $F(0)$ in the limit of zero symmetry breaking in the masses. Without trying further to justify this prescription on theoretical grounds, we shall now briefly examine its performance from the experimental point of view.

According to the results and conventions described in this section, the $K^*K\pi$ coupling is given by

$$f_{K^*K\pi}\overline{K}\tau_a K^*_\mu\, \partial_\mu\pi^a + \text{h.c.}, \qquad (2.42)$$

where

$$f_{K^*K\pi} = (2m_{K^*})\, m_k^{-1} f_q \sqrt{m_K/m_\pi}\,.$$

This gives

$$\Gamma_{K^*K\pi} \approx 75 \text{ MeV},$$

compared with its experimental value of $\sim 50$ MeV.[51] Similarly the $\phi K\overline{K}$ coupling constant $f_{\phi K\overline{K}}$, which enters the interaction as

$$f_{\phi K\overline{K}}\overline{K}\overset{\leftrightarrow}{\partial}_\mu K\phi_\mu, \qquad (2.43)$$

is given by

$$f_{\phi K\overline{K}} - (2m_\phi)\, m_K^{-1}f_q, \qquad (2.44)$$

and yields for the $\phi \to K\overline{K}$ width the value[50]

$$\Gamma_{\phi K\overline{K}} \approx 2.4 \text{ MeV}, \qquad (2.45)$$

compared to its experimental magnitude of $2.9 \pm 0.8$ MeV.[51] It appears from these figures that while the agreement with experiment is far from perfect, the V-W prescription for the mass structures of the MMP coupling constants give much better results than without these peculiar symmetry-breaking effects.

As for baryon couplings, the prescription (2.41) does not seem to have any immediate experimental manifestation since decays such as $Y_1^*(1385) \to N\overline{K}$

are not energetically favoured. The $\Lambda KN$ coupling constant which is defined by the interaction

$$iG_{NK\Lambda}\overline{N}\gamma_5 K\Lambda + \text{h.c.},$$

is now given by

$$G_{NK\Lambda} = \sqrt{3}\,(m_N + m_\Lambda)\,m_K^{-1}\,f_q\,\sqrt{m_K/m_\pi}, \qquad (2.46)$$

so that

$$G_{NK\Lambda}^2/4\pi \approx 5.1, \qquad (2.47)$$

compared with the estimate $6.8 \pm 2.9$ by ZOVKO.[52]

For baryon decays, there is no simple counterpart to the V-W prescription (2.29) and (2.23), since the relativistic $\overline{B}BP$ and $\overline{B}B^*P$ coupling constants are already dimensionless quantities, unlike the relativistic MMP coupling constants. However, if the physical content of the V-W prescription (2.33) is taken seriously, one must pretend that for the decay $\Delta \to N\pi$ in the PCAC limit, $\Delta$ and $N$ are "equally massive" states, a feature not as yet included in the description (2.19). Remembering the form of normalization of the relativistic baryon states, this effect leads to a "compensation factor" $\sqrt{M/m}$ where, $M, m$ are the masses of the parent and daughter baryons respectively.[50] Taking this factor into account, gives for the decuplet decay widths, the results (in MeV)[50]

$$\Gamma_{\Delta \to N\pi} = 101\ (120 \pm 2), \quad \Gamma_{Y_1^* \to \Lambda\pi} = 36.4\ (34 \pm 3),$$
$$\Gamma_{Y_1^* \to \Sigma\pi} = 4.5\ (3.5 \pm 1), \qquad \Gamma_{\Xi^* \to \Xi\pi} = 12.4\ (7.3 \pm 1.7), \qquad (2.48)$$

where the numbers in parentheses denote the experimental values.[51] It is interesting to compare the figures (2.48) with those obtained without the $\sqrt{M/m}$ correction factor, viz., 77.0, 28.8 and 3.9 respectively. This comparison brings out the usefulness of this ad hoc factor depending on the baryon masses, analogously to the V-W factor (2.41) depending on the mass of the radiation quantum. It may be noted, however, that for the $\Xi^* \to \Xi\pi$ case this correction factor acts in the wrong direction, since it gives 12.4 MeV, compared to the experimental figure of $7 \cdot 3 \pm 1 \cdot 7$ MeV, and the value 10.0 MeV obtained without this correction.

Similar results are obtainable for decays within higher baryon or meson supermultiplets, which do not depend on any extra parameters other than the quark coupling constant $f_q$ and the masses of the involved hadrons. Physically such a simplification is due to the fact that the orbital wave functions are the *same* for the initial and final hadron states, so that the evaluation of the form factor reduces essentially to a normalization effect, apart from a slow dependence on the emitted momentum $\mathbf{k}$. While the construction of the states of $L^P \neq 0^+$, with the inclusion of spin and $SU(3)$ wave functions, is described in a later section, it is convenient to record here the results of some decays within given hadron supermultiplets of $L^P \neq 0^+$ for which the above prescriptions are applicable.

The $D$-meson at 1285 MeV which is regarded as an $I = 0$ member of the $J^P = 1^+$ meson nonet[27] has a well-defined decay mode[51] into $\pi_V(1016) + \pi$, where $\pi_V(1016)$ is supposed to be the $I = 1$ member of the $J^P = 0^+$ nonet, or at least to have a strong channel to the $\delta$-meson (962 MeV) of the same quantum numbers.[53,54] Both the states $D(1285)$ and $\pi_V(1016)$ correspond to the same $L$-excitation, viz. $L = 1$. Using the $V$-$W$ prescription (2.33) for the relativistic normalization of the meson states, one obtains[50]

$$\Gamma_{D \to \pi_v \pi} = 23.2 \text{ MeV}, \tag{2.49}$$

on the assumption that $D(1285)$ is in the $SU(3)$ octet. This is a rather clean prediction which checks fairly well with the experimental value of $32 \pm 8$ MeV[51] for the same mode.

Among baryons there is the fairly established decay mode $Y_1^*(1767)$ $\to Y_0^*(1520) + \pi$, where the initial baryon is the well-known $I = 1$ member of the $J^P = 5/2^-$ octet and the final particle is the equally known $SU(3)$ singlet at $J^P = 3/2^-$. If both these particles are supposed to belong to the supermultiplet $L^P = 1^-$ in an $SU(6) \times O(3)$ classification, the corresponding decay width is predicted to be[50]

$$\Gamma_{Y_1^* \to Y_0^* \pi} = 14.4 \text{ MeV}, \tag{2.50}$$

as against the experimental value[51] of $14 \pm 2$ MeV. Note that the correction factor $\sqrt{M/m}$ arising from the relativistic baryon state normalization discussed earlier in this section, has proved helpful in bringing about a better accord with experiment.

Effects of these mass corrections to the couplings between different supermultiplets are described later.


## 3  SYMMETRY CLASSIFICATION OF HADRON RESONANCES

The symmetry classification of hadron resonances has been the subject of periodical reviews[18,27] since the quark model and $SU(6)$ were proposed. The picture which has been slowly building up as a result of interaction between theory and experiment since 1965, is one of $SU(6) \times O(3)$, with or without the quark model. This picture was rather incomplete till the time of the 1966 Berkeley conference because of many important missing links, especially in respect of the strange components of $SU(3)$. However, the discoveries of many more resonances during the last two years have greatly helped in filling the gaps.[53] For the baryons, most of the resonances have been established through phase shift analyses by various groups, especially by Lovelace and Collaborators.[54] For meson states, on the other hand, phase shift methods are not so easily available. Nevertheless, the experimental successes in directly detecting many of the hadron resonances (especially mesons) through peaks in mass plots, have been quite impressive.[55,56] One of the most striking experimental features for both meson and baryon resonances is the continued validity of the simple quark model which predicts only the **1, 8, 10** multiplets for baryons ($QQQ$), and nonets for mesons ($Q\bar{Q}$). Indeed, searches for "exotic" ($Y = 2$) states have not so far proved very fruitful[53], inspite of fluctuating evidences for the so-called $Z$-bumps in $K^+N$ reactions.[57,58]

For the supermultiplet structures of these hadrons based on $SU(6) \times O(3)$ we shall refrain from presenting detailed evidences for such assignments through the alternative of referring to the appropriate reviews,[53,59] including those by Dalitz for baryons[60] and mesons.[61] Instead we shall try to present a connected picture in terms of the quark model which greatly facilitates an $SU(6) \times O(3)$ description through the assumption of supermultiplet two-body forces (in $QQ$ or $Q\bar{Q}$ pairs) which preserve $SU(6) \times O(3)$ symmetry. The simplest assumption is to invoke harmonic oscillator potentials which lead to straight-line behaviour of energy with orbital excitation. This simple result finds strikingly good experimental evidence for both mesons and baryons if the square of the energy is plotted against the $L$-value of the corresponding state. We now summarize separately the status of meson and baryon supermultiplets.

### 3.1  Meson states

Because of their two-body structures, the description of meson states is considerably simpler than for baryons. The various resonances are most naturally interpreted as the result of successively higher excitations corresponding to both rotational and vibrational (radial) motions, in close analogy to the idea of similar mechanisms for the excitations of diatomic molecules.[61] For a given $Q\bar{Q}$ state specified by the orbital quantum number $L$, spin $S$, total angular momentum $J$, and radial quantum number $n$ we use the spectroscopic notation $(n)^{2S+1}L_J$. These are nine $(3 \times 3)$ possible $SU(3)$ states associated with each of these sets of quantum numbers. If the forces are $SU(3)$-*invariant*, these $SU(3)$ states separate into **8** and **1** multiplets. With $SU(3)$-breaking forces, the isosinglet part of **8** can mix strongly with the $SU(3)$-singlet, such as is the case for $\omega - \phi$ mixing, in which situation the totality of the **8** and **1** states is termed a meson *nonet*. The parity $(P)$ of the $(n)^{2S+1}L_J$ state is

$$P = (-1)^{L-1} \tag{3.1}$$

Similarly the charge conjugate parity $C$ of the neutral $Y = 0$ states, and the $G$-parity of the $Y = 0$ substates with isospin $I$ are respectively given by

$$C = (-1)^{L+S}, \quad G = C(-1)^I. \tag{3.2}$$

The ground states $1^1S_0$ and $1^3S_1$ of $L = 0$ and $n = 1$ already account nicely for the $P$-meson $(\pi K \bar{K} \eta X^0)$ and $V$-meson $(\varrho K^* \bar{K}^* \omega \phi)$ nonets respectively.

For $L = 1$, $n = 1$, there are four distinct nonets given by

$$[1^1P_1]; \quad [1^3P_0, 1^3P_1, 1^3P_2]. \tag{3.3}$$

The two groups, both of which have $P$-parity $= (+1)$, have been separated according to their $C$-parities which are $(-1)$ and $(+1)$ respectively. The normal charge conjugation states in (3.3) are those with $P = C = +1$, while the states $1_{P_1}$ which have $C = -P = -1$, are said to have "abnormal" charge conjugation. The $B$-meson at 1220 MeV, which is now fairly well established,[62] is supposed to be of the $1_{P_1}$-type (abnormal) while the other low lying positive parity mesons, such as $\delta(962)$, $A_1(1070)$ and $A_2(1320)$, are respectively of the $3_{P_0}$, $3_{P_1}$ and $3_{P_2}$ types. Table 1 summarises the present states of the comparatively low lying meson states[53] up to $L = 1$, together with their possible quantum number assignments $(L, S, J; P, C; I)$. All these states presumably correspond to $n = 1$ (no radial excitation).

While referring to the Harari[53] and Dalitz[61] reports for details, it may be mentioned that Table 1 accounts for most of the meson states in the region 1000–1500 MeV. As for $SU(3)$ mixing effects, the idea of the "ideal mixing angle" between $\omega$ and $\phi$, may well extend to the $L = 1$ mesons, e.g., between $\sigma(750)$ and $S^*(1070)$ or between $f_0(1260)$ and $f_0'(1515)$. These questions are discussed further on in connection with the decay modes of these states which provide a very useful check on such hypotheses.

The situation is much more confused for higher lying mesons which presumably have $L > 1$, and/or radial excitations associated with $L \geqslant 1$.

Most of the higher lying states have been discovered through the techniques of the "missing mass spectrometer" which studies the reactions

$$\pi^\pm + p \to p + M^\pm,$$

and makes momentum measurements only of the recoil proton[63], to deduce peaks in the "missing mass" distribution for the states $M^\pm$. The peaks beyond

Table 1   Meson states for $L = 0$ and $1$[a, b]

| $L$ | $S$ | $J^{PC}$ | $I = 1$ | $I = \frac{1}{2}$ | $I = 0$ | $I = 0$ |
|---|---|---|---|---|---|---|
| 0 | 0 | $0^{-+}$ | $\pi$ | $K$ | $\eta$ | $X^0$ |
|  | 1 | $1^{--}$ | $\varrho$ | $K^*$ (890) | $\omega$ | $\phi$ |
| $\sim 1$ | 1 | $0^{++}$ | $\delta(960)$ | $K\pi(1100)$[a] | $\sigma(750)$[a] | $S^*(1070)$[a] |
|  | 1 | $1^{++}$ | $A_1(1070)$ | $K^*(1230)$? | $D(1280)$ | ? |
|  | 1 | $2^{++}$ | $A_2(1310)$ | $K_V(1420)$ | $f_0(1260)$ | $f_0'(1515)$ |
|  | 0 | $1^{+-}$ | $B(1220)$ | $K_A^*(1320)$? | ? | ? |

[a] The assignments marked [a] are "good" guesses.
[b] The assignments marked (?) are tentative.

the familiar $\pi$, $\varrho$ and $A_2(1310)$ are the $R(\sim 1700)$, $S(\sim 1929)$, $T(2195)$ and $U(2885)$. The (mass)$^2$ plot of the sequence, $A_2, R, S, T, U$ is surprisingly linear as a function of their sequence number $N$ (e.g. $N = 1$ for $\varrho$, etc). It is tempting to associate this straight line with the Regge trajectory for $\varrho$[61], through the identification $N = J$.*

Among the established higher lying states one should mention the (i) $\varrho_V(1650)$, or the $g$-meson for which probably $J^P = 3^-$ or $1^-$ and (ii) the $\pi A(1640)$ which may be a combination of $(2)\,1_{S_0}$ and $(1)^1\,D_2$. These are discussed further on, in connection with their decay widths.

The problem of radial excitations is much less understood at the present moment. Among the lower lying states the $E(1410)$ meson of $J^P = 0^-$ and $I = 0$ probably represents a radial excitation of the $\eta$-meson. The appearance of certain additional levels in the $A_1 - A_2$ region, such as the $A_{1.5}$-meson at 1170 MeV and the $A_2^L$-meson at 1290 MeV both of which undergo $\varrho\pi$ decay, has considerably complicated the otherwise "clean" picture afforded by the old-fashioned quark model. These states do not seem to admit of any simple

---

* The near coincidence of the $\varrho$ ($P = -1$) and $A_2$ ($P = +1$) trajectories may be interpretated as due to the much smaller effect of exchange of the (massive) diquarks, compared with that of mesons between quarks and antiquarks.

interpretation through the use of the radial quantum number, so one should perhaps wait till these are further confirmed or get subsequently rejected.

To accommodate such states, an ingenuous extension of the quark model has recently been given by Gell–Mann and Zweig[64], through the introduction of a four-dimensional oscillator scheme in which both $L$ and $S$ are considered as $0(4)$ (rather than $0(3)$) quantum numbers. With this extension, $S = 1$ is now accompanied by a fourth component with the same "$C$" but opposite "$P$", so that the $L = 0$, $J^{PC} = 1^{--}$ nonet of normal "$C$" has now a "daughter" $0^{+-}$ of abnormal $C$. Note that this "daughter" has no place in the conventional quark model. Similarly $L = 1$ now acquires an extra component via the $0(4)$ scheme and gives rises to many more states through the coupling with $S$. The essential result of this extension is to introduce a great proliferation of states for higher $L$-values. For $L = 1$, the extra states introduced account for $A_2^L(1^{-+})$ and $A_{1.5}(1^{-+})$. However the scheme also predicts a $0^{+-}$ nonet around the $B$-meson mass, and a $0^{++}$ nonet around the $2^{++}$ nonet mass. The experimental situation at this stage is too premature to confirm or reject this scheme.

## 3.2  Baryon states

The $QQQ$ structure of baryons makes them much less amenable to theoretical techniques than is the case with meson states. In order to understand the sequence of their mass levels, it is necessary to make several simplifying assumptions. First, the non-relativistic model has little alternative to it at this poor state of development of relativistic theories for more than two particles. Secondly, one must keep a more open mind on the question of quark statistics in order to reconcile the **56** representation with a symmetric wave function, the most natural condition not only for obtaining the lowest energy value for this $L^P = 0^+$ supermultiplet, but for exhibiting the absence of nodes in the nucleon form factors as well. The observed sequence of levels for the successively higher supermultiplets $(\mathbf{70}, 1^-), (\mathbf{56}, 2^+)$ etc., also seems to find a natural explanation in the assumption of symmetric wave functions, together with some further assumptions of a simple dynamical nature. The symmetric structure of the $QQQ$ wave function was first suggested by Greenberg[33] in order to obtain a systematic classification of supermultiplet states according to energy levels. The supermultiplet classification depends of course on the assumption of $SU(6)$ symmetry for the $Q-Q$ force, which may be termed as "super strong"[60]. The splitting of the various states into $SU(3)$ and lower multiplets depends in turn on the structure of the symmetry breaking forces which can be of various types, including the aspects of (i) spin-orbit, (ii) spin-spin and (iii) $SU(3)$-breaking, separately and/or in different combinations. Unfortunately the pattern of mass levels of the identified resonances is still so incomplete that it is almost impossible at this stage to obtain much insight into the structures of the various $SU(6)$-breaking forces. On the other hand, the assignments of the observed resonances in appropriate multiplets, are facilitated through their decay characteristics much more reliably than through their observed level sequence, since the former are extremely sensitive to these supermultiplet assignments while the latter would fit many different alternatives for the symmetry-breaking terms in the mass formulae via corresponding assumptions on the two-body potentials. We refer to the detailed reviews by Dalitz[60] and Harari[53] on the question of various types symmetry-breaking terms in the mass formulae and their limitations in fixing the supermultiplet assignments. Instead we choose to give a brief review of the supermultiplet status of various resonances, and such a discussion is possible even with $SU(6)$-symmetric forces. As already stated, all successful classifications within the simplest quark model seem to depend on the "symmetric" assumption for the $QQQ$ function.

The first dynamical classification of supermultiplets through symmetric wave function was given by Greenberg[33] harmonic oscillator well for the quarks. However, since such a picture has the (slight) disadvantage of depending on the effect of the c.m. motion of the three quarks, the Greenberg classification was unfortunately saddled with certain spurious states. A more careful classification was recently given by Karl and Obryk[65] who considered harmonic potentials to be operative in $Q$–$Q$ pairs, and were thus able to eliminate the spurious states. Elimination of the c.m. motion leads to a description of shell model states in terms of only *two* relative coordinates, or sub-angular momenta $l_1$ and $l_{23}$, etc. [see Eq. (1.4)]. Thus the **56** states which have $l_1 = l_{23} = 0$, should be designated by the spectroscopic symbols $(1S)^2\,(0^+)$, where the notation is $[(n_1 + 1)\,l_1]\,[(n_{23} + 1)\,l_{23}]\,(L^P)$ and $n_1$, $n_{23}$ are the radial quantum numbers associated with the sub-angular momenta $l_1$ and $l_{23}$ respectively. It may however be convenient to give a separate Shell-configuration to *each* of the *three* quarks, and designate the **56** states by the notation $(1S)^3\,(0^+)$, emphasizing the fact that each of the three quark pairs are in relative $s$-states.* In this extended notation, the $(\mathbf{70}, 1^-)$ states have the con-

---

* This notation is not of course intended to give the impression of three independent angular momenta.

figuration $(1S)^2 (1p) (1^-)$. Note that in this notation, the parity is reflected in the number of sub-shells containing the $p, f$, etc., states, and *not* in the total $L$-value of the resultant configuration.

The various shell model configurations can be described by the quantum numbers representing the total excitation $(N)$, orbital momentum $(L)$, $SU(6)$ multiplicity (which may be **56**, **70** or **20**), and the symbols $(nl)$ for the individual quark states. Since, according to the naive shell-model picture, the vibrational excitations are expected to be much further apart than their rotational counterparts, the classifications may be conveniently given in order of increasing $N$-values. The shell model configurations of a $QQQ$ system for $N$-values ranging from 0 to 3 are listed in table 2.[60]

It is apparent from this table that while the first two states are fairly unique, number and variety of higher states increases very rapidly with $N$. The experimental status of these states, for whose discussion we refer to Dalitz[60] and Harari[53], suggests that while the two $N = 0$ and 1 supermultiplets are very satisfactorily realized, only a few of the higher $N$-supermultiplets seem to be observed. Thus for $N = 2$, the only successful candidates are $(\mathbf{56}, 2^+)$ and $(\mathbf{56}, 0^+)$. While the former, which may be interpreted as the first Regge recurrence of the **56** states, has had most of its members satisfactorily identified with the observed positive parity resonances in the energy range $(1600-2000)$, the $(\mathbf{56}, 0^+)$ component of $N = 2$ has been gaining increasing experimental support through the interpretation of the $N(1470)$ and $\Delta(1690)$ states as the first *radial* excitations of the $N(938)$ and $\Delta(1236)$ respectively. As for the other $[SU(6); L^P]$ components of $N = 2$, there seems at the moment little or no experimental support, though the harmonic oscillator prediction puts them rougly in an overlapping energy range. Finally, for $N = 3$, the few resonances identified in the range 2100–2500 MeV seem to favour only the $(\mathbf{70}, 3^-)$ supermultiplet, to the exclusion of the rest of the long list, though this assignment must await a much longer scrutiny.

The apparent realization of such a selected list of states as indicated above seems to suggest the occurrence in nature of only the following types of supermultiplets:

$$[\mathbf{56}, (2l)^+], \quad [\mathbf{70}, (2l + 1)^-], \quad l = 0, 1, 2, \ldots \tag{3.4}$$

together with their radial excitations. Such a simple picture is most unlikely to be obtained from considerations of pure group theory, without some additional dynamical assumptions. Some insight into the possible mechanisms for the occurrence of such states may be obtained through an angular-momentum analysis of the $SU(6)$ invariant potentials, into $s, p, d$, etc., waves. A general theory on these lines was developed by the author [22,23], by considering separately the effects of $SU(6)$—invariant potentials in even and odd partial waves, on the energy level structure of the $QQQ$ supermultiplets, in the spirit of Faddeev's three-body theory.[66] This theory has the advantage that the sign of the Faddeev Kernel for the connected three-body system, which directly reflects the energy-level status of a given $[SU(6); L^P]$ state depends crucially on the *spatial* symmetry associated with that supermultiplet. Such an analysis in general leads to a sub-set of states listed in Table 2, as long as only a single partial wave in the $Q$–$Q$ force is considered. In particular, the assumption of a pure $s$-wave force, leads precisely to the states (3.4) as the only ones showing attractive Faddeev Kernels,[24,25] while the other two types of realizable states, viz.,

$$[56, (2l + 1)^-] \quad \text{and} \quad [70, (2l)^+] \tag{3.5}$$

show repulsive Kernels.*

Table 2   Shell model states for $N \leqslant 3, L \leqslant 3$

| $N$-value | $(nl)$ configuration | $[SU(6); L^P]$ assignments |
|---|---|---|
| $N = 0$ | $(1s)^3$ | $(\mathbf{56}, 0^+)$ |
| $N = 1$ | $(1s)^2 (1p)$ | $(\mathbf{70}, 1^-)$ |
| $N = 2$ | $(1s) (1p)^2$ $(1s)^2 (1d)$ $(1s)^2 (2s)$ | $(\mathbf{56}, 2^+), (\mathbf{56}, 0^+), (\mathbf{70}, 2^+)$ $(\mathbf{70}, 0^+), (\mathbf{20}, 1^+)$ |
| $N = 3$ | $(1s)^2 (1f)$ $(1s)^2 (2p)$ $(1s) (1p) (1d)$ $(1s) (1p) (2s)$ $(1p)^3$ | $(\mathbf{56}, 3^-), (\mathbf{56}, 1^-), (\mathbf{70}, 3^-)$ $(\mathbf{70}, 2^-), (\mathbf{70}, 1^-)$ $(\mathbf{20},3^-), (\mathbf{20}, 1^-)$   twice |

We summarise in Tables 3 the currently known baryon resonances, in terms of their $J^P$ values and supermultiplet assignments up to $L = 2^{53,60}$. Tables 3 (a) and 3 (b) which show the octet and decuplet states respectively, do not include the two famous (singlet) states $\Lambda(1405)$ of $J^P = 1/2^-$ and

---

* States with $\mathbf{20}$ representations, or those with $L^P = (2l + 1)^+$ or $(2l)^-$ are not of course realized with pure $s$-wave forces.

$\Lambda(1520)$ of $J^P = 3/2^-$, both of which are important constituents of the $L^P = 1^-$ supermultiplet. These tables do not also include a few higher lying negative parity states, presumably of $L^P = 3^-$, which will be mentioned again in Section 7. Similarly certain well-defined $\Delta$-states of still higher $J^P$ values

**Table 3(a)**   Some of the better known baryon octet states.
For the cases marked (?) the assignments are tentative

| $L$ | $S$ (spin) | $J^P$ | Particle symbols | | | |
|---|---|---|---|---|---|---|
| 0 | 1/2 | $1/2^+$ | $N\,(938)$ | $\Lambda\,(1115)$ | $\Sigma\,(1190)$ | $\Xi\,(1318)$ |
| 1 | 1/2 | $1/2^-$ | $N\,(1710)$ | | | |
| 1 | 3/2 | $1/2^-$ | $N\,(1550)$ | $\Lambda\,(1670)$ | $\Sigma\,(1670)?$ | |
| 1 | 1/2 | $3/2^-$ | $N\,(1518)$ | $\Lambda\,(1690)?$ | $\Sigma\,(1660)?$ | |
| 1 | 3/2 | $3/2^-$ | $N\,(1675)$ | $\Lambda\,(1690)?$ | $\Sigma\,(1660)?$ | |
| 1 | 3/2 | $5/2^-$ | $N\,(1680)$ | $\Lambda\,(1830)$ | $\Sigma\,(1770)$ | $\Xi\,(1930)$ |
| 2 | 1/2 | $5/2^+$ | $N\,(1690)$ | $\Lambda\,(1816)$ | $\Sigma\,(1910)$ | $\Xi\,(2020)?$ |
| 2 | 1/2 | $3/2^+$ | $N\,(1863)$ | | | |
| 3 | 1/2 | $7/2^-$ | $N\,(2190)?$ | | $\Sigma\,(2260)?$ | |
| 3 | 3/2 | $7/2^-$ | $N\,(2190)?$ | | $\Sigma\,(2260)?$ | |

**Table 3(b)**   Decuplet states up to $J^P = 7/2^+$

| $L$ | $S$ (spin) | $J^P$ | Particle symbols | | | |
|---|---|---|---|---|---|---|
| 0 | 3/2 | $3/2^+$ | $\Delta\,(1238)$ | $\Sigma^*\,(1385)$ | $\Xi^*\,(1530)$ | $\Omega\,(1675)$ |
| 1 | 1/2 | $1/2^-$ | $\Delta\,(1640)$ | $\Sigma^*\,(1770)$ | | |
| 1 | 1/2 | $3/2^-$ | $\Delta\,(1691)$ | | $\Xi^*\,(1816)$ | |
| 2 | 3/2 | $1/2^+$ | $\Delta\,(1934)$ | | | |
| 2 | 3/2 | $5/2^+$ | $\Delta\,(1913)$ | | | |
| 2 | 3/2 | $7/2^+$ | $\Delta\,(1950)$ | $\Sigma^*\,(2030)$ | | |

are excluded from Table 3(b). Many of the $SU(3)$ assignments are still tentative, and may well admit of mixing effects, especially between $\Lambda$-states of **1** and **8**, and $\Sigma$-states of **8** and **10**. Among the states of $L^P = 1^-$, there is the further possibility of mixing between certain octets of *intrinsic* spins $S = 1/2$ and $S = 3/2$. These questions are discussed more fully in the later sections in connection with the decays of the relevant states.

While the general pattern exhibited by these states is broadly consistent with (3.4), am ore severe test will be shown to be provided by a specific model of decays to be discussed in Section 6. An alternative pattern is provided by the new Gell–Mann–Zweig (GMZ) theory[64] of parity doubling of baryons on the basis of $O(4)$ symmetry for the spin $(S)$ and orbital $(L)$ degrees of freedom. As in the meson case, this scheme results in a large proliferation of states many of which must await experimental discovery. One important difference between the new GMZ scheme and the "old" $SU(6) \times O(3)$ scheme is the apparent absence of radial excitations in the former. For example the state $N(1470)$ of $J^P = 1/2^+$ which is believed to be a radial excitation of the familiar nucleon in the $SU(6) \times O(3)$ scheme, must be interpreted as the $J^P = 1/2^+$ partner of the $J^P = 1/2^-$ state $N(1550)$ in the GMZ scheme. However, since the quantitative aspects of the GMZ scheme are not yet available, it will not be possible to discuss its implications on decays at this stage.

## 4   COUPLINGS BETWEEN DIFFERENT SUPERMULTIPLETS

This section is devoted to a phenomenological formalism for the evaluation of couplings for hadrons in different multiplets. Specifically, we shall be interested in the couplings of baryons $(B_L)$ or mesons $(M_L)$ corresponding to internal orbital excitations $L$, to the more familiar **56** baryons $(B)$ and **36** mesons $(P$ or $V)$ respectively of $L = 0$, together with the emission of a pseudo-scalar meson $(P)$ in each case. As noted in Section 1, while the parity of the meson state $M_L$ is $(-1)^{L-1}$, the $L$-value by itself does not determine the parity of the baryon state $B_L$. However, if we confine our discussion only to baryon states of the type (3.4), the parity of $B_L$ relative to a **56** baryon $(B)$ is given by $(-1)^L$. In what follows we shall consider only baryon states of the kind (3.4), and not the other (more general) types listed in Table 2. The baryon couplings $\bar{B}_L B P$ will be obtained by taking the matrix elements of the operator (2.2) between the $B$-states of $L = 0$ and the $B_L$-states of $L \neq 0$. Similarly the meson couplings $M_L M P$ $(M = P$ or $V)$ will be obtained by taking the matrix elements of the *sum* of the two operators (2.2) and (2.3) between the $M$-states of $L = 0$ and the $M_L$-states of $L \neq 0$. We now consider the two cases separately.

### 4.1   $M_L M P$ couplings

The complete wave function of a given $M_L$-state of zero c.m. momentum and a specified $(J, m)$ value is of the form

$$\Psi(M_L) = \delta(\mathbf{P}) \, [\psi_L \otimes \chi_S]_m^J \, \phi, \tag{4.1}$$

where $\delta(\mathbf{P}) \, \psi_{Lm_L}(\mathbf{p})$ is the spatial wave function of $Q\bar{Q}$ in terms of the relative momentum $\mathbf{p}$ and total momentum $\mathbf{P}$, $\phi$ is its $SU(3)$ function, $\chi_{Sm_S}$ its spin function ($S = 0$ or $1$), and the notation

$$[\psi_L \otimes \chi_S]_m^J = \sum_{m_L} C(LSJ; m_L m_S m) \, \psi_{Lm_L} \chi_{Sm_S}, \qquad (4.2)$$

represents the irreducible spherical tensor with eigenvalues $(J, m)$, constructed in the standard manner out of the orbital and spin functions.[67] As discussed in sec 3.1., for the singlet state, one has $J = L$ and $C = (-1)^L$. For the triplet states, the possible values of $J$ are $L - 1, L$ and $L + 1$, and $C = (-1)^{L-1}$. If $SU(6)$ symmetry were valid, all these four nonets would have the *same* orbital function $\psi_L$. In practice, however, this symmetry would be broken, so that these functions would no longer be the same.

The $M_L M P$ couplings can now be expressed in the standard manner in terms of the reduced matrix elements of the operators (2.2) and (2.3) between the states $\langle M |$ and $| M_L \rangle$. Note that the $SU(3)$ factors would simply get factored out since the operators (2.2)–(2.3), as well as the expression (4.1) for $M_L$ ($L \neq 0$) and a similar one for $M$ ($L = 0$), all involve the $SU(3)$ functions in a simple product form (without any C.G. expansion). The structures of the spin-cum-orbital parts are however involved in a less trivial way. In this respect one must be careful to distinguish the contributions of the "direct" terms (involving $\mathbf{k}$) from those of the "recoil" terms (involving $\mathbf{P}_i$). Thus for transitions from a given $L$-state to $L = 0$ the direct term corresponds to the operator

$$\sum_{\alpha=1}^{8} [\boldsymbol{\sigma}^{(1)} \cdot \mathbf{k}\lambda_\alpha \pi_\alpha \mp \mathbf{k} \cdot \boldsymbol{\sigma}^{(2)} \bar{\lambda}_\alpha \pi_\alpha], \qquad (4.3)$$

according as $L$ is even or odd respectively. Here $\mathbf{k}$ is the momentum of the emitted $P$-meson ($\pi_\alpha$) and ($\lambda_\alpha$, $\bar{\lambda}_\alpha$) are the Gell–Mann matrices[44] associated with $Q$ and $\bar{Q}$ respectively; $\boldsymbol{\sigma}^{(1)}$, $\boldsymbol{\sigma}^{(2)}$ are the spin-operators for $Q$ and $\bar{Q}$ respectively. Similarly the recoil term corresponds to the operator

$$\left(\frac{\omega_k}{M_Q}\right) \sum_{\alpha=1}^{8} [\boldsymbol{\sigma}^{(1)} \cdot \mathbf{p}\lambda_\alpha \pi_\alpha \pm \boldsymbol{\sigma}^{(1)} \cdot \mathbf{p}\bar{\lambda}_\alpha \pi_\alpha], \qquad (4.4)$$

according as $L$ is even or odd respectively. Here $\pm\mathbf{p}$ are the relative momenta of recoil for $Q$ and $\bar{Q}$ respectively.

Clearly the operators (4.3) and (4.4) give null results for $1_{S_0} \to 1_{S_0}$ transitions, as they should. For the transitions $3_{S_1} \to 3_{S_1}$ and $3_{S_1} \to 1_{S_0}$, we have respectively

$$\boldsymbol{\sigma}^{(2)} \equiv \pm \boldsymbol{\sigma}^{(1)} \equiv \mathbf{S}, \tag{4.5}$$

where $\mathbf{S}$ is the spin-operator for the $Q\bar{Q}$ system. This simplification helps in factoring out the $SU(3)$ variables from the spin-cum-angular variables in the transition matrix elements. The structure of the latter is now of the form

$$\langle OS'J'm'| \, \mathbf{S} \cdot \mathbf{q} \, |LSJm\rangle = \delta_{mm'}\delta_{JJ'} \, \langle J \| \mathbf{S} \cdot \mathbf{q} \| J\rangle, \tag{4.6}$$

where $S$, $S'$ are the spin values of the initial and final states respectively, and $\mathbf{q}$ stands for the meson momentum $\mathbf{k}$ or the quark recoil momentum $\mathbf{p}\omega_k M_Q^{-1}$ as the case may be. The spatial part of this matrix element has the structure:

$$\int d\mathbf{p} \, \psi_{LM}^*(\mathbf{p}) \, \mathbf{q}\psi_0 \, (\mathbf{p} - \tfrac{1}{2}\mathbf{k}); \quad \mathbf{q} = \mathbf{k} \ \ \text{or} \ \ \frac{\mathbf{p}\omega_k}{M_Q}. \tag{4.7}$$

The structure of this integral may be visualized through the Taylor expansion

$$\psi_0 \, (\mathbf{p} - \tfrac{1}{2}\mathbf{k}) = 1 + \sum_{l=1}^{\infty} \frac{1}{l!} \, (-\tfrac{1}{2}\mathbf{k} \cdot \mathbf{V}_p)^l \psi_0(p), \tag{4.8}$$

which makes the contributions of all the terms $l < L$ vanish on integration over $\hat{p}$ in (4.7). Using (4.8) the expression (4.7) simplifies, for the "direct" term ($\mathbf{q} = \mathbf{k}$) to

$$f_L(k^2) \, Y_{LM}(\hat{k}) \, \mathbf{k},$$

where $f_L(k^2)$ is some scalar function of the momentum. The last term in turn gives rise to the normalized vector spherical harmonics $\mathbf{T}_{L\pm1,L,M}$ via the relation[67]

$$\mathbf{k}Y_{LM}(\hat{k}) = -\sqrt{\frac{L+1}{2L+1}} \, \mathbf{T}_{L+1,L,M} + \sqrt{\frac{L}{2L+1}} \, \mathbf{T}_{L-1,L,M}. \tag{4.9}$$

This shows that for the direct term, the amplitudes for the $L \pm 1$ wave emissions of a $P$-meson bear a *geometrical* ratio to each other. However, this result is not true for the quark recoil term ($\mathbf{q} = \mathbf{p}\omega_k M_Q^{-1}$) which gives different scalar functions multiplying the harmonics for $L \pm 1$ emissions respectively. Note further that, due to the structure of Eq. (4.8), the "threshold momentum dependence" associated with *both* the $L \pm 1$ wave emissions via the direct term is the same, viz., $\mathbf{k}^{L+1}$. This quantity has the correct number of $k$-factors for $(L + 1)$ emission but an excess factor $k^2$ for $(L - 1)$ emission. The recoil term, on the other hand, gives a more faithful index of the threshold momentum dependence of the amplitudes which have the factors

$k^{L\pm1}$ for $L \pm 1$ wave emissions respectively. It may be remarked further that though the recoil terms are proportional to $M_Q^{-1}$, these are not necessarily small, since the quantity $M_Q$ is *not* the mass of the free quark (which may be heavy), but rather, the *effective mass* of the *bound* quark which may be quite small.[27,31,68] The question of parametrization of the decay amplitudes in terms of direct and/or recoil terms is therefore more involved than would appear at first sight, and perhaps is best answered through a broad phenomenological assessment of the decay characteristics of an $M_L$ resonance. This question is discussed further in Sections 5 and 7.

## 4.2  $\overline{B}_L BP$ couplings

For the baryon state $B_L$, the wave functions are more complicated because of the additional requirement of three-particle symmetry. However, the consideration of only the states (3.4) greatly simplifies the structure of the wave functions since these involve only $s$ or $m$ symmetries. For the **56** states, the symmetrization is a particularly simple process since the orbital function $\psi_L$ must have $s$-symmetry. For zero c.m. momentum this function will be denoted by

$$\delta(\mathbf{P})\, \psi_{Lm_L}^{s} \equiv \psi_{Lm_L}^{s}\,(\mathbf{P}', \mathbf{P}'')\, \delta(\mathbf{P}), \tag{4.10}$$

in the notation of Section 2 for the various momenta. For full $SU(6)$ symmetry, the **8** and **10** $SU(3)$ states of $B_L$ have the respective wave functions

$$\Psi_8(B_L) = \frac{1}{\sqrt{2}}\,([\psi_L^s \otimes \chi_{1/2}']_m^J \phi' + [\psi_L^s \otimes \chi_{1/2}'']_m^J \phi''); \tag{4.11}$$

$$\Psi_{10}(B_L) = [\psi_L^s \otimes \chi_{3/2}^s]_m^J \phi^S; \tag{4.12}$$

where the notation for the spin functions $(\chi', \chi''; \chi^S)$ and the $SU(3)$ functions $(\phi', \phi''; \phi^S)$ are the same as employed in Section 2, and the irreducible tensor products are defined as in Eq. (4.2). For definiteness we have explicitly exhibited the spin values 1/2 and 3/2 as suffixes to the functions $(\chi', \chi'')$ and $\chi^S$ respectively.

For the **70** states, we first note that the $SU(3)$ content for $L = 0$ is given by the list

$$(\mathbf{10};\tfrac{1}{2}); \quad (\mathbf{8};\tfrac{1}{2}); \quad (\mathbf{1};\tfrac{1}{2}); \quad (\mathbf{8};\tfrac{3}{2}). \tag{4.13}$$

The first three states therefore involve only the doublet (mixed symmetric) spin functions $(\chi', \chi'')$, while the last one, an octet, involves only the quartet (symmetric) spin function $\chi^S$. For convenience we designate these states as doublet ($d$) and quartet ($q$) respectively. Note that these alternatives exist only for the octets, and not for singlets and decuplets which always have doublet spin structures in the **70** representation. For the octets, the two possibilities may be distinguished by the symbols $\mathbf{8}_d$ and $\mathbf{8}_q$. With this notation, the $SU(3)$ content of a **70** representation of $L \neq 0$ is

$$(\mathbf{10}; L \pm \tfrac{1}{2}); \quad (\mathbf{1}; L \pm \tfrac{1}{2}); \quad (\mathbf{8}_d; L \pm \tfrac{1}{2}); \tag{4.14}$$

$$(\mathbf{8}_q; L \pm \tfrac{1}{2}, L \pm \tfrac{3}{2}), \tag{4.15}$$

where the second symbol in each bracket stands for the possible $J$-values associated with that $SU(3)$ state. Note that unlike the case of $L = 0$, there are now *two* sets of octet states for all $J$-values except $J = L \pm 3/2$, which correspond to the "stretched" configurations. These two sets which we now distinguish through the symbols $\mathbf{8}_d$ and $\mathbf{8}_q$, have entirely different internal spin structures since $\mathbf{8}_d$ corresponds to $m$-symmetry and $\mathbf{8}_q$ to $s$-symmetry in spin-space. These differences should manifest themselves through the couplings (and hence the decay characteristics) of the resonances which are sensitive to these quantum number assignments.

The complete wave functions of the **70** states are now easily written down through the observation that their spatial parts must have $m$-symmetry. We designate these functions as $\psi'_{LM}$ and $\psi''_{LM}$ and use the normalization

$$\int \psi'^{*}_{LM}\psi'_{LM'} = \int \psi''^{*}_{LM}\psi''_{LM'} = \delta_{MM'}. \tag{4.16}$$

The normalized wave functions for the different $SU(3)$ multiplets (4.14) and (4.15) of a given supermultiplet $B_L$ are now expressible in the notation (4.2) as follows:

$$\Psi_1(B_L) = \frac{1}{\sqrt{2}}\,\phi^a\,([\psi'_L \otimes \chi''_{1/2}]^J_m - [\psi''_L \otimes \chi'_{1/2}]^J_m), \tag{4.17}$$

$$\Psi_{10}(B_L) = \frac{1}{\sqrt{2}}\,\phi^s\,([\psi'_L \otimes \chi'_{1/2}]^J_m + [\psi''_L \otimes \chi''_{1/2}]^J_m), \tag{4.18}$$

$$\Psi_{8_d}(B_L) = \frac{1}{2}\,(\phi'\,[\psi'_L \otimes \chi''_{1/2}]^J_m + \phi'\,[\psi''_L \otimes \chi''_{1/2}]^J_m$$
$$+ \phi''\,[\psi'_L \otimes \chi'_{1/2}]^J_m - \phi''\,[\psi''_L \otimes \chi''_{1/2}]^J_m), \tag{4.19}$$

$$\Psi_{8_q}(B_L) = \frac{1}{\sqrt{2}}\,(\phi'\,[\psi'_L \otimes \chi^S_{3/2}]^J_m + \phi''\,[\psi''_L \otimes \chi^S_{3/2}]^J_m), \tag{4.20}$$

where the other notations are the same as used earlier.

The $\bar{B}_L BP$ couplings can now be evaluated in terms of the reduced matrix elements of the operator (2.2) for quark number 1 (and multiplying the result by *three*) between the states $\langle B|$ and $|B_L\rangle$. Unlike the meson case, the $SU(3)$ factors are now involved in a non-trivial fashion since these have also a part to play in the three-particle symmetrization. The techniques for evaluating the spin-cum-$SU(3)$ matrix elements are the same as described in Section 2.1. In particular Eqs. (2.13)–(2.16) represent the precise correspondence to be used for expressing these elements, defined in the composite $QQQ$ space for baryons, in terms of the corresponding elements between baryon states *regarded as a whole*. The spatial parts of the matrix elements are of the form

$$\iint d\mathbf{P}'\, d\mathbf{P}''\, \psi_{LM}''^* (\mathbf{P}', \mathbf{P}'') \, [\mathbf{k} + \sqrt{\tfrac{2}{3}}\, \omega_k \mathbf{P}'' M_Q^{-1}) \, \psi_0^S (\mathbf{P}', \mathbf{P}'' + \sqrt{\tfrac{2}{3}}\, \mathbf{k}), \quad (4.21)$$

where the direct and recoil contributions are shown together. Using a Taylor expansion similar to (4.8) it is now possible to show in exactly the same way as in section 4.1, that the direct term contributes geometrical ratios for the $L \pm 1$ emissions according to the content of Eq. (4.9), while the recoil term gives different form factors for them. The remarks made earlier about their threshold momentum dependence for the meson case are valid without modification for baryons as well.[42]

### 4.3  Self-couplings among hadrons

As already mentioned in section 2, there are several cases of physical interest where a hadron within a given $[SU(6); L^P]$ supermultiplet decays into another hadron belonging to the same supermultiplet. The couplings involving such states have the remarkable property that they do not depend on any parameters other than the basic $\bar{Q}QP$ coupling constant defined in (2.2) or (2.3). This may be seen as following. The evaluation of the spin and $SU(3)$ matrix elements for such cases follows on the same lines as already outlined in this section. However, the orbital matrix elements for the $\bar{B}_L B_L P$ and $M_L M_L P$ couplings are respectively given by

$$\iint d\mathbf{P}'\, d\mathbf{P}''\, \psi_{LM}^* (\mathbf{P}', \mathbf{P}'') \left[\mathbf{k} + \sqrt{\tfrac{2}{3}}\, \omega_k M_Q^{-1} \mathbf{P}''\right] \psi_{LM'} \left(\mathbf{P}', \mathbf{P}'' + \sqrt{\tfrac{2}{3}}\mathbf{k}\right), \quad (4.22)$$

$$\int d\mathbf{p}\, \psi_{LM}^*(\mathbf{p}) \, [\mathbf{k} \pm \omega_k M_Q^{-1}\mathbf{p}] \, \psi_{LM}, (\mathbf{p} - \tfrac{1}{2}\mathbf{k}), \quad (4.23)$$

where, in (4.23), the symbols $\pm\mathbf{p}$ represent the recoil momenta of the $Q$ or $\bar{Q}$ respectively. From considerations of angular momentum and parity it is immediately clear that the recoil terms in both cases give exactly *zero* contributions to the above matrix elements. The normalizations of both these types of hadron states may be collectively depicted as

$$\int d\tau \, \psi_{LM}^* \psi_{L'M'} = \delta_{MM'} \delta_{LL'}, \tag{4.24}$$

where the volume element $d\tau$ stands for the symbols $d\mathbf{p}$ and $d\mathbf{P}' \, d\mathbf{P}''$ for meson and baryon states respectively. Thus if, as in Sec. 2 (for $L^P = 0^+$) we ignore the $\mathbf{k}$-dependence of the wave functions in (4.22) and (4.23), both these expressions reduce, via (4.24) to the value $\mathbf{k}\delta_{MM'}$, so that the most general form of either of them is

$$\mathbf{k}\delta_{MM'} f(\mathbf{k}^2), \tag{4.25}$$

where $f(0) = 1$. Therefore the only parameters that enter such "self-couplings" of baryon or meson states are (i) the $\bar{Q}QP$ coupling constant $f_q$ and (ii) at most a (gentle) body form factor $f(k^2)$ similar to the (more familiar) $L^P = 0^+$ case of baryons.


## 5  PHENOMENOLOGICAL ANALYSIS OF HADRON DECAYS ($L \neq 0$)

In this section we shall describe the phenomenology of hadron decays on the basis of (i) their $SU(6) \times O(3)$ assignments given in Section 3 and (ii) their couplings to the **56** baryons and the **36** mesons outlined in Section 4. The emphasis will be on certain general results of a qualitative nature concerning the quantum number assignments to these resonances, rather than on more quantitative numerical values for the decay widths. This limitation is the price for the lack of an adequate model for the overlap integrals (described in Section 4) which must therefore be parametrized in a very general manner. Though this parametrization is quite similar for mesons and baryons, it is convenient to discuss the two cases separately. Since the study of baryons decays by this method is historically the earlier[42], we consider this case first. In this section we shall follow closely the method of Mitra and Ross[42], which we refer to as MR.

### 5.1  Parametrization for baryon decays

The reduced matrix element for a transition $B_L \rightarrow B + P$ involves the $SU(3)$ isoscalar factors $(8\alpha \,|\, \alpha_L)$ or $(1\alpha \,|\, \alpha_L)$, where $\alpha_L$ and $\alpha$ are the respective $SU(3)$ multuplicities of $B_L$ and $B$, while the $P$-meson has the corresponding values $1$ or $8$. These isoscalar factors are uniquely determined for all cases, except for $\alpha = \alpha_L = 8$, when one would encounter the problem of a "$D/F$"

ratio. However our $SU(6) \times O(3)$ description provides geometrical numbers for this ratio, through the explicit use of the fully normalized and unambiguous wave functions (4.20) and (4.21) appropriate to the $\mathbf{8}_d$ and $\mathbf{8}_q$ states respectively. If therefore we consider the different resonances to correspond to the $\mathbf{8}_d$ and $\mathbf{8}_q$ break-up for octet states all the isoscalar factors are unambiguous, so that the reduced widths $g_L$ for all the cases are of the form

$$g_L = X_L \, (PB \mid B_L), \qquad (5.1)$$

where the factor $X_L$ is a quantity depending on the spin-cum-orbital structures of $B$ and $B_L$, while the $(I, Y)$ values of the involved states are incorporated in the isoscalar factor. For application to actual decays the quantities (5.1) must be multiplied by some further quantities which should at least incorporate the qualitative features of the body form factors (4.21), without the assumption of a fuller model for these $QQQ$ wave functions, such as the harmonic oscillator model. These qualitative features are (i) threshold dependence on the decay momentum and (ii) some sort of damping factor, only depending on the decay momentum. This second feature which physically represents the effect of a *finite size* for the $QQQ$ wave function is especially important for decays in high partial waves where the threshold factor $(\sim k^{2l+1})$ would otherwise show an extremely sharp rise with large $l$-values. Such damping factors have been known from the very earliest studies of the 3,3-resonance.[69] As already discussed in Section 4, the number of threshold

Table 4  Parametrizations of amplitudes for $L^P = 0^+$ decays<br>via the direct and recoil terms

| Decay | "Direct" | "Recoil" |
|---|---|---|
| $(L - 1)$ wave | $k \, (k/\mu)^L \, (1 + \alpha k^2)^{-\frac{L+1}{2}}$ | $\dfrac{\omega_k}{\mu M_Q} \, (k/\mu)^{L-1} \, (1 + \alpha k^2)^{-\frac{L-1}{2}}$ |
| $(L + 1)$ wave | $k \, (k/\mu)^L \, (1 + \alpha k^2)^{-\frac{L+1}{2}}$ | $\dfrac{\omega_k}{\mu M_Q} \, (k/\mu)^{L+1} \, (1 + \alpha k^2)^{-\frac{L+1}{2}}$ |

factors is dependent on whether the "direct" or the "recoil" term is under consideration. All these features are summarized in Table 4 which exhibits the parametrizations for the decay amplitudes via the direct and recoil terms.

The quantity $\mu$ in Table 4, which is typically of the dimension of a momentum, arises from the dimensional structure of the overlap integral (4.21). While the quantity need not be the same for the direct and recoil terms, its order of magnitude should not differ much for the two cases, since both involve the same wave functions. Qualitatively, $\mu$ may be regarded as a typical order of magnitude for the quark momentum operator, which may even be large, since what is involved in this evaluation is the effective mass of the quark *inside* the baryon and the latter is believed to be small.[68] Thus it is reasonable to assume that $k^2/\mu^2 \ll 1$. The effect of damping for large momenta is represented in Table 4 through the multiplication of each threshold factor $k$ by the quantity[42,70] $(1 + \alpha k^2)^{-1/2}$, where $\alpha$ is a parameter of the dimension of (momentum)$^{-2}$, to be determined from experiment. A more precise interpretation of this damping factor (whose importance increases with large values of $k$) will be given in Section 6.

On the basis of the foregoing it is reasonable to assume that the recoil term dominates $(L - 1)$ wave decays for low $Q$-values (as it has fewer threshold factors) while the direct term above is quite satisfactory for parametrizing $(L + 1)$ wave decays (as it has the same number of threshold factors as the recoil term). Thus for the various decay widths we finally adopt the parametrizations:

$$\Gamma_{L-1} = C_{L-1} g_L^2 \omega_k^2 k \, [k^2 \mu^{-2}/(1 + \alpha k^2)]^{L-1}, \tag{5.2}$$

$$\Gamma_{L+1} = C_{L+1} g_L^2 k \, [k^2 \mu^{-2}/(1 + \alpha k^2)]^{L+1}, \tag{5.3}$$

where $g_L$ is defined through (5.1), the extra factor $k$ in $\Gamma_{L\pm 1}$ comes from phase space, and the quantities $C_{L\pm 1}$ are overall constants governing the two modes of decay from $L^P$ to $0^+$ supermultiplets. Note that in this simple picture we are breaking $SU(6)$ and $SU(3)$ symmetry only in phase space, as long as we do not change the parameters $\mu$, $C_{L\pm 1}$ and $\alpha$ for the different $SU(3)$ or $SU(2)$ modes of emission. On the other hand, breaking of $SU(3)$ symmetry in the *coupling constants* would require suitable modifications in these parameters.

A large number of investigations on the decays of resonances has been carried out by various workers on the basis of such a formalism. We shall refrain from giving a detailed review of the numerical results obtained for a rich variety of decay modes from a large number of resonances (largely on the pattern of the Rosenfeld tables[51]), by giving suitable references to the literature. However, it should be of interest to mention some important qualitative conclusions which have emerged within this broad framework which has the great virtue of not depending on any detailed model, either of interactions, or of the radial functions.

## 5.2 Results on $L^P = 1^-$ baryon decays

The case of negative parity baryon decays was first studied in detail by Mitra and Ross on the basis of available data in 1966. One of the most interesting results of this investigation was the prediction of enhanced decay widths associated with heavy meson modes of decay. These modes, such as $\Lambda K$, $N\eta$ and $\Lambda\eta$ can manifest themselves either directly (if enough phase space is available for decay) or indirectly (if the decay is not energetically favoured) through peaks in $N\pi$ cross sections near the $N\eta$ and $\Lambda K$ thresholds. Experimentally, there is now very good evidence for much larger partial widths for such modes than would be warranted on the grounds of $SU(3)$ and phase space above. This result can be understood very simply through a comparison of the structures of the amplitudes due to the direct and recoil terms given in Table 4 for the $(L - 1)$ wave mode of emission. Because of the extra threshold factor $k^2$ associated with the direct term for the $(L - 1)$ wave emission, the latter would greatly reduce the heavy meson modes for which $k^2$ is necessarily much smaller than for the pion modes. On the other hand, the recoil term involves the factor $\omega_k^2$ (instead of $k^2$) which greatly enhances the heavy meson modes relative to the pion modes, especially at small values of $k^2$. The parametrization (5.2) of the widths for $(L - 1)$ decays by the recoil term therefore incorporates this very important feature of enhancement in the heavy meson modes for $s$-wave decays of the negative parity baryons.*

More recently, Holladay and collaborators[71] have obtained the couplings of the low lying nucleon resonances (of both parities) to the $\Lambda K$ and $N\eta$ channels by parametrizing the observed $\Lambda K$ and $N\eta$ production amplitudes directly through a Breit-Wigner form and making some simple assumptions on the threshold momentum dependence of the partial widths. The result of this analysis is not only consistent with the above feature of heavy meson enhancement but gives much better fits with the $SU(3)$ octet assignments for these resonances than other alternatives like $\overline{\mathbf{10}}$ and $\mathbf{27}$.

---

* For higher $L$-values this feature cannot be so pronounced because the substantially higher masses of the resonances involved with such $L$-values will considerably obscure the sensitivity to $\omega_k^2$ versus $\mathbf{k}^2$.

Another useful feature of the MR type of analysis is the facility it provides for the inclusion of possible mixing effects between $SU(3)$ multiplets within an $[SU(6); L^P]$ supermultiplet. Thus for the $(\mathbf{70}, \overline{1})$ baryons, the physical $N$-states of $J^P = 1/2^-$ and $3/2^-$ could well be mixtures of $\mathbf{8}_d$ and $\mathbf{8}_q$ multiplets. Similarly for the strange particles ($Y = 0, -1$) such mixings could occur between appropriate members of $\mathbf{8}$ and $\mathbf{10}$ multiplets, in addition to those between $\mathbf{8}_d$ and $\mathbf{8}_q$ states. It was found in MR that the following $J^P = 3/2^-$ resonances

$$N^* (1518), \quad \Sigma (1660), \quad \Xi^* (1816)$$

exemplify the possibilities of the above types of mixtures. The remarkable fact that this analysis seemed to indicate was that even *small* admixtures ($\sim 5$–$10\%$) could significantly influence the decay rates so that sensitive adjustments were necessary for obtaining good agreement with experiment. Thus it was found that $\Xi^* (1816)$ should be mostly in the $\mathbf{10}$, $N^* (1710)$ in $\mathbf{8}_d$ and $N^* (1550)$ in $\mathbf{8}_q$, with negligible mixtures from competing states. Indeed the opposite assignments to $N^* (1710)$ and $N^* (1550)$ would lead to a dominant $\eta$-mode for $N^* (1710)$ and a large $\pi$-mode for $N^* (1550)$, results which are exactly the opposite of what are observed.[51] Such criteria are generally more reliable than those based on proximity in mass values of like states, for the purpose of detailed quantum number assignments.

More recently, it has been suggested[72] that even for the well known $\Lambda (1520)$ of $J^P = 3/2^-$, which is generally believed to be a pure $SU(3)$ singlet, a slight mixing with one of the available $\Lambda$-states, viz., $\Lambda (1670)$, can greatly improve the branching ratio of its $N\overline{K}$ to $\Sigma\pi$ decay modes. Other indications of mixing effects arise from the observed decay rates of resonances like $\Lambda (1670)$ which have $\eta$-modes comparable to $\pi$-modes. For, under the assumption of pure $SU(3)$, simple Clebsch-Gordan arguments show that an unmixed $\Sigma$-state of $\mathbf{8}$ should have equal transition matrix elements to $\Sigma\eta$ and $\Lambda\pi$ modes, while a pure $\Lambda$-State of $\mathbf{8}$ should exhibit a $3:1$ ratio between its $\Sigma\pi$ to $\Lambda\eta$ modes. Similarly, the effect of spin-structures on strange particle decays from the octet is such that $\mathbf{8}_q$ states generally lead to strong $\eta$-modes, and the $\mathbf{8}_d$ states to large $\overline{K}N$ modes. Using these broad guidelines, the anomalies in the decay modes of certain strange particles can be understood partly from the effect of the recoil term for $s$-wave emission (heavy meson enhancement) and partly from the effect of spin-cum-$SU(3)$ mixing. Taking the example of $\Lambda (1670)$, an explicit (but ad hoc) construction suggested in MR, is

$$\Lambda (1670) = \sqrt{\tfrac{17}{20}}\, (\mathbf{8}_q) - \sqrt{\tfrac{1}{10}}\, (\mathbf{10}) + \sqrt{\tfrac{1}{20}}\, (\mathbf{8}_d),$$

which gives

$$\Gamma_{\overline{K}N} = 3 \text{ MeV}, \quad \Gamma_{\pi\Sigma} = 12 \text{ MeV}, \quad \Gamma_{\eta\Lambda} = 0.25k.$$

As for self-couplings among $(\mathbf{70}, 1^-)$ states a good example is provided by $\Sigma(1767) \to \Lambda(1520) + \pi$ which has already been treated in Section 2.

### 5.3  Decays of higher baryons $(L > 2)$

For the still higher resonances of positive parity, such as $N^*(1688)$ and $\Delta(1920)$, which may be regarded as Regge recurrences of $N(938)$ and $\Delta(1236)$ respectively, analyses similar to above have been carried out by Lipkin *et al.*[73] as well as by Katyal and Mitra[74] on the assumption that these belong to the supermultiplet $(\mathbf{56}, 2^+)$. Taking the assignments as given in Tables 3, the following types of transition can be evaluated[74]:

$$(\mathbf{56}, 2^+) \to (\mathbf{56}, 2^+) \tag{5.4}$$

$$(\mathbf{56}, 2^+) \to (\mathbf{56}, 0^+) \tag{5.5}$$

$$(\mathbf{56}, 2^+) \to (\mathbf{70}, 1^-). \tag{5.6}$$

Type (5.4), which is again a case of self-coupling, predicts an appreciable width of 53 MeV for the mode $\Delta(1920) \to N^*(1688) + \pi$, to be compared with $\sim 120$ MeV for the width of the 33-resonance $\Delta(1236)$. Similarly the width of $\Sigma(2035) \to \Lambda(1820) + \pi$ is predicted as 19 MeV, to be compared with 34 MeV for the $\Sigma(1385) \to \Lambda + \pi$ decay. It should be of interest to check these modes with experiment, when available.

The modes (5.5) depend of course on parametrizations like (5.2) and (5.3) for the $p$ and $f$ wave emissions respectively. The results for these case are in generally good agreement with the observed modes, even with some variations over (5.2) and (5.3) in the structure of the damping factors[73,74].

Decay modes of the type (5.6) can be calculated by obvious extension of the techniques outlined in Section 4. Unfortunately, at this stage such data which involve sequential decays from $L^P = 2^+$ to $L^P = 0^+$, *via* certain $L^P = 1^-$ states are not yet available. It is however important to note hat the quark model predicts appreciable values for such modes for the following reason. Spatial symmetry in $(\mathbf{56}, 2^+) \to (\mathbf{70}, 1^-)$ transitions changes merely from $s$-type to $m$-type, and such matrix elements which can be brought about through single quark transitions are therefore necessarily large. This case stands in contrast to a transition from an $s$-type to an $a$-type state which corresponds so to say, to *two* units of symmetry change and hence should involve at least two quark transitions. For a mechanism like ours which is based on single quark transition the former matrix elements must be much larger than the latter. This may be termed as a "symmetry selection rule",[75] according to which a symmetry change like $s \to m$ is merely $\Delta Sym = 1$, while one like $s \to a$ is equivalent to $\Delta Sym = 2$ ($s \to m$ *followed by* $m \to a$). Thus the transitions such as (5.6) involving as do, $\Delta Sym = 1$, can give appreciably widths, and hence their detection would be of great significance for the broad validity of the single quark model of transitions.

It was pointed out by Lipkin *et al.*[73] that a rather sensitive test of the quantum number assignments to baryons, can be provided by polarization experiments on the final baryon to check on the interference between two partial waves (where applicable) for the emitted meson. This argument is based essentially on the geometrical relationship, given by (4.9), between the $(L \pm 1)$ wave amplitudes, if only the direct term is considered, and this predicts a definite polarization for the final baryon if the letter is measured in a experiment involving the production and subsequent decay of the relevant resonance. However, if as we have seen for the negative parity resonances, the $(L-1)$ emission is parametrized by the recoil term and $(L+1)$ by the direct term the two amplitudes would become incoherent and hence vitiate the prospects of any simple polarization prediction.

For more detailed predictions one must be prepared to supplement the above frame work by more specific models which facilitate the evaluation of the various parameters in terms of fewer input quantities. For example, if the $QQQ$ wave functions are assumed to be given by harmonic oscillator functions of a common linear dimension $R$, then the parameters $\alpha$, $\mu$ and $C_{L \pm 1}$ of (5.2) or (5.3) are all expressible in terms of $R$, and so also the various types of decay modes considered above. Such a calculation was performed recently by Faiman and Hendry[76] with fairly encouraging results. One important result of this calculation was to confirm the assignment of $N^*(1470)$ to a $(\mathbf{56}, 0^+)$ state which presumably represents the first radial excitation of the $N(938)$. These authors found that the assignment of $N^*(1470)$ to other supermultiplets gave decay widths in complete discord with experiment.

The calculations of $L^P = 3^-$ decays have been given by Katyal and Mitra[74]. These cases are much more tentative because of the present limitations on their experimental details. Such calculations do however provide a broad frame work against which it should be possible to check on the spin-parity and $SU(3)$ multiplet assignments for the observed resonances above 2000 MeV.

## 5.4 Similarity to $SU(6)_W$

It may be appropriate at this stage to discuss briefly the scope and predictions of such calculations in relation to those of certain allied investigations on the basis of more formal group structures, especially $SU(6)_W$. Since these groups allow the baryon to retain its "elementary" character, it is possible to give a relativistic shape to the relevant form factors. The general assumption in such approaches is to use the *physical* masses of the particles involved in the vertex functions. The damping factors in these approaches are now more natural functions of the momenta and masses of the decay products, facilitated by the relativistic normalizations of the various wave functions.

The broad similarity of $SU(6)_W$ to the $SU(6) \times O(3)$ quark model has been emphasized by Lipkin[77]. This is based on the observation that the basic $\bar{Q}QP$ interaction operator $\boldsymbol{\sigma} \cdot \mathbf{k}$ transforms under $W$-spin rotations, line the component of a vector in the direction $\mathbf{k}$ of the emitted meson in a collinear coordinate system in which all momenta are in the $Z$-direction. Thus the matrix element for any particular transition satisfy the same relations as required by $W$-spin conservation. The same argument is applicable with the inclusion of the isospin and hypercharge degrees of freedom for the quarks. Now for transitions between two *like* baryon states belonging to the same **56** but different $L$-excitations (e.g., $L = 0$ and 2), the transformation operator involves only $L_Z = 0$ (collinear transformations) so that even a non-zero $L$-excitation of the internal baryon state does not play much of a role in an $SU(6)_W$-invariant interaction. Therefore for cases such as $N^*(1688) \to N(938) + \pi$ involving $J^P = 5/2^+$ to $J^P = 1/2^+$ transitions, a strong similarity of $SU(6)_W$ to the $SU(6) \times O(3)$ quark model results should be expected. However, the situation is some what different for transition between *unlike* baryon states, e.g., $N^*_{5/2}(1688) \to \Delta(1236) + \pi$, when the intrinsic spin functions in the initial state have $S = \frac{1}{2}$, while those in the final state have $S = \frac{3}{2}$. Thus a transition from an initial $L = 2$ state of $J_Z = \frac{5}{2}$ to a final $L = 0$ state of $J_Z = S_Z = \frac{3}{2}$ necessarily requires $L_Z = +1$ in the initial state. Such a requirement quark makes the $L$-excitations play a more active role in the $SU(6) \times O(3)$ quark model than envisaged in an $SU(6)_W$ theory. One should therefore expect differences in the decay predictions between a quark model based on $SU(6) \times O(3)$ and an $SU(6)_W$ theory, especially when unlike baryons are involved, and these differences should increase with the $L$-excitation of the initial state.

Calculations of decay widths of various resonances on the $SU(6)_W$ model have been made by Dobson using separate parameters for successively higher $L$-excitations.[78] The differences in the predictions of $SU(6) \times O(3)$ and those of $SU(6)_W$ do indeed show the trend discussed above, being much more marked for $L^P = 3^-$ than for $L^P = 2^+$ or lower values. The two predictions are of course identical for the $L^P = 0^+$ (the **56** baryons themselves). Another calculation of decay widths based on the colinear, $U(6)_W \times O(2)_W$ group, which is similar in spirit to $SU(6)_W$, has been reported by Freund et al.[79] again with results very similar to $SU(6)_W$. As to the comparison with experiment, the main disadvantage of such models lies in the nature of their predictions for $(L-1)$ wave with the *direct* term. In other words, the $(L-1)$ decays in these models have an extra threshold factor $k^2$ in the meson momentum, which as we have seen greatly suppresses the heavy meson modes of decay, contrary to observation. The recoil term in the quark model which is so successful in explaining the enhanced heavy meson modes has no counterpart in these relativistic models, since the former is a necessary consequence of the *compositeness* assumption on the hadron structures in contrast to their "elementarity" assumption in a relativistic group theory. In the next Section we shall try to formulate a somewhat improved form of parametrization in the $SU(6) \times O(3)$ model which incorporates relativistic kinematics in the form factors and therefore shows a much closer resemblance to relativistic models like $SU(6)_W$, without giving up the advantage of the "recoil term".*

## 5.5  Meson decays

Calculation of meson decays follows a pattern very similar to the baryon case. The coupling scheme is already described in Section 4.1. The parametrization of the widths for the $(L+1)$ wave modes is adequately described by Eq. (5.3), though a different shape for the damping factor is not ruled out. For the $(L-1)$ wave emission on the other hand, the distinction between the direct and recoil terms is not so pronounced as in the baryon case, since unlike the $s$-wave heavy meson modes in $L^P = 1^-$ baryon decays, the corre-

---

* Note that any theory which does not have the recoil term must depend entirely on the possibility of "representation mixing" for bringing about the observed enhancement in heavy meson modes of decay. On the other hand, no amount of representation mixing can overcome the disadvantage arising from the extra threshold factor in the $(L-1)$ mode of emission.

sponding modes in the $L^P = 1^+$ meson decays are appreciably above thesh-old for the two final mesons.

Calculations of $L^P = 1^+$ meson decays have been performed by several authors within or without the quark model. Within $SU(3)$, the decays of tensor mesons into two pseudoscalars[80] can provide information on the mixing angle between the $SU(3)$ singlet and octet members of $I = 0$, such as between $f_0(1250)$ and $f_0'(1514)$, or between $\eta$ and $X^0$ mesons, and this can be compared with similar information obtainable from the quadratic Gell-Mann-Okubo mass formula. An interesting possibility is the extension of the concept of the ideal mixing angle beyond the $(\omega, \phi)$ pair. Thus the latest analysis of tensor meson decays[53] suggests that the $f_0(1250)$ and $f_0'(1514)$ arc probably exact counterparts of $\omega$ and $\phi$ respectively, since $f_0'(1514)$ hardly decays into two pions and $f_0(1250)$ has a very small $K\bar{K}$ mode. For the $(\eta, X^0)$ pair, the picture is not so clear. For, an ideal mixing angle which makes $\eta$ like $\phi$ would greatly suppress the $A_2 \to \eta\pi$ with respect to the prediction of $SU(3)$ plus phase space, a result partly conforming to experiment. However such an assumption would simultaneously inhibit $N\eta$ modes of certain well-known negative parity mesons especially $N^*(1550)$, which is quite the contrary of what is observed.

As we have already seen in Section 2, the quark model gives a simple raison d'être for the ideal mixing angle, since the mixed states, can now be interpreted as made up of strange and non-strange quarks respectively exactly as in Eq. (2.36) for $\omega$ and $\phi$. Within the quark model it is of course possible to make many more predictions on the decay rates in a unified fashion. Thus for the tensor mesons, both their $PV$ and $PP$ modes of decay are obtainable in the same scheme. Such a procedure was adopted by Lipkin et al.[81] using a quark rearrangement model in which the $\eta$ and $f_0'$ were considered to have $\lambda\bar{\lambda}$ structures. Actually the predictive powers of the quark model extend to the decays of not only the tensor meson nonet ($J^{PC} = 2^{++}$), but the entire set of all the *four* nonets ($2^{++}, 1^{++}, 0^{++}, 1^{+-}$) listed in Eq. (3.3), since the decays of all of them must be governed by the same spatial form factor, multiplied by appropriate Clebsch-Gordan factors. Calculations of $L^P = 1^+$ decays on such lines were reported by Mitra and Srivastava[82], and by Uretsky[83]. Mitra and Srivastava used a general form of parametrization based on the direct term for both $(L \pm 1)$ wave emissions. However, instead of taking the geometrical relationship (4.9) between the corresponding amplitudes, they used two different parameters $g_S$ and $g_T$ for the coupling constants governing the $s$-wave (scalar) and $d$-wave (tensor) emissions respectively. Some of their typical extra-$SU(3)$ relations among the coupling constants for important meson states are

$$\tfrac{1}{2}g_S\,(A_1\varrho\pi) = g_S\,(B\omega\pi) = -\sqrt{2}\,g_S\,(K_A^*K\omega); \tag{5.7}$$

$$\tfrac{3}{10}\,\sqrt{5}\,g_T\,(A_1\varrho\pi) = -\tfrac{3}{10}\,\sqrt{5}\,g_T\,(B\omega\pi) = \tfrac{1}{2}g_T\,(A_2\varrho\pi). \tag{5.8}$$

The parametrization of the decay widths based on the use of the direct term for both $L \pm 1$ wave emissions is of the general form[82]

$$g_{IY}^2\,[\lambda_S(S_{LSJ})^2 + \lambda_T(T_{LSJ})^2]\,k^5F^2\,(k), \tag{5.9}$$

where $g_{IY}$ is the unitary spin factor, $(S, T)$ are the reduced matrix elements for $s$- and $d$-wave emissions respectively, and $(\lambda_S, \lambda_T)$ are the squares of the overall scalar and tensor coupling constants. The quantum numbers of the different $L^P = 0^-$ and $L^P = 1^+$ states may be taken as in Table I.

A considerable role can be played by the choice of form factors in $L^P = 1^+$ meson decays since the availability of much larger phase space for these decays makes the results more crucially dependent on the variation of the form factors with energy, than in the baryon case. Mitra and Srivastava used a Gaussian form

$$F(k) = e^{-\frac{1}{2}\alpha k^2}; \quad \left(\sqrt{\alpha} = (250\text{ MeV})^{-1}\right), \tag{5.10}$$

which leads to a heavy damping for large $k$-values and therefore automatically suppresses the corresponding widths, such as $f_0' \to \pi\pi$, even without a mixing angle between $f_0$ and $f_0'$. This would seem to suggest that the exact role of mixing between $SU(3)$ multiplets is considerably linked to the shape of the form factor.

Uretsky[83] used a (more specific) model of square well potential for the calculation of decays of $L^P = 1^+$ as well as still higher mesons. However, the energy dependence of the widths does not seem to be as well reproduced as with the Gaussian form (5.10). More recently, Choudhury[84] has obtained a much better quality of fits to the decay widths of the various $L^P = 1^+$ nonets through the use of the (i) Gaussian form factors and (ii) the recoil term for $s$-wave decays.

While the recoil term is not so helpful for the $1^+$ and $2^+$ meson decays (which already exhibit appreciable phase space), it seems to be very useful for the heavy meson modes (e.g., $K\bar{K}$) of certain scalar mesons for which evidence has been gaining a lot of ground.[53] Thus the $\delta(962)$ meson of $J^PI^G = 0^+1^-$, observed in $K^-p$ and $\bar{p}p$ reactions[55], and the $\sigma$-meson ($J^PI^G$

$= 0^+0^+$), deduced from $\pi$–$\pi$ phase shift analysis[85] seem to be fairly well established. Since the $\delta$-meson is very near threshold for an $s$-wave $K\bar{K}$ state of $I = 1$, this "pole" is expected to give a peak in the $K\bar{K}$ system[61], thus providing a reasonable interpretation for the $K\bar{K}$ enhancement $\pi_v(1016)$[51] of $J^P I^G = 0^+1^-$ which does not then have to be listed as a separate particle. There is also some evidence[53] for a second scalar meson $S^*(1070)$ of $J^P I^G = 0^+0^+$, which together with the $\sigma$-meson, probably forms the singlet-octet pair of $I = 0$. While the small width of $S^* \to \pi\pi$ and the large width of $\sigma \to \pi\pi$, again suggests their mixing in the "ideal" manner, at least a partial explanation within the quark model lies in the decay mechanism provided by the recoil term.

More sensitive tests of the quark model can be provided by polarization measurements such as those now available in $B \to \pi\omega$ decay.[62] This process which may be visualized as an $^1P_1 \to {}^3S_1$ transition in the $\bar{Q}Q$ state, with the emission of a pion, is desribed by a (non-relativistic) coupling of the form

$$F(k^2)\,(B_i^a k_i)\,(\omega_j k_j)\,\pi_a, \tag{5.11}$$

which is due entirely to the *direct* term in $\bar{Q}QP$ coupling. Here $a$ is the isospin index, and $(i, j)$ represent the (three-dimensional) cartesian indices. As was pointed out by Uretsky[83] and Lipkin[77], the interaction (5.11) may be seen to lead to a $\cos^2 \chi$ distribution in the angle $\chi$ between the normal to the decay plane of $\omega \to 3\pi$, and the direction of the outgoing pion in $B \to \omega\pi$. However, the recent experimental distribution in this angle[62] seems to suggest an almost pure $\sin^2 \chi$-law. The difficulty may be remedied, at least in principle, through the use of the recoil term; for the latter, by itself, gives the coupling

$$F_1(k^2)B_i^a \omega_i \pi_a, \tag{5.12}$$

where the polarization indices of $B$ and $\omega$ are *coupled* to each other, unlike the direct term coupling (5.11) where these indices are not so coupled. It is now clearly possible to exploit this last feature to get a $\sin^2 \chi$-law through a suitable mixture of (5.11) and (5.12), since the two form factors $F$ and $F_1$ are now completely independent.

Similar remarks apply to $A_1 \to \varrho\pi$ decay for which the direct term gives the coupling

$$k_i k_j A_i^a \varrho_j^a - k^2 A_i^a \varrho_i^a \tag{5.13}$$

and the recoil term simply

$$A_i^a \varrho_i^a, \tag{5.14}$$

where each expression is multiplied by an independent form factor. The direct term would now predict the angular distribution $\sin^2 \theta$ in the angle $\theta$ between $\varrho \to 2\pi$ decay direction and that of the pion in $A_1 \to \varrho\pi$. On the other hand, the recoil term would predict isotropy in this angle. At this stage the experimental information on this question is extremely meagre.

With the exception of $B$ and $A_1$, and perhaps $D(1285)$, the present status of the $1^+$ mesons, especially their strange components, is much less satisfactory than their $J^P = 2^+$, counterparts. The situation has been further complicated through the conjectured existence of particles like $A_{1.5}(1170)$ and $A_{2_L}(1290)$ which do not even fall under the simple quark model pattern and have necessitated the new Gell-Mann–Zweig scheme[64]. However, the decay properties of these particles are hardly known at the present time. The $D$-meson of $J^{PC} = 1^{++}$ and $I^G = 0^+$ decays mainly through $D \to \delta\pi$[55] and is a good example of self-coupling for which the prediction of the quark model is clean and in good accord with experiment (see Eq. (2.49)).

For the higher mesons of $L > 2$ and/or radial excitations of $L^P = 0^-$, $1^+$ mesons, very few cases have been properly studied so far, to warrant a meaningful discussion on the basis of this general analysis. The cases of the $g$-meson of $\varrho_V(1650)$ which decays into $\pi\pi$, and the $\varrho(1700)$ meson which decays into $\pi\omega$, will be discussed briefly in Section 7 in connection with an improved model based on fewer parameters.

We close this section with the remark that the "ideal mixing angle" between single and octet members of $I = 0$ may well form the basis of a more rational classification of isosinglet mesons than pure $SU(3)$ assignments.


## 6   IMPROVED (RELATIVISTIC) MODEL OF HADRON COUPLINGS

So far we have been considering the problem of hadron couplings and decays in a phenomenological manner through the use of a suitable number of parameters to describe the coupling constants and the structure of the various overlap integrals. While this approach is capable of yielding several results of a qualitative nature, their quantitative aspects are clearly of much less significance than those of a more formal theory. However some of the qualitative features of the experimental decay characteristics deduced from such an analysis can be profitably employed to build a more complete (hence less parametric) theory. One such feature is represented by role of the recoil term which, as we have seen, accounts for the observed enhancement in heavy meson modes for low partial wave ($s$-wave) decays from relatively low energy hadron resonances. We have also seen that the special tensorial structure of the couplings due to the recoil term can even help in understanding certain experimental angular correlations among the decay products (e.g., in $B \to \pi \omega$ decay). These features may or may not be present in many of the formal relativistic theories of hadron couplings that have been proposed from time to time, and unless some conscious effort is made to ensure such conditions more explicitly, these theories would to that extent fail to satisfy the observations.

In this section we shall outline a scheme of hadron couplings[87] in a unified manner designed to give a much clearer look to the parametric studies discussed in the earlier sections through a significant reduction in the number of parameters. Such a reduction will be achieved first by making certain plausible guesses on the structures of the overlap integrals (form factors) in terms of the *masses* of the decay products, and secondly, through a semi relativistic extension of the structure of the interactions so as to incorporate the desirable features of the direct and recoil terms, without the need for their separate parametrizations. It turns out that such a program is indeed possible and it provides a rather wide range of agreement with experiment with a surprisingly small number of parameters. In this last respect it has not only a good deal of resemblance to a more formal theory but in addition gives sufficient indications on the structure of $SU(3)$ symmetry breaking (due to masses) that a more formal theory should preferably possess. As we shall see in what follows, the derivation of such couplings will depend partly on the quark model and partly on certain intuitive guesses on the structure of the form factors. While discussing the baryon and meson cases separately for convenience, we shall keep in view the essential of approach for both the cases.

### 6.1 Baryon couplings ($L > 0$)

The initial procedure for this evaluation, in particular for expressing the spin-cum-$SU(3)$-matrix elements in terms of "elementary" baryon structures is the same as for the quark model described in Section 2 [Eqs. (2.13)–(2.16)]. Thus the spin matrix elements (between states of $S = 1/2$ and/or $S = 3/2$) are all deducible from any one of the following invariant couplings (see Section 2 for detailed notations)

$$\chi^\dagger \sigma_i K_i \chi, \quad \chi_i^\dagger K_i \chi, \quad \chi_i^\dagger \sigma_i K_j \chi_i. \tag{6.1}$$

Here $\chi$, $\chi_i$ are Pauli and Rarita–Schwinger (non relativistic) spinors for $S = 1/2$ and $3/2$ respectively and $(i, j)$ are 3-vector indices. In deriving these structures only the direct term $\sigma_i K_i$ in the basic $\bar{Q}QP$ coupling has been considered. Likewise a few $SU(2)$ invariant forms for the couplings are

$$N^\dagger \tau_a \pi_a N, \quad \Delta_a^\dagger \pi_a N, \quad \Delta_a^\dagger \tau_b \pi_b \Delta_a, \quad N^\dagger \eta N;$$

$$\Sigma_a^\dagger \pi_a \Lambda, \quad \Lambda^\dagger \bar{K} N, \quad \Sigma_a^\dagger \bar{K} \tau_a N, \quad \Lambda^\dagger \eta \Lambda; \tag{6.2}$$

where the symbols (in obvious notation) designate the types of states involved. The $SU(6)$ matrix elements will then be given by various products obtained from (6.1) and (6.2), multiplied with appropriate C.G. coefficients. This completes the scheme for couplings among $(56, 0^+)$ baryons whose relativistic extensions have already been described in Section 2. For couplings between $L^P = 0^+$ and $L^P \neq 0^+$ baryons, the invariant quantities (6.1) must be further multiplied by the overlap integrals (4.21) involving the radial functions. For the *direct* term a typical overlap integral (remembering that the $\sigma_i K_i$ factor has already been absorbed in (6.1)) is

$$\iint d\mathbf{P}' \, d\mathbf{P}'' \, \psi_{LM}^* (\mathbf{P}', \mathbf{P}'') \, \psi_0 \left(\mathbf{P}', \mathbf{P}'' + \sqrt{\tfrac{2}{3}}\mathbf{k}\right). \tag{6.3}$$

On general grounds of three-dimensional covariance, this expression is equivalent to the matrix element, between states of $|LM\rangle$ and $|00\rangle$, of an invariant coupling of the form

$$f_L(k^2) \, B_{\alpha_1\alpha_2\cdots\alpha_L}^{LM*} k_{\alpha_1} k_{\alpha_2} \cdots k_{\alpha_L}, \tag{6.4}$$

where $f_L(k^2)$ is a (scalar) form factor in $k^2$ and $B_{\alpha_1\cdots\alpha_L}^{LM}$ is an irreducible spherical tensor of rank $L$ involving the tensor indices $(\alpha_1, \ldots, \alpha_L)$. This tensor which is symmetric and traceless in all its indices, is normalized by the condition[88]

$$\sqrt{p_L} \, k_{\alpha_1} k_{\alpha_2} \cdots k_{\alpha_L} B_{\alpha_1\alpha_2\cdots\alpha_L}^{LM} = \sqrt{4\pi} \, k^L Y_{LM}(\hat{k}), \tag{6.5}$$

where

$$p_L = (2L + 1)!!/L!. \tag{6.6}$$

To calculate the couplings for states of definite $J$ from the product of factors like (6.1) and (6.4), one encounters products of the form

$$B_{\alpha_1\cdots\alpha_L}^L \otimes \chi \quad \text{or} \quad B_{\alpha_1\cdots\alpha_L}^L \otimes \chi_\alpha, \tag{6.7}$$

which must be expressed as suitable Clebsch–Gordan series. A very suggestive form of such an expansion in a relativistic manner was given by Fronsdal several years ago,[89] using the techniques of generalized Rarita–Schwinger (R-S) fields.[47,90] We shall merely require the non-relativistic version of the

Fronsdal technique through the simple replacement of the Dirac matrices $\gamma_\mu$ by the Pauli matrices $\sigma_\alpha$. We define the normalized R-S spinor $(\chi^{L+1/2}_{\alpha_1\alpha_2\cdots\alpha_L})$ for spin-$(L+\frac{1}{2})$ which is symmetric and traceless in all its (3-dimensional) indices and also satisfies the condition

$$\sigma_\alpha \chi^{L+1/2}_{\alpha\alpha_2\cdots\alpha_L} = 0. \tag{6.8}$$

Then following Fronsdal it is possible to obtain the result

$$B^L_{\alpha_1\cdots\alpha_L} \otimes \chi = \chi^{L+1/2}_{\alpha_1\cdots\alpha_L} + \sqrt{\frac{L}{2L+1}}\, S_L \sigma_{\alpha_1} \chi^{L-1/2}_{\alpha_2\cdots\alpha_L}, \tag{6.9}$$

where

$$\chi^{L+1/2}_{\alpha_1\cdots\alpha_L} = B^L_{\alpha_1\cdots\alpha_L} \otimes \chi - \frac{L}{2L+1}\, S_L \sigma_{\alpha_1}\sigma_\alpha B_{\alpha\alpha_2\cdots\alpha_L} \otimes \chi \tag{6.10}$$

and

$$\chi^{L-1/2}_{\alpha_2\cdots\alpha_L} = \sqrt{\frac{L}{2L+1}}\, \sigma_\alpha B^L_{\alpha\alpha_2\cdots\alpha_L} \otimes \chi \tag{6.11}$$

are both normalized R-S spins of $J = L \pm \frac{1}{2}$ respectively and $S_L$ is a symmetrizer for the indices $(\alpha_1 \cdots \alpha_L)$. In a similar way, we have the more involved Clebsch–Gordan series

$$B^L_{\alpha_1\cdots\alpha_L} \otimes \chi_\alpha = \chi^{L+3/2}_{\alpha\alpha_1\cdots\alpha_L} + \sqrt{\frac{L+1}{2L+3}}\, S_{L+1}\sigma_\alpha \chi^{L+1/2}_{\alpha_1\cdots\alpha_L}$$

$$+ \sqrt{\frac{L(L+1)}{2(2L+1)}}\, S_{L+1}\delta_{\alpha\alpha_1}\chi^{L-1/2}_{\alpha_2\cdots\alpha_L}$$

$$+ \sqrt{\frac{L(L^2-1)}{2(4L^2-1)}}\, S_{L+1}\delta_{\alpha\alpha_1}[S_{L-1}\sigma_{\alpha_2}\chi^{L-3/2}_{\alpha_3\cdots\alpha_L}], \tag{6.12}$$

where all the functions $\chi^J$ are normalized R-S spinors. The evaluation of (6.12), together with the normalizations of the various spinors, is outlined in Appendix I. The C.G. expansions (6.9) and (6.12) lead directly to the geometrical ratios for the various couplings generated from the direct product of the spin and orbital factors. Thus we have

$$[B^L_{\alpha_1\cdots\alpha_L} \otimes \chi]^\dagger \boldsymbol{\sigma}\cdot\mathbf{k}\chi\,(k_{\alpha_1}k_{\alpha_2}\cdots k_{\alpha_L})$$

$$= \left[ \chi^{L+1/2\dagger}_{\alpha_1\alpha_2\cdot\alpha_L} k_{\alpha_1}\boldsymbol{\sigma}\cdot\mathbf{k}\chi + \sqrt{\frac{L}{2L+1}} \cdot \mathbf{k}^2 \chi^{L-1/2\dagger}_{\alpha_2\cdots\alpha_L}\chi \right] k_{\alpha_2}\cdots k_{\alpha_L} \tag{6.13}$$

and

$$[B^L_{\alpha_1\cdots\alpha_L} \otimes \chi_a]^\dagger \, \boldsymbol{\sigma} \cdot \mathbf{k}\, k_{\alpha_1} \cdots k_{\alpha_L}$$

$$= \chi^{L+3/2\dagger}_{\alpha\alpha_1\cdots\alpha_L}\, k_\alpha k_{\alpha_1} \cdots k_{\alpha_L} + \sqrt{\frac{L+1}{2L+3}}\, \chi^{L+1/2\dagger}_{\alpha_1\cdots\alpha_L}\, \boldsymbol{\sigma} \cdot \mathbf{k}\, k_{\alpha_1} k_{\alpha_2} \cdots k_{\alpha_L}\, \chi$$

$$+ \sqrt{\frac{L(L+1)}{2(2L+1)}}\, \chi^{L-1/2\dagger}_{\alpha_2\cdots\alpha_L}\, \mathbf{k}^2\, (k_{\alpha_2} \cdots k_{\alpha_L})\, \chi$$

$$+ \sqrt{\frac{L(L^2-1)}{2(4L^2-1)}}\, \chi^{L-3/2\dagger}_{\alpha_3\cdots\alpha_L}\, \mathbf{k}^2\, \boldsymbol{\sigma} \cdot \mathbf{k}\, k_{\alpha_3} \cdots k_{\alpha_L}\, \chi. \tag{6.14}$$

These couplings are if course multiplied by the $SU(3)$ factors with appropriate geometrical coefficients necessary to ensure $SU(6)$ invariance in the limit of small momentum transfer ($\mathbf{k}$) to the $P$-meson. Finally, for a given supermultiplet transition $L^P \to 0^+$, there is an overall form factor $f_L(k^2)$ multiplying all the above couplings.

## 6.2 Relativistic extension

So far the physical content of this formulation is identical with that outlined in the previous sections. However it is clear that this new form holds out hopes for at least a partial relativistic generalization. It may also be noted that we have so far considered only the role of the direct term in the basic $\bar{Q}QP$ coupling for deriving (6.13) and (6.14). An obvious relativistic extension is now suggested if the non-relativistic spinors $\chi^J_{(\alpha)}$ are replaced by the full-fledged R-S spinors $\psi^J_{\mu_1\mu_2\cdots\mu_J-1/2}$ which include in the usual way *both* the negative and positive energy components for each 4-vector index $\mu$[90]. Simultaneosly the three-vector indices ($\alpha_i$) appearing in the momenta must be replaced by the corresponding 4-vector indices $\mu_i$. For $\boldsymbol{\sigma} \cdot \mathbf{k}$ in (6.13) and (6.14), we should further make replacements of the type

$$\chi^\dagger_B\, \boldsymbol{\sigma} \cdot \mathbf{k}\, \chi_A \Rightarrow \bar{\psi}_A \gamma_\alpha \gamma_5 k_\alpha \psi_B \Rightarrow i\,(m+M)\, \bar{\psi}_B \gamma_5 \psi_A, \tag{6.15}$$

through the standard application of the Dirac equation for the initial state $A$ (mass $M$) and the final state $B$ (mass $m$). These modifications take care of most of the factors in the various terms of (6.13) and (6.14) except for the factor $\mathbf{k}^2$ which appears at certain places. Indeed the factor $\mathbf{k}^2$ appears in precisely those terms which are associated with **strongly antiparallel alignments of the vectors L and S** to form a total **J**. Strictly speaking these factors should best be left alone, since in the non-relativistic limit, they are necessary for obtaining the correct threshold dependence of the corresponding matrix elements, such as we have seen in the earlier sections. Such an attitude would however be too conservative for yielding new results, for we have already seen that the extra threshold dependence for the $(L-1)$ wave decay from the direct term is a great drawback for heavy meson decay. On the other hand, since the recoil term shows the right threshold dependence in this regard, we must somehow be prepared to incorporate it in the present analysis. If as in the previous analysis we add the recoil term only for $(L-1)$ wave couplings it will have the effect of making extra contributions precisely to those terms in (6.13) and (6.14) which have the extra factor $k^2$. These additional contributions will of course be multiplied by a second, independent form factor parametrizing the overlap integral for the recoil term, which would *not* show the unwanted threshold factor $k^2$. We now see the possibility of including the recoil term in a *special combination* if we make the replacement $\mathbf{k}^2 \to K_\mu K_\mu$ whose value on the energy shell is $(-\mu^2)$, $\mu$ being the mass of the emitted meson. This special combination corresponds to taking the ratio $1 : (-\omega_k^2)$ for the form factors associated with the direct and recoil terms respectively (since $k_\mu K_\mu = \mathbf{k}^2 - \omega_k^2$). Note that the replacement $\mathbf{k}^2 \to K_\mu K_\mu$ destroys the geometrical relationship (4.9) between the $(L \pm 1)$ wave amplitudes, a property which is shared by the couplings (6.13) and (6.14). It also violates $SU(6)$ invariance for the resultant coupling structures. However, it has the great advantage of exhibiting enhanced heavy meson modes without having to introduce a fresh parametrization for the recoil term. This is the crucial step in this new (phenomenological) formulation of baryon couplings in terms of R-S fields, one which simultaneously allows the possibility of a relativistic extension and incorporation of recoil-like effects without proliferation of parameters. The price for such a modification is of course the loss of $SU(6)$ invariance for the couplings. This should not be hard to tolerate when it is remembered that apart from the incompatibility of any relativistic generalization with $SU(6)$ invariance, the latter is in any case broken through the mass differences among the hadrons in the coupling terms—an effect which we would like to study.

With the above modifications, the couplings (6.13) and (6.14) are now respectively expressible as

$$(M + m)\, \overline{\psi}_{\mu_1 \cdots \mu_L}^{L+1/2}\, i\gamma_5 k_{\mu_1} \cdots k_{\mu_L} \psi$$

$$+ (-\mu^2) \sqrt{\frac{L}{2L + 1}}\, \overline{\psi}_{\mu_2 \cdots \mu_L}^{L-1/2}\, k_{\mu_2} \cdots k_{\mu_L} \psi; \qquad (6.16)$$

$$\bar{\psi}_{\mu\mu_1\cdots\mu_L}^{L+3/2}\, k_\mu k_{\mu_1}\cdots k_{\mu_L}\,\psi + (m+M)\sqrt{\frac{L+1}{2L+3}}\;\bar{\psi}_{\mu_1\cdots\mu_L}^{L+1/2}\, i\gamma_5 k_{\mu_1}\cdots k_{\mu_L}\,\psi$$

$$+\sqrt{\frac{L(L+1)}{2(2L+1)}}\,(-\mu^2)\,\bar{\psi}_{\mu_2\cdots\mu_L}^{L-1/2}\, k_{\mu_2}\cdots k_{\mu_L}\,\psi + (m+M)(-\mu^2)\times$$

$$\times\sqrt{\frac{L(L^2-1)}{2(4L^2-1)}}\,\bar{\psi}_{\mu_3\cdots\mu_L}^{L-3/2}\, i\gamma_5\, k_{\mu_3}\cdots k_{\mu_L}\,\psi. \qquad (6.17)$$

### 6.3  Structure of form factors

These couplings must now be multiplied by suitable form factors just as in
the more phenomenological analysis in the earlier sections. First we note
that the occurrence of multiple derivative terms in these couplings (represented by the $K_\mu$-factors) necessitates that the form factors be heavily dimensional. It is most natural to demand that the mass ($\mu$) of the emitted meson
be the main carrier of this dimensionality, and on this basis we must multiply each term by $\mu^{-L-1}$. Note that since $\mu$ differs widely from the pion to
the heavy mesons ($K, \eta$), the factor $\mu^{-L-1}$ which predicts a strong violation
of $SU(3)$ is a very sensitive function of this parameter, so that even a relatively crude experimental test should suffice to determine the validity of its
occurrence in the couplings. To this factor we must add the "Weisskopf-
correction factor" $\sqrt{\mu/m_\pi}$ as well as another empirical factor $\sqrt{M/m}$ representing the baryons masses, the usefulness of both of which has already been
noted in Section 2 for the $0^+ \to 0^+$ decays. Actually if these were the only
factors to be added to the structures (6.16) and (6.17), these would remain
fully Lorentz—invariant and show strong resemblance to the theories of
effective Lagrangians based on relativistic groups[91] and chiral dynamics.[92-94]
However, such a simple prescription would give too strong a dependence on
the momentum and would greatly enhance the decay rates from high mass
resonances. A more reasonable theory would, on the other hand, be expected
to predict a more gentle dependence on the momentum. A considerable
amount of physical insight for obtaining bounded form factors with momentum is provided by the work of Barut and collaborators on the basis of
irreducible representations of the group $O(3, 1)$[95] and group $O(4, 2)$[96] and
also by the work of Nambu[97] on the basis of infinite component wave equations with $O(4, 2)$. While it is not in the scope of this article to go into the
technical details of these theories, the essential fact that emerges from them
is that the form factors are certain complicated functions of *velocities* and

*not* of the momenta. These formal theories have met with considerable success in the evaluation of decay widths for both baryon[98] and meson resonances.[99]

A simple way to incorporate this feature of boundedness with momentum phenomenologically in our couplings is to multiply each 4-momentum factor $k_\mu$ in (6.16) or (6.17) with the (dimensionless) Lorentz—contraction factor $(\mu\omega_k^{-1})$, which would essentially have the effect of replacing the momenta (unbounded for $\mathbf{k} \to \infty$) by the velocities (bounded for $\mathbf{k} \to \infty$). As we shall see in Section 7, this simple prescription which now plays the role of the form factors (5.2) and (5.3) seems to work remarkably well over a wide range of decays, without having to introduce any new parameters. Unfortunately this factor violates formal relativistic covariance for the couplings, but this is apparently the price for agreement with experiment.

Collecting the results of the foregoing discussion the phenomenological form factors multiplying the couplings (6.16) or (6.17) are now given by

$$f_L(k^2) = g_L \mu^{-L-1} \sqrt{\mu/m_\pi}\, (\mu/\omega_k)^{L\pm 1} \sqrt{M/m}, \qquad (6.18)$$

where the dimensionless constant $g_L$ is expected to be almost a constant, since we have (hopefully) extracted all the rapidly varying quantities. The indices $(L\pm 1)$ in (6.18) are appropriate for the decays in the corresponding waves.

It may be noticed that the quantity $g_L$ is the only free parameter governing the coupling of the entire $L^P$-supermultiplet to the **56** baryons. In this respect it strongly resembles a more formal theory like $SU(6)_W$, though its theoretical basis is by no means comparable to the latter.

We do note however that each factor in (6.18) has its roots in some sort of intuitive physical consideration. Moreover, the complete form factor $f_L$, being a rather rapidly varying function of $\mu^2$ and $k^2$, should not need very precise experimental data for testing its general validity.

### 6.4  Meson couplings $(L > 0)$

The derivation of meson couplings is very similar to that of baryon couplings except that the $Q\bar{Q}$ spin functions are now scalars and vectors rather than Pauli and R-S spinors. The spin matrix elements are now obtainable from either of the following two invariant forms (see also Section 2.2) which can be derived by taking the matrix elements of the operator $\sigma^{(1)} \cdot \mathbf{k}$ between (i) triplet and singlet, or (ii) triplet and triplet spin states.*

$$\varepsilon_i k_i; \quad \varepsilon_i^{(1)} \varepsilon_j^{(2)} \varepsilon_{ijk} k_k. \qquad (6.19)$$

---

* Singlet to singlet transitions for $\boldsymbol{\sigma} \cdot \mathbf{K}$ are of course forbidden.

Here $\varepsilon_i$, $\varepsilon_i^{(1)}$ and $\varepsilon_i^{(2)}$ are unit 3-vectors representing polarization states of the relevant mesons, and $\mathbf{k}$ is the momentum of the emitted $P$-meson. The spatial matrix elements are similarly obtainable from the same invariant form (6.4) as for baryons. The Clebsch–Gordan reduction is now achieved through the relation (see Appendix I)

$$\varepsilon_\alpha^1 \otimes B_{\alpha_1 \alpha_2 \cdots \alpha_L}^L = A_{\alpha\alpha_1 \cdots \alpha_L}^{L+1} + i \sqrt{\frac{L}{L+1}}\, S_L \varepsilon_{\alpha\alpha_1\beta_1} A_{\beta_1 \alpha_2 \cdots \alpha_L}^L$$

$$+ \sqrt{\frac{L(L+1)}{2(2L+1)}}\, S_{L+1} \delta_{\alpha\alpha_1} A_{\alpha_2 \cdots \alpha_L}^{L-1}, \tag{6.20}$$

where $A_{\alpha_1\alpha_2 \cdots \alpha_J}^J$ is a normalized tensor of rank $J$, symmetric and traceless in any two indices and has the $C$-parity $(-1)^{L+1}$ (being a $Q\bar{Q}$ triplet). On the other hand, the tensor $B_{\alpha_1 \cdots \alpha_L}^L$ of rank $L$ occurring on the left hand side of (6.20) has the $C$-parity $(-1)^L$ (being a $Q\bar{Q}$ singlet). The tensors $A^J$ are appropriate for the description of mesons like $A_1$, $A_2$ etc., while $B^L$ is good for state like the $B$-meson. Contraction with the $k_\alpha$-factors in (6.4) now leads to the following couplings of $A$-mesons with $PP$ states*

$$A_{\alpha\alpha_1 \cdots \alpha_L}^{L+1} k_\alpha k_{\alpha_1} \cdots k_{\alpha_L} + \sqrt{\frac{L(L+1)}{2(2L+1)}}\, \mathbf{k}^2 A_{\alpha_2 \cdots \alpha_L}^{L-1} k_{\alpha_2} \cdots k_{\alpha_L}. \tag{6.21}$$

For couplings to $PV$ states, we must distinguish between $B_L PV$ and $A_J PV$. For $B_L PV$ couplings, such as $B\omega\pi$, we have immediately from (6.4) and the $\varepsilon \cdot \mathbf{k}$ term in (6.19) the following expression

$$B_{\alpha_1 \cdots \alpha_L}^L k_{\alpha_1} \cdots k_{\alpha_L} V_\alpha k_\alpha. \tag{6.22}$$

For the $A_J PV$ couplings on the other hand we must consider the expression (6.4) and the $\varepsilon^{(1)} \times \varepsilon^{(2)} \cdot \mathbf{k}$ term in (6.19) and carry out the necessary contraction of indices with those in the C.G. series (6.20). This gives the result

$$A_{\alpha\alpha_1 \cdots \alpha_L}^{L+1} k_{\alpha_1} \cdots k_{\alpha_L} i\varepsilon_{\alpha\beta\gamma} k_\beta V_\gamma$$

$$+ \sqrt{\frac{L}{L+1}}\, (k_\alpha k_{\alpha_1} V_\alpha A_{\alpha_1 \cdots \alpha_L}^L - \mathbf{k}^2 V_{\alpha_1} A_{\alpha_1 \cdots \alpha_L}^L)\, k_{\alpha_2} \cdots k_{\alpha_L}. \tag{6.23}$$

Note that $A^{L-1}$ does *not* couple to $PV$ states.

So far the physical content of the formulation is the same as outlined in Section 4 in terms of the direct term alone. The relativistic extension of the

---

* Note that the $A^L PP$ and the $B^L PP$ couplings vanish identically.

couplings, which simultaneously entails at least a partial inclusion of the recoil term in a special combination, is almost identical to the baryon case which amounts to a replacement of the isolated factors $\mathbf{k}^2$ in (6.21)–(6.23) by the invariant quantity $k_\mu k_\mu$ which equals $(-\mu^2)$ on the energy shell. As to the extension of the tensor indices from three to four dimensions, this is immediately achieved in (6.21), (6.22) and the second term of (6.23). For the first term of (6.23), we first note the correspondence

$$\varepsilon_{ijk} \Rightarrow \varepsilon_{ijk4}; \quad A^{L+1} \Rightarrow iM^{-1}\partial_4 A^{L+1},$$

which immediately leads to a Lorentz invariant $A^{L+1}PV$ coupling. The necessary generalizations of (6.21)–(6.23) are now respectively

$$\left[ A^{L+1}_{\mu\mu_1\cdots\mu_L} k_\mu k_{\mu_1} + (-\mu^2) \sqrt{\frac{L(L+1)}{2(2L+1)}}\, A^{L-1}_{\mu_2\cdots\mu_L} \right] k_{\mu_2} \cdots k_{\mu_L} P; \quad (6.24)$$

$$B^L_{\mu_1\cdots\mu_L} k_\mu W_\mu k_{\mu_1} \cdots k_{\mu_L}; \quad (6.25)$$

$$-M^{-1}\partial_\varrho A^{L+1}_{\mu\mu_1\cdots\mu_L} k_{\mu_1} \cdots k_{\mu_L} \varepsilon_{\mu\nu\lambda\varrho} k_\nu V_\lambda$$

$$+ \sqrt{\frac{L}{L+1}}\,(k_\mu V_\mu k_{\mu_1} + \mu^2 V_{\mu_1}) A^L_{\mu_1\cdots\mu_L} k_{\mu_2} \cdots k_{\mu_L}. \quad (6.26)$$

These expressions must now be multiplied by a form factor whose structure is again governed by considerations similar to the baryon case, except for certain additional remarks necessary for the $PP$ modes of decay of these mesons. These remarks concern the choice of the final meson which must be considered "elementary". As we have seen in the baryon case the determination of the compensating dimension to be associated with the various momentum factors in the interactions (6.24)–(6.26), depends sensitively on this choice, since the mass differences are large. We choose the convention that the *heavier* of the two $P$-mesons be considered elementary, as the opposite choice would give unreasonably large values for the modes $\eta\pi$, $K\pi$ etc., compared with experiment.*

This provides us with the factor $\mu^{-L-1}$, where $\mu$ is the mass of the $P$-meson for $VP$ modes, and that of the *heavier $P$-meson* for $PP$ modes of decay. This should be multiplied by the Weisskopf factor $\sqrt{\mu/m\pi}$ as in other hadronic

---

* In an earlier work, Mitra and Srivastava had taken the opposite convention. But this effect was considerably masked, partly by the strong Gaussian dependence of the form factors, and partly by certain ad hoc mass suppression factors used in that paper.

modes. Further, as was seen in Section 2.2, the relativistic normalization for the $Q\bar{Q}$ mesons in the initial and final states provides the additional factor $(2M)$ where $M$ is the mass of the initial (decaying) meson. Finally, the replacement of "momenta" by the "velocities" as in the baryon case provides the necessary mechanism for toning down the otherwise strong momentum dependence of the decay widths. Collecting all these points, the form factor to multiply the meson couplings (6.24)–(6.26) has the structure

$$f_L'(k^2) = g_L' \mu^{-L-1} \sqrt{\mu/m_\pi} \cdot (\mu/\omega_k)^{L\pm 1} (2M), \qquad (6.27)$$

where $g_L'$ is a dimensionless quantity, representing the only adjustable parameter governing the magnitudes of the supermultiplet transitions. Note that except for the difference represented by the last factors in (6.18) and (6.27), which arise necessarily from the basic difference in relativistic normalizations for fermion and boson fields, these empirical formulae for the form factors have almost identical structures for the two cases. There is no obvious connection between the quantities $g_L$ and $g_L'$ in (6.18) and (6.27), both of which must be regarded as free parameters in the absence of a more specific model. It is interesting to note however, that for $L^P = 0^+ \to 0^+$ transitions, the quantities $g_0$ and $g_0'$ *are equal*, as has already been seen in Section 2. Indeed, the results of Section 2 for the **56** baryon and **36** meson couplings can be summarised in a unified manner in the following structures for the respective form factors:

$$f_0(k^2); f_0'(k^2) = [g_0; g_0'] \mu^{-1} \sqrt{\mu/m_\pi} \left[ \sqrt{M/m}; 2M \right], \qquad (6.28)$$

where the constants are directly expressible in terms of the $Q\bar{Q}P$ coupling constant $f_q$ as:

$$g_0 = g_0' = f_q \approx \sqrt{4\pi \times 0.03}, \qquad (6.29)$$

since for these "self-coupling" cases the form factor reduces only to a normalization integral. The experimental tests of (6.28) have already been discussed in Section 2. The tests of (6.18) and (6.27) which are discussed in the next section for $L > 0$ hadron resonances will be seen to bring out the essential unity of description for the two cases in terms of a common form of parametrization.

## 6.5  Self-couplings

For completeness we list some examples of self-couplings among hadron states of the same $L^P$ in terms of the tensor formalism outlines in this section. From (4.21) and (4.22), the *orbital* structures of the matrix elements for both meson and baryon cases, are described by the coupling:

$$B^{LM*}_{\alpha_1\cdots\alpha_L} B^{LM'}_{\alpha_1\cdots\alpha_L} = \delta_{MM'}, \tag{6.30}$$

where the $B^L$-field is normalized by (6.5) and (6.6). These matrix elements are multiplied by spin matrix elements (6.1) or (6.19) for baryons or mesons respectively. Proceeding as in (6.19) and (6.20) for the C.G. reductions, the couplings between $(B^L, A^L)$ or $(A^{L+1}, A^{L-1})$ etc. fields are easily written down together with their relativistic extension. For example the $(A^L, A^{L-1})$ coupling which is appropriate for $D \to \pi_V + \pi$ (or $\delta + \pi$), is of the form

$$\sqrt{\frac{L^2}{2\,(2L+1)}}\; k_{\mu_1} A^L_{\mu_1\cdots\mu_L} A^{L-1}_{\mu_2\cdots\mu_L}. \tag{6.31}$$

Similar expressions hold for the couplings between baryons within the same supermultiplets to simulate decays like $\Sigma(1767) \to \Lambda(1520) + \pi$. For all these cases the form factors are given by (6.28), which is free from parameters.


## 7  COMPARISON OF IMPROVED MODEL WITH EXPERIMENT

The improved form of parametrization proposed in Section 6 stands a much better comparison with experiment than would have been worth while with the earlier analysis of Section 5, since the former depends on appreciably fewer parameters than the latter. For the baryons, the cases of partciular interest are the decays

$$(\mathbf{70}, 1^-) \to (\mathbf{56}, 0^+) \quad \text{and} \quad (\mathbf{56}, 2^+) \to (\mathbf{56}, 0^+),$$

since these are replete with a large number and variety of decay modes. Several interesting cases of self couplings within $(\mathbf{70}, 1^-)$ and $(\mathbf{56}, 0^+)$ have already been discussed in Section 2. For mesons, the only transitions for which enough data are available are those involving the nonets $L^P = 1^+$ to $L^P = 0^-$ for the higher nonets, the data are too few to be meaningfully systematized.

### 7.1  Baryon decays from $(\mathbf{70}, 1^-)$

Table 5 is a sufficiently representative list of decays from the negative parity resonances whose quantum number assignments are indicated in Table 3. The list includes several modes involving a wide range of masses which should provide a fairly sensitive test of the form factor structure (6.18),

**Table 5** $(70, 1^-)$ decays from some familiar states. The experimental data are partly from Rosenfeld *et al.* (Ref. 51) and partly from Tripp (Ref. 101)

| $SU(3)$ decay | Mode | $\Gamma$ (Theory) (MeV) | | $\Gamma$ (Expt.) (MeV) | $\Gamma$ (Dobson) (MeV) |
| --- | --- | --- | --- | --- | --- |
| | | $8q$ | $8d$ | | |
| $(8)_{5/2^-} \to (8)_{1/2^+}$ | $N(1680) \to N\pi$ | 20.8 | | 68.0 | |
| | $\to N\eta$ | 6.1 | | $\lesssim 4.0$ | |
| | $\to \Lambda K$ | 0 | | $< 2.7$ | |
| | $\Lambda(1830) \to \Sigma\pi$ | 56.7 | | $33 \pm 6$ | |
| | $\to N\overline{K}$ | | | $6 \pm 1$ | |
| | $\Sigma(1770) \to N\overline{K}$ | 48.3 | | $44 \pm 8$ | 23.1 |
| | $\to \Lambda\pi$ | 17.8 | | $14 \pm 6$ | 50.2 |
| | $\to \Sigma\eta$ | 0.1 | | 0.5 | 0.08 |
| | $\to \Sigma\pi$ | 10.0 | | 2.0 | 20.9 |
| $(8)_{3/2^-} \to (8)_{1/2^+}$ | $N(1518) \to N\pi$ | 24.7 | 164.1 | 62.0 | 116.0 |
| | $\to N\eta$ | 0.3 | 0.12 | $\sim 0.6$ | 0.02 |
| | $\Lambda(1690) \to \Sigma\pi$ | 52.7 | 21.9 | $35 \pm 15$ | |
| | $\to N\overline{K}$ | 0 | 153.8 | $11 \pm 4$ | |
| | $\Sigma(1660) \to N\overline{K}$ | 42.0 | 4.4 | 5.0 | 1.5 |
| | $\to \Sigma\pi$ | 10.7 | 111.3 | 25.0 | 15.9 |
| | $\to \Lambda\pi$ | 20.9 | 8.7 | $14 \pm 2$ | 12.4 |
| $(10)_{3/2^-} \to (8)_{1/2^+}$ | $\Delta(1691) \to N\pi$ | | 31.6 | 37.0 | 158.3 |
| | $\Xi^*(1816) \to \Xi\pi$ | | 7.1 | 1.6 | 1.9 |
| | $\to \Lambda\overline{K}$ | | 5.7 | 10.4 | 1.6 |
| | $\to \Sigma\overline{K}$ | | 1.9 | $\sim 0.3$ | 11.6 |
| $(1)_{3/2^-} \to (8)_{1/2^+}$ | $\Lambda(1520) \to N\overline{K}$ | | 10.0 | 7.2 | 3.2 |
| | $\to \Sigma\pi$ | | 85.4 | 7.2 | 3.1 |
| $(8)_{1/2^-} \to (8)_{1/2^+}$ | $N(1710) \to N\pi$ | 56.0 | 159.3 | 240.0 | |
| | $N(1550) \to N\pi$ | 47.3 | | 52.0 | |
| | $\to N\eta$ | 80.6 | | 91.0 | |
| | $\Lambda(1670) \to N\overline{K}$ | 0 | 434.3 | $3 \pm 0.5$ | |
| | $\to \Sigma\pi$ | 45.2 | 18.8 | $10 \pm 2$ | |
| $(10)_{1/2^-} \to (8)_{1/2^+}$ | $\Delta(1640) \to N\pi$ | | 18.4 | 54.0 | |
| $(1)_{1/2^-} \to (8)_{1/2^+}$ | $\Lambda(1405) \to \Sigma\pi$ | | 59.1 | 50.0 | |
| $(8)_{5/2^-} \to (10)_{3/2^+}$ | $\Sigma(1770) \to \Sigma^*\pi$ | 25.2 | | 13.3 | 3.1 |
| $(8)_{3/2^-} \to (10)_{3/2^+}$ | $N(1518) \to \Delta\pi$ | 54.6 | 91.0 | 50.0 | 20.1 |
| | $\Sigma(1660) \to \Sigma^*\pi$ | 8.8 | 14.7 | 8.0 | 3.4 |

whose variation with the meson masses and momenta is particularly rapid. The only free parameter is the quantity $g_1$ of (6.18) whose input value has been chosen as

$$g_1^2/4\pi \approx 0.14 \qquad\qquad (7.1)$$

to give the best overall fit to some of the well-known modes. Since there is no other adjustable parameter, the quality and range of agreement with experiment must be considered surprisingly good, though there exist certain notarious cases of disagreement as well. The comparison also warrants some general conclusions.

First the model breaks $SU(3)$ not only in phase space, but in the couplings themselves, mainly through the factor $\mu^{-2}$ in the limit $\mathbf{k} \to 0$. This is particularly evident for $d$-wave decays where this factor is very explicitly present in the corresponding couplings (apart from the momentum factors). The breaking is not so strong for $s$-wave decays, since these amplitudes involve a compensating factor $(-\mu^2)$ in the numerators due to our special choice of the recoil contribution which manifests itself through the replacement $\mathbf{k}^2 \to (-\mu^2)$. Thus for $s$-wave decays, the $SU(3)$ breaking in the couplings is limited only to the factor $\sqrt{\mu/m\pi}$ while for the $d$-wave decays it is proportional to $\mu^{-3/2}$ (in the limit $\mathbf{k} \to 0$). Thus the model predicts *different degrees of SU(3) breaking for emissions in different waves*, the higher the wave, the bigger the breaking in the limit $\mathbf{k} \to 0$. This broad mass dependence has the consequence that for all $d$-wave decays the $\pi$-modes are large compared with the $K$ and $\eta$ modes, while the opposite is true of $s$-wave decays.

The prediction of enhanced $\eta$, $K$ modes in $s$-wave decays is of course borne out well by the experimental results. The best example is perhaps that of $N^*(1550) \to N\eta$ which is particularly large. Note that such a large width is totally incompatible with the assignment of $\eta$ as a $\lambda\bar\lambda$ composite, which would predict zero $N\eta$ widths for all $N^*$ decays. This fact would thus seem to rule out the ideal mixing angle between $\eta$ and $X^0$, though some mixing need not be excluded.

An interesting conclusion from Table 5 is that $N^*(1710)$ is likely be mainly in $\mathbf{8}_d$ and $N^*(1550)$ in $\mathbf{8}_q$, as the opposite assignments would predict quite the contrary magnitudes for the principal modes ($N\pi$ and $N\eta$) of their decays. This confirms the result of the earlier Mitra–Ross[42] analysis but is the opposite of the Dalitz conclusion on the basis of proximity of mass splittings due to a spin-orbit potential.[60] As pointed out earlier, it is reasonable to suppose that decays provide a more sensitive test of quantum number assignments than do the mass patterns due to simple symmetry breaking effects. A further test of this assignment would be furnished by the observation of the (energetically allowed) $\Lambda K$ mode from $N^*(1710)$ decay, since the quark model predicts a zero $\Lambda K$ width from an $\mathbf{8}_q$ state, and an appreciable width for the same mode from an $\mathbf{8}_d$ state. As yet experiment seems inconclusive on this point, but it is signifiacnt (from Table 5) that the $N\pi$ mode fails to saturate the total width of $N^*(1710)$. Such a selection rule is similar to that of Moorhouse[100] which forbids photoproduction of spin quartet states in $(\mathbf{70}, 1^-)$, through the $E1$ transition, since a spin-flip is involved in the process. On the whole, the $s$-wave decay predictions are extremely good on this model wish furnishes strong support for the ($SU(3)$-breaking) structure of the form factor for such cases. Even the decay $\Lambda(1405) \to \Sigma\pi$ is rather well reproduced, considering the fact that the model predicts unusually large values for most other $\Sigma\pi$ modes.

For the $d$-wave decays, the model predicts a stronger $SU(3)$ breaking effect, which is proportional to $\mu^{-3/2}$ in the $\mathbf{k} \to 0$ limit. This mass dependence seems to be well-confirmed for the decays of $J^P = 5/2^-$ resonances whose widths are very well reproduced (except for $\Sigma\pi$ modes). However, the decays of $J^P = 3/2^-$ states are not in such good accord. This is due partly to the structure of the couplings for "unstretched" states for which $J^P \leqslant 3/2$ and partly to a lack of proper knowledge of configuration mixing for such states. As to the structure of the couplings for unstretched states, these involve the factor $\boldsymbol{\sigma} \cdot \mathbf{k}$ in the non-relativistic form [see Eqs. (6.13) and (6.14)], unlike the couplings of "stretched" states like $J^P = 5/2^-$ which do not involve this factor. Now the factor $\boldsymbol{\sigma} \cdot \mathbf{k}$ in its relativistic ($i\gamma_5$) form, according to (6.15), brings along the unusually large factor $(M + m)$ on the mass shell in comparison to merely an extra factor $k_\mu$ for the couplings of states like $J^P = 5/2^-$. Thus the discrepancy for $J^P = 3/2^-$ cases is at least partly traceable to the very prescription for the relativistic generalization of a $\boldsymbol{\sigma} \cdot \mathbf{k}$ term. This limitation can be partially overcome through the recognition of configuration mixing possibilities for unstretched states. As noted in Section 5.2, these mixing effects can occur between $SU(3)$ structures, or spin structures, or both. For $N$-type states, $SU(3)$ mixing is forbidden, so one must depend only on spin-mixing. A good example is provided by the unusually large and small widths for the $N\pi$ modes of $N^*(1518)$ decay, on the $\mathbf{8}_d$ and $\mathbf{8}_q$ assignments respectively. It is therefore reasonable to suppose that at least a part of the discrepancy should be ascribed to a mixing with its more recent $J^P = 3/2^-$ counterpart at $N^*(1675)$ whose detailed properties are still largely unknown. For the strange members, $\Sigma$, $\Lambda$ and $\Xi$, these mixing effects can occur between different $SU(3)$ multiplets as well. Thus, as already noted in Section 5.2, the large $\Lambda(1520) \to \Sigma\pi$ decay could be partly due to its mixing with $\Lambda(1690)$. Such a mixing also helps the latter which otherwise shows a large $N\bar{K}$ mode. Unfortunately the model outlined in Section 6 does not give any concrete recipe for configuration mixing, so that these conclusions are largely qualitative. However, if these mixing effects are regarded as small, an examination of Table 5 would suggest that the model is capable of making fairly unambiguous assignments of most resonances to well-defined multiplets. We note in passing that $SU(3)$ mixing effects of the order of 15–20° have been suggested by various authors in recent times[101,102] to narrow down the discrepancies of the principal decay modes of $\Lambda(1520)$, $\Lambda(1690)$, etc., between their $SU(3)$ predictions and experimental values.

## 7.2   Higher baryon decays

The most important states in this category are the positive parity ones belonging to the $(\mathbf{56}, 2^+)$ supermultiplet. Note that in this case $SU(3)$ mixing in much less feasible than for $(\mathbf{70}, 1^-)$ states, since such mixing is now confined only to $\Sigma$-type states. Further, since the $\mathbf{56}$ representation gives only spin-doublet states for $\mathbf{8}$ and quartet ones for $\mathbf{10}$, its very structure, unlike that of $\mathbf{70}$, is incompatible with any important spin-mixing effects. Thus the predictions in this case are expected to be cleaner than for the previous case, so that a comparison with data would provide a more direct test of the model. This comparison is shown in Table 6 *for* a typical set of decay modes, designed to bring out the mass and momentum dependence of the particles involved, similar to the $(\mathbf{70}, 1^-)$ case. The input value of the coupling constant $g_2$ is taken as

$$g_2^2/4\pi \approx 0.04 \pm 0.01, \tag{7.2}$$

which is designed to give a fit to some of the principal $5/2^+$ modes, especially $N^*(1688) \to N\pi$, to within about 10 per cent. As in Table 5 the overall fit again looks surprisingly good, and indeed better than the $SU(6)_W$ predictions. The agreement again brings out the general validity of the mass-effect on $SU(3)$ breaking which is now proportional to $\mu^{-1/2}$ and $\mu^{-5/2}$ for $p$ and $f$ wave decays respectively. There are however, certain discrepancies similar to the $(\mathbf{70}, 1^-)$ cases. Thus the predicted rates of $J^P = 7/2^+$ decays are geerally smaller than those of lower $J^P$ states for the same reason as mentioned earlier, viz., the effect of the baryons mass factor $(M + m)$ in the couplings of $J \leqslant 5/2$ states, and its absence for the stretched ($J^P = 7/2^+$) configuration. However, the *relative* widths even within the "stretched" category are well reproduced by the model. Moreover, the discrepancies for $\Sigma\pi$ modes are now much less than their ($\mathbf{70}$, $1^-$) counterparts.

The cases of the still higher baryon states of negative parity are too few for proper enumeration. However a reasonably good comparison seems to result from the ($\mathbf{70}$, $3^-$) if $N^*(2190)$ and $\Lambda(2100)$, each of $J^P = 7/2^-$, are both put in $\mathbf{8}_d$, and if the coupling constant $g_3$ is taken as

$$g_3^2/4\pi \approx 0.11. \tag{7.3}$$

**Table 6**   ($\mathbf{56}$, $2^+$) decays from some familiar states. The experimental data are mostly from Rosenfeld *et al.* (Ref. 51)

| $SU(3)$ decay | Mode | $\Gamma$ (Theory) (MeV) | $\Gamma$ (Expt) (MeV) | $\Gamma$ (Dobson) (MeV) |
|---|---|---|---|---|
| $(\mathbf{10})_{7/2^+} \to (\mathbf{8})_{1/2^+}$ | $\Delta(1950) \to N\pi$ | 28.8 | 88.0 | 146.6 |
| | $\to \Sigma K$ | 5.9 | seen | 4.1 |
| | $\Sigma^*(2020) \to N\overline{K}$ | 10.5 | 14.2 | 38.0 |
| | $\to \Lambda\pi$ | 12.6 | 43.2 | 65.8 |
| | $\to \Sigma\pi$ | 7.5 | 10.8 | 29.2 |
| | $\to \Xi K$ | 1.3 | 2.4 | 1.4 |
| $(\mathbf{8})_{5/2^+} \to (\mathbf{8})_{1/2^+}$ | $N(1690) \to N\pi$ | 76.3 | 84.5 | 96.6 |
| | $\to \Lambda K$ | 0.25 | 0.17 | 0.05 |
| | $\to N\eta$ | 1.6 | 1.95 | 0.3 |
| | $\Lambda(1816) \to N\overline{K}$ | 38.4 | $46.6 \pm 3.2$ | 42.1 |
| | $\to \Sigma\pi$ | 6.5 | $8.1 \pm 0.6$ | 14.4 |
| | $\to \Lambda\eta$ | 0.6 | $0.74 \pm 0.07$ | 0.4 |
| | $\Sigma(1910) \to N\overline{K}$ | 2.4 | 4.8 | 3.8 |
| | $\to \Lambda\pi$ | 3.7 | 6.0 | 21.0 |
| | $\to \Sigma\pi$ | 28.8 | 1.8 | 30.7 |
| | $\to \Sigma\eta$ | 0.85 | | 0.6 |
| | $\Xi(2020) \to \Sigma\overline{K}$ | 42.5 | | |
| | $\to \Lambda\overline{K}$ | 10.5 | | |
| | $\to \Sigma\pi$ | 2.6 | | |
| $(\mathbf{10})_{5/2^+} \to (\mathbf{8})_{1/2^+}$ | $\Delta(1913) \to N\pi$ | 46.3 | 56.0 | |
| $(\mathbf{10})_{1/2^+} \to (\mathbf{8})_{1/2^+}$ | $\Delta(1934) \to N\pi$ | 113.8 | 10.2 | |
| $(\mathbf{8})_{3/2^+} \to (\mathbf{8})_{1/2^+}$ | $N(1863) \to N\pi$ | 64.7 | 63 | |
| $(\mathbf{8})_{5/2^+} \to (\mathbf{10})_{3/2^+}$ | $\Lambda(1816) \to \Sigma^*\pi$ | 103 | $8.1 \pm 0.6$ | 7.4 |

## 7.3   Universal coupling of Regge recurrences

A very interesting consequence of this model relates to the magnitudes of the coupling constants governing the various supermultiplet transitions. Thus a comparison of the magnitudes (6.29) and (7.2) for $g_0$ and $g_2$ respectively shows such a close proximity that it is most unlikely to be a mere chance coincidence. On the other hand, since according to our traditional ideas, the $(\mathbf{56}, 2^+)$ states represent the first Regge recurrence of $(\mathbf{56}, 0^+)$, it is more reasonable to interpret the near equality of $g_0$ and $g_2$ in terms of a universal coupling for the Regge trajectory which contains the $\mathbf{56}$ baryons of positive parity. To pursue this point further, it is interesting to compare the predictions of this model for the still higher resonance or positive parity. In this regard, the only available data are in the $\Delta$-type states $\Delta(1920)$, $\Delta(2420)$, $\Delta(2850)$ and $\Delta(3230)$ of successive $J^P$ values $7/2^+$, $11/2^+$, $15/2^+$ and $19/2^+$, respectively each of which has a well-defined $N\pi$ mode. If these states are regarded as successive Regge recurrences of $\Delta(1236)$, with $L = 2^+$, $4^+$, $6^+$ and $8^+$ respectively, the model predicts a very simple result for their $N\pi$ decay widths. For this purpose it is easy to derive the following general formula for $\Delta \to N\pi$ decay!

$$\Gamma_{\Delta \to N\pi} \approx (g_L^2/4\pi)\,[(L + 1)!/(2L + 3)!!]\,8k\,(M + m - \omega_k)/m, \quad (7.4)$$

where the "stretched" value $J = L + 3/2$ has been used, and certain obvious simplifications based on the ultra relativistic approximation for the pion energy, have been performed. If now the same value (7.2) is used for all the coupling constants $g_L\,(L = 2, 4, 6, 8)$ in (7.4), this formula predicts the following values for the widths of these $\Delta$-type states of $L = 2, 4, 6, 8$ respectively (in MeV)

$$\Gamma_{\Delta \to N\pi} = 28.8,\ 9.2,\ 2.8,\ 0.75 \quad (7.5)$$

to be compared with the corresponding experimental values

$$88.8,\ 34.1,\ 12.5,\ 2.4. \quad (7.6)$$

These figures show that, apart from an over all reduction factor of $0.3 \pm 0.04$, the relative magnitudes are in excellent agreement with experiment, thus strengthening the assertion of a uinversal coupling of these Regge recurrences.*

---

* As noted earlier, the decau modes of stretched states are somewhat underestimated in relation to these of unstretched states. The reduction factor between (7.5) and (7.6) merely represents the effect of parametrizing the coupling constant $g_2$ by tuning it to the decay modes of unstretched states of $L^P = 2^+$.

For negative parity states, the data are too inadequate for a corresponding conclusion, although the proximity of (7.1) and (7.3) may not again be a coincidence.

### 7.4   Meson decays from $L > 0$ nonets

The data on higher mesons are limited mostly to the $L^P = 1^+$ states which are listed in Table 1. The information on the $K^*$ states is still very incomplete except for $K_V(1420)$ of $J^P = 2^+$. For the non-strange mesons, the $SU(3)$ mixing effects are confined mainly to the isosinglet members. In Section 5.5 we have already noted certain evidences in favour of the ideal mixing angle between $(f_0, f_0')$ and perhaps also between $\sigma$ and $S^*(1070)$. For $(\eta, X^0)$ on the other hand, it is unlikely that $\eta$ is made up of $\lambda\bar{\lambda}$, as the evidences on certain $N\eta$ modes from $N$-type resonances suggest. Table 7 is a representative collection of results on several decay modes of the better known $L^P = 1^+$ mesons into $PP$ and $PV$ systems according to the coupling scheme outlined in Section (6.3). It is not a complete list but includes a sufficient varity of decay modes designed to bring out the rather sensitive mass and momentum structures of the form factor (6.27). The hypothesis of the ideal mixing angle for $(\eta, X^0)$ is tested through a comparison of results of pure $SU(3)$ assignments with those of the mixed assignment. The only adjustable parameter is $g_1'$ which may be estimated from the rate for a well-defined mode, such as $A_2 \to \varrho\pi$ to give

$$g_1'^2/4\pi \approx 0.08. \tag{7.7}$$

The overall agreement may be seen to be at least as good as, or perhaps somewhat better than, the baryon case. Here $f_0$ and $f_0'$ are assumed to be made up of non-strange and strange quarks respectively, and likewise for $\sigma$, $S^*(1070)$. For the $(\eta, X^0)$ states the table shows that, apart from evidences due to the baryonic modes, the singlet-octet assignments actually works better than the ideal mixing angle. This result is in apparent discord with a pure $SU(3)$ analysis[53], where the only phase space correction is taken in the factor $k^{2l+1}$. The reason for this difference from the results of $SU(3)$ lies in the recognition that the present model of meson couplings (i) breaks $SU(3)$ in the $k \to 0$ limit, according to the $\mu^{1/2}$ and $\mu^{-3/2}$ laws for $s$- and $d$-wave couplings respectively, and (ii) has a strong damping factor with momentum to tone down the effect of $k^{2l+1}$. It is reasonable to assume that the couplings of both types of hadrons should exhibit qualitatively similar features, so that if these features are more or less valid for baryons, they are,

**Table 7** Decays from $L = 1$ to $L = 0$ states for mesons. The ideal mixing angle is used for $f_0(1260)$ and $f_0'(1514)$. For $(\eta, \chi^0)$, both the ideal mixing angle as well as a pure octet assignment of $\eta$ are used. $S^*(1070)$ is assumed to be in the octet

| Particle | $I^G(J^P)$ | Mode | $\Gamma_{\text{th}}$ (MeV) | $\Gamma_{\text{expt}}$ (MeV) |
|---|---|---|---|---|
| $\pi_V(1016)$ | $1^-(0^+)$ | $K\bar{K}$ | 40 | $25 \pm 5$  (only mode seen) |
| | | $\eta\pi; \eta_8\pi$ | $0; 24.3$ | $\lesssim 20$ |
| $S^*(1070)$ | $0^+(0^+)$ | $\pi\pi$ | 55.0 | $\lesssim 56$ |
| | | $K\bar{K}$ | 24.8 | $\gtrsim 24.0$ |
| $A_1(1070)$ | $1^-(1^+)$ | $\varrho\pi$ | 65.0 | $\sim (80 \pm 35)$ |
| $B(1220)$ | $1^+(1^+)$ | $\omega\pi$ | 129.0 | $123 \pm 16$ |
| $f_0(1260)$ | $0^+(2^+)$ | $\pi\pi$ | 75.1 | large |
| | | $K\bar{K}$ | 8.1 | $\sim 3.6$ |
| $f_0'(1514)$ | $0^+(2^+)$ | $K\bar{K}$ | 76.2 | $52.5 \pm 16.5$ |
| | | $K^*\bar{K} + \bar{K}^*K$ | 17.6 | $7.3 \pm 2.3$ |
| | | $\pi\pi$ | 0 | $\lesssim 10$ |
| $A_2(1300)$ | $1^-(2^+)$ | $\varrho\pi$ | 82 | $77 \pm 8.5$ |
| | | $K\bar{K}$ | 11.2 | $2.2 \pm 1.0$ |
| | | $\eta\pi; \eta_8\pi$ | $0; 8.8$ | $10.9 \pm 1.2$ |
| | | $X^0\pi; \eta_0\pi$ | $1; 9.0$ | $\sim 1$ |
| $K_V(1420)$ | $\frac{1}{2}(2^+)$ | $K\pi$ | 42.3 | $45 \pm 3$ |
| | | $K^*\pi$ | 30.4 | $29.3 \pm 2.0$ |
| | | $K\varrho$ | 9.4 | $10 \pm 1$ |
| | | $K\omega$ | 2.7 | $3.0 \pm 1.0$ |
| | | $K\eta; K\eta_8$ | $17.2; 2.8$ | $1.8 \pm 0.1$ |
| $\delta(960)$ | $1^-(0^+)$ | $\eta\pi; \eta_8\pi$ | $0; 22.7$ | $\sim 5$ |

unlikely to be violated in the meson case. On the other hand, the equally good agreement shown with the data, using exactly similar form factors for baryon and meson cases, should perhaps be interpreted as "experimental evidence" of this phenomenological coupling scheme.

The only "bad" case in Table 7 is the $\delta \to \pi\eta$ mode which is predicted to be 0 MeV in the model, though it is very small experimentally. A possible way out might lie in the recognition that the proximity to $\delta(960)$, of the neighbouring $K\bar{K}$ channel, called $\pi_V(1003)$[51], could strongly affect the decay rate of $\delta \to \eta\pi$.

Polarization of the $B$-meson in $B \to \pi\omega$ decay is still a problem in this model which claims to take at least partial account of the recoil effect. For, the $B\pi\omega$ coupling (6.25) still predicts the angular distribution $\cos^2 \chi$, which, as already noted in Section 5.5, is in disagreement with experiment. This should be interpreted as necessitating the addition of an "extra recoil term" having the structure (5.12). This would seem to suggest that the partial inclusion of the recoil contribution through the replacement $\mathbf{k}^2 \to (-\mu^2)$ in some of the "direct term" couplings is probably not enough, and should be supplemented by additional prescriptions.

For the higher meson states the only cases where some definite prediction is possible in this model are (i) the $\varrho_V(1650)$ meson[86] whose total width of 120 MeV is mainly in the $\pi$–$\pi$ mode, and (ii) the $\varrho(1700)$ meson[103] with a well-established $\pi\omega$ mode representing about 20 per cent of a total width of 110 MeV. According to Johnston $et$ $al.$[103], these two mesons are essentially the same. A recent analysis by Dalitz[61] of the angular distributions of these modes puts $\varrho(1700)$ at $J^P = 3^-$ and $\varrho_V(1650)$ at $J^P = 3^-$ or $1^-$. If this meson is regarded as the first Regge recurrence of the $\varrho$-meson, it is reasonable to take its $L$-excitation as $L^P = 2^-$, in which case the model predicts the ratio of the $\pi\omega$ mode of $\varrho(1700)$ to the $\pi\pi$ mode of $\varrho_V(1650)$ as

$$\Gamma(\pi\omega)\,\Gamma(\pi\pi) = 0.2 \text{ or } 0.0, \qquad (7.8)$$

according as $\varrho_V$ has $J^P = 1^-$ or $3^-$ respectively. The experimental value is about 0.2, which somewhat favours the assignment of $\varrho_V(1650)$ to $J^P = 1^-$ in this model. The absolute rate for $\varrho(1700) \to \pi\pi$ leads to the following estimate for the coupling constant $g_2'$

$$g_2'^{2}/4\pi \approx 0.08, \qquad (7.9)$$

under the assumption of $J^P = 1^-$. This estimate of $g_2'$ compares unfavourably with the value (6.29) of $g_0'$, though this would not be expected on the assumption of universal coupling of the Regge trajectory for the $\varrho$-meson. A part of this discrepancy under the $J^P = 1^-$ assignment could perhaps be ascribed to the possible effect of $radial$ excitations whoch $may$ compete with the effect of the orbital excitation $L^P = 2^-$. Such a possibility cannot however be discussed quantitatively within this model. Freund[104] has recently suggested a theory of radially excited mesons which predicts the ratios of $1 : \frac{18}{35} : \frac{2}{3}$ for the decay rates of $\varrho_V(1650)$ or $\varrho(1700)$ under the respective assumptions of radial excitation, and the assignments $3_{D_3}$ and $3_{D_1}$ for this state.

We conclude this section with the remark that the model of Section 6 works rather well for both mesons and baryons up to $L = 1$ and $L = 2$ respectively. Comparisons for higher modes of excitation are limited by lack of knowledge about the decay characteristics of the higher energy resonance of both types of hadrons. However, within this limited range of $L$-excitations, the results are very suggestive of universal couplings of Regge trajectories, *separately* for "even" and "odd" parity baryons. The model also makes specific predictions for the mode of $SU(3)$ breaking due to the mass of the emitted meson for individual waves of emission, and these predictions are in fairly good accord with experiment.

## 8  SUMMARY AND CONCLUSION

We have tried to outline the theory and application of a semi-quantitative phenomenological model of hadron couplings with special reference to the type of information that it bears on the decay widths of various resonances. The most important use of this study lies in its great sensitivity to the quantum number assignments for the large variety of resonance that are being discovered almost daily. While mass formulae are practically insensitive to differences in $SU(3)$ and/or $J^P$ assignments, the decay modes can vary drastically from one assignment to another. Even small mixing effects between different multiplets (spin or $SU(3)$) can make big differences to the decay widths.

An important consequence of this general analysis is the more or less firm establishment of the supermultiplet classification based on $SU(6) \times O(3)$ group, which seems to accommodate a large number of baryons and mesons in quite a convincing manner, as judged from the harmony of the predicted decay patterns with experiment. Such a framework is therefore reliable for the assignments of quantum numbers for subsequent resonances on similar lines. The fairly economical size of the $SU(6) \times O(3)$ helps in keeping open the possibility of its imbedding in a bigger (presumably non-compact) group, without having to make any early choice in this regard at the present stage when the data are still too inadequate to warrant a more complete theory.

In more specific terms, the coupling scheme yields a number of semi-quantitative results. Being a model of broken $SU(6) \times O(3)$, where the breaking occurs mostly in phase space, it finds a ready formulation in terms of quarks. However it hardly has to make use of any dynamical property of quarks, since the only ingradients which go into the formulation are the quark labels and at most certain overlap integrals over the quark wave functions which are best parametrized as such. Even without going into any detailed model for evaluating these overlap integrals, the coupling scheme (which is based on the additivity property for individual quark transitions), can distinguish between the algebraic structures of terms arising from the "direct" and "recoil" transitions respectively. The "recoil" transition has the great advantage of exhibiting the correct threshold behaviour for emission in a given partial wave, a feature which makes it particularly useful for the description of *low* partial wave emissions. Thus the heavy meson $(\eta, K)$ enhancements in $s$-wave decays of hadrons finds a very convincing explanation in terms of the recoil term which is essential for reproducing the desired momentum behaviour. For example, the large $N\eta$ mode from $N(1550)$, or similar $\bar{K}K$ modes from $\pi_V(1003)$ and $S^*(1070)$ would not be possible to understand even qualitatively without such a mechanism. It is doubtful if a formal group theory would automatically explain these features without making a conscious effort to include such an input information. The recoil term is also helpful for understanding certain polarization phenomena such as in the process $B \to \pi\omega$ and $\omega \to 3\pi$.

The scheme leads to certain natural mixing effects which are mainly of two types (i) mixing among hadron $SU(3)$ multiplets and (ii) mixing among baryon multiplets with different internal spin structures (doublet and quartet). The role of the recoil term overlaps somewhat with that of mixing between multiplets, but cannot be completely substituted by the latter, especially for low wave decays of low energy resonances. Thus no amount of mixing can, *by itself*, explain the large $N\eta$ mode of $N(1550)$. For high partial waves, on the other hand, the role of the recoil term is more limited, and one must depend on mixing alone. A particularly interesting possibility for mesons is the ideal mixing angle which seems to work beautifully for several pairs like $(\omega, \phi)$, $(f_0, f_0')$, $(\sigma, S^*(1070))$ and possibly others in $J^P = 1^+$ states.

While $SU(3)$ breaking in such a scheme is considered mostly in phase space, a considerable insight into the structure of this breaking in the couplings themselves can be achieved through a more careful choice of the form factors on the basis of simple physical considerations. Thus if the basic picture of meson emission by individual quarks is taken to be correct, then the mass $(\mu)$ of the meson appears as a natural dimension in the coupling scheme which is characterised by the same degree of momentum ($k$) dependence as the partial wave of emission. This suggests a dimensionless combination ($k/\mu$) for each of the $k$-factors appearing in a certain coupling, which thus provides a definite mechanism for the breaking of $SU(3)$ in the limit of $\mathbf{k} \to 0$, since the C.G. coefficients in the couplings would now be multiplied by a certain power of ($\mu$) depending on the partial wave involved. A similar consideration provides some understanding of the structure of form factors if the parameter for damping the momentum dependence for large $k$ is again taken as $\mu$. For, it is reasonable to suppose that the form factors should be functions of "velocities" rather than momenta, as more formal group theories for couplings would seem to suggest. Such general considerations can be of great help not only in reducing the number of free parameters for a comparison of the general theory with experiment, but in deducing certain definite mechanisms for $SU(3)$ breaking due to the masses involved.

A specific model based on such considerations has been discussed in some detail in the article. The model is by no means unique, but nevertheless has some very desirable features. Thus it not only admits of a relativistically invariant generalization for a significant portion of the coupling, but is also able to incorporate the effect of the recoil term (which is supposed to provide for enhanced heavy meson modes) through the simple replacement of $\mathbf{k}^2$ in certain low wave terms by its relativistically invariant form $k_\mu k_\mu$ which equals $(-\mu^2)$ on the mass shell. The only non-Lorentz invariant (yet relativistic) feature of the model is in the structure of its form factor which is governed by considerations of effectively replacing the momenta by the velocities. With all these features the model shows a good deal of similarity with the more formal theories like $SU(6)_W$ though it lacks a complete theoretical basis possessed by the later. However, such a model, which must at most be regarded as a good guess for the parametrization of data, with a single adjustable parameter for an entire supermultiplet transition, seems to make for its lack of formal theoretical support, by providing very good agreement with the data for hadron resonances, the agreement being definitely better than those of more formal theories at the present stage. A more interesting fact about this phenomenology is that is provides a unified description for *both* types of hadron decays with almost identical parametrizations of the two form factors. While this phenomenology should by no means be claimed as a substitute for a more formal theory, it should perhaps be emphasized that such considerations may well be useful for laying the groundwork for a future theory which must stay closer to experiment in order to be acceptable.

A very interesting consequence of the semi-quantitative model described above is that the magnitudes of the "coupling constants" describing super-multiplet transitions were strongly reminiscent of the idea of the universal couplings of Regge trajectories for hadrons. This seems to be well substantiated for positive parity baryons which have enough data for $L^P = 0^+$ and $L^P = 2^+$ states to check the numerical accuracy of the above statement. And since this conclusion is based on the use of the *concrete* couplings (including the form factors), it provides some confidence in the basic validity of such a form of parametrization. Even for negative parity baryons, the limited data for $L^P = 3^-$ states are in conformity with the idea of equality of the coupling constants for $L^P = 1^-$ and $3^-$ transitions to $L^P = 0^+$ states. As for mesons, the data for $L > 1$ are too few and too inaccurate to check on the veracity of such a hypothesis. However, the fact that two sets of coupling constants are indicated for baryons, *one* for positive and a different *one* for negative parity states, would seem to suggest that the same would probably be true of mesons as well, when adequate data become available.

This last feature brings us somewhat closer to facing dynamical questions concerning the nature of quarks. For the experimental success of the above coupling scheme, which indicates two sets of universal baryon couplings for possitive and negative parity decays respectively, also indicates the *existence* of only two sets of baryon supermultiplets, viz $(\mathbf{56}, 2^+)$ and $(\mathbf{70}, (2 + 1)^-)$, together with their possible radial excitations. This list is considerably smaller than, e.g., one based on the harmonic oscillator scheme (see Table 2). It would be most interesting to check the assignments of future resonances on the basis of this simpler classification together with the decay patterns that this scheme provides for them, since it is at least not in disagreement with experiment so far. A remarkable feature of this scheme is that it provides only baryon states of "natural" parity viz., that the total $L$-value essentially carries the parity label. It seems unlikely that any simple group theory would give rise to only such representations for a $QQQ$ system, without being supplemented by some dynamical assumptions. On the other hand, as noted in Section 3, a simple dynamical model based only on *s-wave* $Q$–$Q$ interactions leads precisely to the existence of only such states, and none else, as those exhibiting *attractive Faddeev Kernels,* and hence of comparatively lower energy. Whether or not such a naive model is eventually found to have a deeper theoretical basis for the quark system may be an open question but it is clear that the simplification it provides in the classification of baryon systems by ruling out unnatural parity states, is non-trivial.* Even more important is the fact that this simplified picture has its roots, *not* so much in any theoretical model, as in the *experimental* results for decay widths.

Another dynamical manifestation, though less important than the above, lies in the important role of the quark recoil term, which would seem to suggest that the quarks do indeed take a more active part in the transitions than is merely indicated by their quantum number labels. In this spirit it becomes meaningful to speak of scattering and re-scattering of the emitted mesons by the other quarks of the hadron system, and thus obtain corrections to the above couplings which based on the addition of individual quark transition amplitudes. While any detailed theory in this regard may yet be a long way to come, it has been found that even a naive application of the Glauber theory[108] based on the eikonal approximation to high energy scattering, seems to yield better results for the couplings among the **8** and **10** multiplets of the **56** baryons,[109] without having to make use of any new parameters other than those required to fit $\pi - N$ scattering in the Glauber model.[110-111]**

A different picture which pushes the quark dynamics considerably to the background, is provided by the chiral model of $O(4)$ symmetry proposed recently by Gell–Mann and Zweig[64] as an extension of the more orthodox quark model to accommodate certain new states like $A_{1.5}$ and $A_{2l}$. In this model the quarks completely lose their individuality and give way to certain collective indices labelling the chiral representations of the $O(4)$ augmented by the $SU(3)$ labels. The consequences of this model, which gives rise to a large proliferation of hadron states rapidly increasing with $L$, do not seem to have been worked out in any great details as yet. Perhaps the future experimental status of the new states like $A_{1.5}$ and $A_{2l}$ would have a considerable bearing on the physical urgency of this model.

At the present state of experimental knowledge on hadrons, the quark

---

* This picture is consistent with the classification of states predicted by a quark-diquark ($Q$–$D$) model, advocated by Lichtenberg and collaborators[105,107], provided the diquark is considered to be an object of positive parity, along with the quark. However, one needs to postulate extra symmetric conditions to bring the results of the $Q$-D model on par with those of the $QQQ$ model.

** This is not to suggest that the high energy approximation for quark-meson scattering is literally valid for this application. However, the extensive parametrization of the form factors necessitated in any concrete application of the Glauber theory, reduces this question to one of not more than academic interest. (See Ref. (109).)

philosophy seems to be at the cross roads between two entirely uncorrelated schools of thought. If quarks are indeed dynamical objects, whether or not they exhibit entirely different properties in "free" and "bound" states, these would presumably have to be governed by a full-fledged dynamical theory of two and three particle systems. However, the answer to this basic question should preferably come from experiment. In this respect, indirect evidences such as the role of the "recoil term" and the "existence" of natural parity baryons, must be supplemented by many others of a similar nature, before the claims of the "dynamical" idea can be taken seriously. On the other hand, if the "mathematical" quark idea gains more ground through more convincing experimental evidences on the existence of unnatural charge conjugation states, group theory would presumably play a bigger role than formal dynamics in the formulation of a more complete theory. Perhaps experiments of a more discriminating character than available at present can hold the answer to this basic question.

## Acknowledgement

This review was undertaken at the suggestion of Professor Hadi Aly to whom the author expresses his grateful thanks. He is also indebted to D. K. Choudhury and D. L. Katyal for help in the preparation of this manuscript.

## References

1. H. Yukawa, *Proc. Phys. Math. Soc. Japan.* **17, 48,** (1935).
2. S. Sakata, *Prog. Theo. Phys.,* **16,** 686 (1956).
3. M. Gell–Mann, *Phys. Rev.,* **125,** 1067 (1962).
4. Y. Nee'man, *Nucl. Phys.,* **26,** 22 (1961).
5. F. Gursey and L. Radicati, *Phys. Rev. Letters,* **13,** 173 (1964); A. Pais, *Phys. Rev. Letters,* **13,** 175 (1964).
6. E. P. Wigner, *Phys. Rev.,* **51,** 106 (1937).
7. K. T. Mahanthappa and E. C. G. Sudarshan, *Phys. Rev. Letters,* **14,** 163 (1965).
8. A. Salam, R. Delbourgo and J. Strathdee, *Proc. Roy. Soc. A.,* **248,** 146 (1965).
9. P. Budini and C. Fronsdal, *Phys. Rev. Letters,* **14,** 968 (1965).
10. H. J. Lipkin and S. Meshkov, *Phys. Rev. Letters,* **14,** 670 (1965).
11. M. Gell-Mann and R. F. Dashen, *Phys. Letters,* **17,** 142, 145 (1965); K. Bardakci, J. M. Cornwall, P. G. O. Freund and B. W. Lee, *Phys. Rev. Letters,* **14,** 48, 264 (1965).
12. L. Michel, *Phys. Rev.,* **137,** B405 (1965); L. O'Raifertaigh, *Phys. Rev.,* **139,** B1052 (1965).
13. G. Feldman and P. T. Mathews, "Unitary, Causality and Statistics".
14. $\tilde{U}(12)$: R. Delbourgo, M. A. Rashid and J. Strathdee, *Phys. Rev. Letters,* **14,** 719 (1965); *SL* (6, *C*): C. Frousdal and R. White, *Phys. Rev.,* **151,** 1287 (1966); $SU(6)_W$: D. Horn, M. Kugler, H. J. Lipkin, S. Meshkov, J. C. Carter and J. Coyne, *Phys. Rev. Letters,* **14,** 717 (1965).

15. M.Gell–Mann, *Physics*, **1**, 63 (1964).
16. M.Gell–Mann, *Phys. Letters*, **8**, 214 (1964).
17. G.Zweig, C.E.R.N. preprint (1964).
18. R.H.Dalitz, in the *Proceedings of the Oxford Conference on Elementary Particles*, Rutherford Laboratory, Harwell, 1966.
19. R.H.Dalitz, in *High Energy Physics*, edited by M.Jacob and C.Dewitt, Gordon and Breach, New York, 1965.
20. G.Morpurgo, *Physics*, **2**, 95 (1965).
21. T.K.Kuo and L.Radicati, *Phys. Rev.*, **139**, B746 (1965).
22. A.N.Mitra, *Phys. Rev.*, **151**, 1168 (1966).
23. A.N.Mitra, *Ann. Phys. (N.Y).*, **43**, 126 (1967).
24. A.N.Mitra and D.L.Katyal, *Nucl. Phys.*, **B5**, 308 (1968).
25. A.N.Mitra, *Nuovo Cimento*, **56A**, 1164 (1968).
26. A.N.Mitra and R.Majumdar, *Phys. Rev.*, **150**, 1194 (1966).
27. R.H.Dalitz, Rapporteur Talk at the XIII International Conference on High Energy Physics at Berkeley (1966); Univ. of California Press, Berkeley (1967).
28. H.Bacry, J.Nuyts and L. van Hove, *Phys. Letters*, **9**, 279 (1964); Y.Hara, *Phys. Rev.*, **134**, B701 (1964); Z.Maki, *Prog. Theoret. Phys. (Kyoto)*, **31**, 331 (1964).
29. Y.Nambu in *Preludes in Theoretical Physics* edited by A. de Shalit, H.Feshbach and L. van Hove, North Holland Publishing Company Amsterdam, 1966, p.133.
30. M.Y.Hau and Y.Nambu, *Phys. Rev.*, **139**, B1006 (1965).
31. A.N.Tavkhelidze in *High Energy Physics and Elementary Particles*. Internation. Atomic Energy Agency, Vienna, 1965; p.763.
32. Y.Katayama, I.Umemura and E.Yamada, *Prog. Theoret. Phys. Kyoto Suppl.*, Yukawa No., 1965.
33. O.W.Greenberg, *Phys. Rev. Letters*, **13**, 598 (1964).
34. H.S.Green, *Phys. Rev.*, **90**, 270 (1953).
35. O.W.Greenberg and D.Zwanziger, *Phys. Rev.*, **150**, 1177 (1966).
36. A.N.Mitra and S.A.Moskowski, *Phys. Rev.*, **172**, 1474 (1968).
37. G.C.Joshi, V.S.Bhasin and A.N.Mitra, *Phys. Rev.*, **156**, 1572 (1967).
38. S.Das Gupta and A.N.Mitra, *Phys. Rev.*, **156**, 158 (1967).
39. C.Becchi and G.Morpurgo, *Phys. Rev.*, **149**, 1284 (1966); referred to as BM, in Section 2.
40. E.M.Levin and L.L.Frankfurt, *JETP Letters*, **2**, 65 (1965); H.J.Lipkin and F.Scheck *Phys. Rev. Letters*, **16**, 71 (1966).
41. J.D.Jackson, in *High Energy Physics*, edited by C.Dewitt and M.Jacob, Gordon and Breach, New York, 1965.
42. A.N.Mitra and M.Ross, *Phys. Rev.*, **158**, 1630 (1967).
43. R. van Royen and V.F.Weisskopf, *Nuovo Cimento*, **50A**, 617 (1967); referred to as V.W. in Section 2.
44. M.Gell–Mann, in *The Eightfold Way*, edited by M.Gell–Mann and Y.Nee'man, W.A.Benjamin and Sons, Inc, New York 1964.
45. M.Goldhaber, *Phys. Rev.*, **100**, 433 (1956).
46. See, e.g., M.Verde, in the *Handbuch der Physik*, Vol. **39**, Ed. by S.Flugge, Springer-Verlag, Berlin (1957); p.170.
47. M.A.B.Beg and A.Pais, *Phys. Rev.*, **137**, B1514 (1965).

48. W. Rarita and J. Schwinger, *Phys. Rev.*, **61**, 61 (1941).

49. P. A. Cook, *Nuovo Cimento*, **48A**, 570 (1967).

50. A. N. Mitra, Rutherford Laboratory Preprint, 1968 (August).

51. A. H. Rosenfeld *et al*, *Rev. Mod. Phys.*, **41**, Jan. (1969).

52. N. Zovko, *Phys. Letters*, **23**, 143 (1966); See, however, J. K. Kim, *Phys. Rev. Letters*, **19**, 1079 (1967).

53. H. Harari, Rapporteur talk on Resonances (theoretical) at the XIV International Conference on High Energy Physics, Vienna, 1968.

54. A Donnachie, Rapporteur Talk, loc. cit.

55. B. Rench, Rapporteur Talk, loc. cit.

56. R. D. Tripp, Rapporteur Talk, loc. cit.

57. R. Cool, G. Giacomelli, T. Kycia, B. Leontic, K. Li and A. Lundby, *Phys. Rev. Letters*, **17**, 102 (1966).

58. R. Abrams, R. Cool, G. Giacomelli, T. Kycia, B. Leontic, K. Li and D. Michel, *Phys. Rev. Letters*, **19**, 259 (1967).

59. C. Lovelace, in the *Proc. Heidelberg Conf. on Elementary Particles (Sept. 1967)*, North Holland Publishing Co, Amsterdam, 1968.

60. R. H. Dalitz, Introductory talk on Baryon Supermultiplets at the Irvine (California) Conference on $\pi$–$N$ Scattering. December, 1967.

61. R. H. Dalitz, at the Topical Conference on Meson Spectroscopy at the University of Philadelphia in May, 1968.

62. G. Ascoli, H. Crawley, D. Mortara and A. Shapiro, *Phys. Rev. Letters*, **20**, 1411 (1968).

63. M. Focacci, W. Kienzle, B. Levrat, B. Maglic and M. Martin, *Phys. Rev. Letters*, **17**, 890 (1966).

64. M. Gell–Mann and G. Zweig, at the XIV International Conf. on High Energy Physics, Vienna, 1968.

65. G. Karl and E. Obryk, Oxford Univ. Preprint (1968).

66. L. D. Faddeev, *Soviet Physics JETP*, **12**, 1014 (1961).

67. M. E. Rose, *Elementary Theory of Angular Momentum*, John Wiley and Sons Inc., New York, 1957.

68. H. J. Lipkin, in the *Proceedings of the Heidelberg International Conf. on Elementary Particles (Sept. 1967)*, North Holland Publishing Co, Amsterdam, 1968.

69. M. Gell-Mann and K. M. Watson, *Annual Reviews of Nuclear Science*, **4**, 219 (1954).

70. cf., S. L. Glashaw and A. H. Rosenfeld, *Phys. Rev. Letters*, **10**, 192 (1963).

71. S. R. Deans, W. G. Holladay and J. E. Rush, *Phys. Rev.*, (1969); to be published.

72. D. R. Divgi, *Phys. Rev.* (1969), in press.

73. H. J. Lipkin, H. R. Rubinstein and H. Stern, *Phys. Rev.* **161**, 1502 (1967).

74. D. L. Katyal and A. N. Mitra, *Phys. Rev.*, **169**, 1322 (1968).

75. S. Das Gupta and A. N. Mitra, *Phys. Rev.*, **159**, 1285 (1967).

76. D. Faiman and A. W. Hendry, *Phys. Rev.*, **173**, 1720 (1968).

77. H. J. Lipkin, *Phys. Rev.*, **159**, 1303 (1967).

78. P. N. Dobson, Jr., *Phys. Rev.*, **160**, 1501 (1967).

79. P. G. O. Freund, A. N. Maheshwari and E. Schonberg, *Phys. Rev.*, **159**, 1232 (1967).

80. S. L. Glashaw and R. Socolow, *Phys. Rev. Letters*, **15**, 329 (1965).

81. M. Elitzur, H. R. Rubinstein, H. Stern and H. J. Lipkin, *Phys. Rev. Letters*, **17**, 420 (1966).

82. A.N.Mitra and P.P.Srivastava, *Phys. Rev.*, **164**, 1803 (1967).

83. J.Uretsky, in *Lectures in Theoretical High Energy Physics*, John Wiley–Interscience, London, (1968), H.H.Aly, Ed., p. 285.

84. D.K.Choudhury, *Phys. Rev.*, **174**, 2101 (1968).

85. See, e.g. E.Malamud and P.E.Schlein, *Phys. Rev. Letters*, **19**, 1056 (1967). W.D. Walker *et al.*, *Phys. Rev. Letters*, **18**, 630 (1967).

86. e.g., D.Crenell, P.Hough, G.Kalbfleisch, K.Lai, J.Scarr, T.Schumann, I.Skillicorn, R.Strand, M.Webster, P.Baumel, A.Bachman and R.Lea, *Phys. Rev. Letters*, **18**, 323 (1967).

87. A.N.Mitra, Rutherford Laboratory Preprint RPP/A 44 (1968); also *Nuovo Cimento*. (in press).

88. cf., R.Blankenbecler and R.Sugar, *Phys. Rev.*, **168**, 1597 (1968).

89. C.Fronsdal, *Nuovo Cimento Suppl.*, **9**, 416 (1958).

90. H.Umezawa, *Theory of Quantized Fields*, North Holland Publishing Co, Amsterdam, 1956.

91. See, e.g., F.J.Dyson, *Symmetry Groups in Elementary Particles*, W.A.Benjamin and Sons, Inc., New York, 1966.

92. J.Schwinger, *Phys. Letters*, **24B**, 473 (1967).

93. B.W.Lee and H.T.Nieh, *Phys. Rev.*, **166**, 1507 (1968); (other references therein).

94. S.Weinberg and H.J.Schnitzer, *Phys. Rev.*, **164**, 1828 (1967); S.Weinberg, *Phys. Rev. Letters*, **18**, 507 (1967).

95. A.O.Barut and H.Kleinert, *Phys. Rev.*, **156**, 1546 (1967).

96. A.O.Barut and H.Kleinert, *Phys. Rev.*, **157**, 1180 (1967).

97. Y.Nambu, *Phys. Rev.*, **160**, 1171 (1967).

98. A.O.Barut and H.Kleinert, *Phys. Rev. Letters*, **18**, 754 (1967); A.O.Barut and K. C.Tripathi, *Phys. Rev. Letters*, **19**, 1081 (1967).

99. A.O.Barut and K.C.Tripathy, *Phys. Rev. Letters*, **19**, 918 (1967).

100. R.G.Moorhouse, *Phys. Rev. Letters*, **16**, 771 (1966).

101. R.D.Tripp, *Phys. Rev. Letters*, **21**, 1721 (1968).

102. R.H.Capps, *Phys. Rev. Letters*, **22**, 215 (1969).

103. T.Johnston, J.Prentice, N.Steenberg, T.Yoon, A.Garfinkel, R.Morse, B.Oh and W.Walker, *Phys. Rev. Letters*, **20**, 1414 (1968).

104. P.G.O.Freund, Radially excited mesons; I.C.T.P. preprint Trieste (1968).

105. D.B.Lichtenberg and L.J.Tassie, *Phys. Rev.*, **155**, 1601 (1967).

106. D.B.Lichtenberg, L.J.Tassie and P.J.Kaleman, *Phys. Rev.*, **167**, 1535 (1968).

107. D.B.Lichtenberg, Rutherford Laboratory Preprint (1968).

108. R.J.Glauber, in *Lectures in Theoretical Physics*, Vol. I, Ed. by W.E.Brittin and L. G.Dunham, Interscience Publishers, New York, 1958.

109. J.D.Anand, V.S.Bhasin and A.N.Mitra, *Phys. Rev.*, **172**, No. 5 (1968).

110. A.Deloff, *Nucl. Phys.*, **B2**, 597 (1967).

111. V.Franco and R.J.Glauber, *Phys. Rev.*, **142**, 1195 (1966).

## APPENDIX

In this Appendix we briefly outline the main steps necessary for obtaining some results quoted in Section 6. This method is a non-relativistic adaptation of Fronsdal's corresponding procedure for relativistic Rarita–Schwinger (R–S) fields. Consider first the simpler case of a three-dimensional tensor $B^L_{(\alpha)}$, $((\alpha) \equiv \alpha_1 \cdots \alpha_L)$ and a Pauli spinor $\chi$, whose properties are described in Section 6. We write the obvious identity

$$B^L_{(\alpha)} \otimes \chi = \chi^L_{\alpha_1 \cdots \alpha_L} + A S_L \sigma_{\alpha_1} \sigma_\alpha B^L_{\alpha \alpha_2 \cdots \alpha_L} \chi, \tag{A.1}$$

where

$$\chi^L_{\alpha_1 \cdots \alpha_L} = B^L_{(\alpha)} \otimes \chi - A S_L \sigma_{\alpha_1} \sigma_\alpha B^L_{\alpha \alpha_2 \cdots \alpha_L} \chi, \tag{A.2}$$

and $S_L$ is a symmetrizer for the indices $(\alpha_1, \ldots, \alpha_L)$. This decomposition is motivated by the requirement that the first and second terms of (A.1) should represent R–S fields of spins $J = L \pm \frac{1}{2}$ respectively. That the second term is an $L - \frac{1}{2}$ R–S field is clear from the fact that one of the indices of has been contracted with a Pauli spin operator. Therefore a suitable choice of the constant $A$ would ensure that the first term is a pure $(L + \frac{1}{2})$ R–S field. This is achieved by imposing the usual R–S condition

$$\sigma_{\alpha_1} \chi^L_{\alpha_1 \cdots \alpha_L} = 0 \tag{A.3}$$

on (A.2), which leads in a simple way to

$$A = L (2L + 1)^{-1}. \tag{A.4}$$

As for the normalizations of these $J = L \pm \frac{1}{2}$ fields, it may be seen directly that (A.2) is already normalised, since the projection operator

$$\Theta \equiv 1 - L (2L + 1)^{-1} S_L \sigma_{\alpha_1} \sigma_\alpha$$

necessary to extract this state from the full direct product $B^L_{(\alpha)} \otimes \chi$ already has its first term as unity. The normalization of the $J = L - \frac{1}{2}$ state is then obtained as follows. Define the normalized $J = L - \frac{1}{2}$ state $\chi^{L-1/2}_{\alpha_2 \cdots \alpha_L}$ which is explicitly dependent on $(L - 1)$ indices, as

$$\chi^{L-1/2}_{\alpha_2 \cdots \alpha_L} = N_L \sigma_\alpha B^L_{\alpha \alpha_2 \cdots \alpha_L} \chi. \tag{A.5}$$

The constant $N_L$ must now be determined from the condition that the second term of (A.1), viz.,

$$\phi^L_{(\alpha)} \equiv N_L^{-1} L (2L + 1)^{-1} S_L \sigma_{\alpha_1} \chi^{L-1/2}_{\alpha_2 \cdots \alpha_L}, \tag{A.6}$$

has the same normalization as (A.5), i.e.,

$$\| \phi^L_{(\alpha)} \|^2 = \| \chi^{L-1/2}_{\alpha_2 \cdots \alpha_L} \|^2. \tag{A.7}$$

A little algebra then yields

$$N_L^2 = L\,(2L + 1)^{-1}, \tag{A.8}$$

which is the result used in the explicit reduction (6.9) of $B_{(\alpha)}^L \otimes \chi$. The main point to be noted in connection with this derivation is that though an R–S spinor of $J = L - \tfrac{1}{2}$ only needs $(L - 1)$ tensor indices, as in (A.5), the same is being expressed in a higher dimensional space of $L$ indices, as in (A.6). Such an extended representation for an unstretched state like $J = L - \tfrac{1}{2}$ necessitates a non-trivial normalization like (A.8), unlike the case of the stretched state $J = L + \tfrac{1}{2}$ for which all the $L$-indices are required and the normalization is a trivial unity factor.

The reduction (6.12) of the product of a tensor $B_{(\alpha)}^L$ and R S spinor $\chi_\alpha$ can be achieved as a two-step process. The first step consists in the reduction of the product of two tensors $B_{(\alpha)}^L$ and $V_\alpha$ of ranks $L$ and 1 respectively, so as to give rise to a linear combination of tensors of ranks $L + 1$, $L$ and $L - 1$, which we denote by $A_{(\alpha)}^J$, $J = L, L \pm 1$. The second step is merely a repetition of the method outlined above for the reduction of the product $A_{(\alpha)}^J \otimes \chi$ ($\chi$ being a Pauli spinor), for each of the three values $0 = L \pm 1, L$. Incidentally, the first step already suffices for the problem of meson couplings, where the spin functions are merely scalars and vectors, in contrast to the Pauli and R–S spinors characterising baryon spin-functions.

The reduction of $B_{(\alpha)}^L \otimes V_\alpha$ is expressed by the identity

$$V_\alpha \otimes B_{(\alpha)}^L \equiv \frac{1}{L + 1}\,(V_\alpha B_{\alpha_1 \alpha_2 \cdots \alpha_L} + L S_L V_{\alpha_1} B_{\alpha \alpha_2 \cdots \alpha_L})$$

$$+ \frac{L}{L + 1}\,(V_\alpha B_{\alpha_1 \alpha_2 \cdots \alpha_L} - S_L V_{\alpha_1} B_{\alpha \alpha_2 \cdots \alpha_L}), \tag{A.9}$$

where the first term is symmetrical in all the $(L + 1)$ indices $\alpha, \alpha_1, \alpha_2, \ldots, \alpha_L$, while the second term is anti-symmetric (in turn) between $\alpha$ and the members of the aggregate $(\alpha_1, \alpha_2, \ldots \alpha_L)$. Since these two terms have opposite parities, it is clear that the first term corresponds to $J = L \pm 1$ and the second one to $J = L$. To project out the $(L \pm 1)$ states explicitly, we note that while the $J = (L + 1)$ state must be traceless in all the $(L + 1)$ indices $(\alpha, \alpha_1, \alpha_2, \ldots \alpha_L)$, the $J = L - 1$ state needs to be traceless in only $(L - 1)$ indices, so that the two additional indices must be involved only in an isotropic form (like $\delta_{\alpha \alpha_1}$). The equivalent statement for the spinor case is that the additional index for $J = L - \tfrac{1}{2}$ is involved only in the factor $\sigma_{\alpha_1}$.

The first term of (A.9) is re-expressible as $S_{L+1}V_\alpha B_{\alpha_1\alpha_2\cdots\alpha_L}$ where $S_{L+1}$ is the symmetrizer in all the $(L+1)$ indices. We now write

$$A^{L+1}_{\alpha\alpha_1\cdots\alpha_L} = S_{L+1}V_\alpha B_{\alpha_1\cdots\alpha_L} - S_{L+1}\left(c\delta_{\alpha\alpha_1}V_\beta B_{\beta\alpha_2\cdots\alpha_L} + d\delta_{\alpha_1\alpha_2}V_\beta B_{\beta\alpha\alpha_3\cdots\alpha_L}\right),$$

$$(A.10)$$

and determine the constants $c$ and $d$ from the requirement that $A^{L+1}$ be traceless in all its $(L+1)$ indices. This readily gives

$$c = d = L(2L+1)^{-1}. \qquad (A.11)$$

As in the case of $B^L_{(\alpha)}\otimes\chi$, this stretched state $A^{L+1}$ is clearly a normalized one. A normalized state of $J = L - 1$ is represented by the contraction

$$A^{L-1}_{\alpha_2\cdots\alpha_L} = N_S V_\beta B^L_{\beta\alpha_2\cdots\alpha_L}, \qquad (A.12)$$

where $N_S$ must be determined from a condition similar to one employed for (A.5). Now the terms of rank $J = L - 1$ in the reduction of the symmetric part of (A.9) are just the last two terms of (A.10), viz.,

$$\frac{L}{2L+1}\, S_{L+1}\delta_{\alpha\alpha_1}V_\beta B_{\beta\alpha_2\cdots\alpha_L}, \qquad (A.13)$$

which according to (A.12) is re-expressible as

$$N_S^{-1}\,\frac{L}{2L+1}\,S_{L+1}\delta_{\alpha\alpha_1}A^{L-1}_{\alpha_2\cdots\alpha_L}. \qquad (A.14)$$

We now use the condition analogous to (A.7) viz., that (A.12) and (A.14) have the same normalization. This gives with a little algebra

$$N_S^2 = 2L(2L+1)^{-1}(L+1)^{-1}. \qquad (A.15)$$

Finally, the state of $J = L$ is given by the antisymmetric term in (A.9) which may be re-expressed as

$$L(L+1)^{-1}S_L\varepsilon_{\alpha\alpha_1\beta}\varepsilon_{\beta\gamma\delta}V_\gamma B_{\delta\alpha_2\cdots\alpha_L}. \qquad (A.16)$$

Now define the normalized tensor $A^L_{(\alpha)}$ as

$$A^L_{\alpha_1\cdots\alpha_L} = -iN_A S_L\varepsilon_{\alpha_1\beta\gamma}V_\beta B_{\gamma\alpha_2\cdots\alpha_L}, \qquad (A.17)$$

where $N_A$ must be determined from the condition that (A.17) has the same normalization as the tensor (A.16) of rank $L$, but defined in the extended space of $(L + 1)$ indices. This gives in a similar way

$$N_A^2 = L (L + 1)^{-1}. \tag{A.18}$$

Collecting all the results from (A.9) to (A.18), the C.G. reduction of the direct product of $V_\alpha$ and $B_{(\alpha)}^L$ is given in terms of normalized tensors $A^J$ of $J = L \pm 1, L$ as

$$V_\alpha \otimes B_{(\alpha)}^L = A_{\alpha\alpha_1\cdots\alpha_L}^{L+1} + i \sqrt{\frac{L}{L + 1}} \, S_L \varepsilon_{\alpha\alpha_1\beta} A_{\beta\alpha_2\cdots\alpha_L}^{L}$$

$$+ \sqrt{\frac{L (L + 1)}{2 (2L + 1)}} \, S_{L+1} \delta_{\alpha\alpha_1} A_{\alpha_2\cdots\alpha_L}^{L-1}. \tag{A.19}$$

The reduction of the direct product $B_{(\alpha)}^L \otimes \chi_\alpha$ where $\chi_\alpha$ is an R–S spinor now follows by repeated use of Eq. (6.9) for each term of (A.19), and this leads directly to Eq. (6.12) of the text.

# Field Theories with Indefinite Metric

J.G.TAYLOR*

*Queen Mary College, London, England*

## Contents



## 1 INTRODUCTION

The quantum theory of fields has had a chequered history. After the initial introduction by Dirac[1] of a theory of light quantum emission and absorption by atomic systems which corresponded to quantisation of the radiation field, it was soon realised[2] that the quantisation of the complete electromagnetic field generated divergent quantities. These divergent quantities have a

* Present address: Department of Physics, The University, Southampton, England.

classical ancestry; whilst in the classical theory the divergent self-energy of a point charge can be avoided in a suitable fashion[3] the divergences of quantum electrodynamics posed a much more difficult question. It was not until 1947 that the work of Feynman, Schwinger, Tomonaga and of Dyson[4] showed how these divergences could be eliminated by means of renormalisation. This renormalisation allowed the absorption of all the divergent quantities into the mass and charge of the electron. The resulting mass and charge are then set equal to the experimentally measured values of these parameters; this renormalisation means that the original "bare" mass and charge are not measurable or even calculable in terms of their physical values.

Besides this drawback the renormalisation program seemed closely related to the perturbation expansion used to solve the equations of motion. Whilst such a perturbation expansion is satisfactory for quantum electrodynamics, where the perturbation parameter is equal to the fine structure constant, with the value of 1/137, this is not true for the strong interactions, when the corresponding parameter is about 14. Nor does the renormalisation program apply to weak interactions which are non-renormalisable, in the sense that not all of the divergences can be absorbed into the masses and charges of the weakly interacting particles.

It was later shown[5] how the renormalisation program can be divorced from the perturbation expansion; it is possible to set up coupled non-linear integro-differential equations for the renormalised Green's functions of the Yukawa type of strong interactions whose perturbation expansion in powers of the renormalised coupling constant involves no divergences. This non-perturbative approach has not proved successful for weak interactions, unless highly non-perturbative methods are used.[6] Even then the problem is very difficult, and not too much progress along such non-perturbative lines has been made. Attempts are presently being made to remove divergences from the self-coupled Yang–Mills theory of vector mesons with mass[7] and extension of methods successful in this case to weak or gravitational interactions is hoped for.

In spite of this the underlying divergences of all these theories, including that of quantum electrodynamics, leave a bad taste in the mouth. It would appear more reasonable to attempt to set up a theory which has no divergences in it from the beginning, instead of removing them by a suitable fudge after the event of their appearance. For many years this has appeared possible in a theory in which there is a cancellation of the divergences by other divergences. This cancellation can only be achieved by means of states which have a negative length in the vector space $\mathscr{H}$ of states of the theory. Such states prevent $\mathscr{H}$ from being a Hilbert space, but require it to be an inner product space which is indefinite in sign. In other words the state space has an indefinite metric. The use of an indefinite metric in quantum mechanics was first suggested by Dirac[8]. Since that time its use to remove the divergences in quantum field theory has been studied very often[9]. It has also been noted as essential in the Lee model without a cut-off[10], though its appearance there is not related to the removal of divergences. More recently Lee and Wick[11] have attempted to use an indefinite metric to give a finite theory of quantum electrodynamics and weak interactions. The main emphasis in that work was to preserve unitarity, which is usually violated in the presence of an indefinite metric. This is achieved by requiring that all eigenstates of the Hamiltonian of the system with real energy have positive length; the other eigenstates of the system must have complex energy but since they cannot be reached from the eigenstates of real energy then the scattering matrix $S$ will be truly unitary. Lee and Wick confined their analysis mainly to the Lee model, both in its original version and in an extended form. We will see in Section 2 that it is quite possible that many theories with indefinite metric have complex energy levels, so that a theory of system with indefinite metric must take such complex energies into account. These complex energies will evidently cause difficulties in defining Fourier transforms of Green's functions, since these functions will have exponential increase in their variables. Further the use of a non-hermitian Hamiltonian and possible non-locality of the theory[9] may cause a breakdown of causality, Lorentz invariance, time reversal invariance, the TCP theorem, etc. It also makes the calculation of $S$-matrix elements non-trivial, especially if the interaction representation is used[11].

It is the purpose of this article to describe how a general field theory may be set up in the presence of an indefinite metric when complex energies arise. We will give a set of axioms which replace the usual axioms of quantum field theory[12]. We will especially consider the asymptotic condition which is vital in determining the validity of dispersion relations, invariance under the full Lorentz group and under the TCP operation. We will also show how the general theory we set up may be used to obtain the results of reference (11) for the Lee model.

We will start by showing how complex eigenvalues *always* arise in a simple theory with an indefinite metric, that of the static Lee model. Following that we give in Section 3 a general discussion of quantum mechanics with an indefinite metric, and then in section 4 we extend our discussion to quantum field theory. Having done that we consider in Section 5 how our general approach may be used to solve the $N$–$\theta$ sector of the Lee model with indefinite metric; that is extended in Section 6 to a modified Lee model. Finally in Section 7 we return to discuss the implications of our general results.


## 2   THE STATIC LEE MODEL

This model contains static $N$, $V$ and $\theta$ particles, which are respectively described by the annihilation and creation operators $c, c^+$; $b, b^+$; $a, a^+$. These satisfy the commutation relations

$$[a, a^+]_- = [b, b^+]_- = [c, c^+]_- = 1 \tag{2.1}$$

with all other commutator brackets being zero (where $a^+$ is the hermitian conjugate of $a$, etc.). The Hamiltonian for the system is

$$H = mb^+b + wc^+c + g\,(b^+ac - c^+a^+b) \tag{2.2}$$

so describes the process $V \to N + \theta$ and its converse. We see that since $g$ is assumed to be a real constant then $H$ is not a hermitian operator; it does however, satisfy the property:

$$H^+ = \eta H \eta \tag{2.3}$$

where $\eta = (-1)^{b^+b}$. If we use $\eta$ as the metric operator, so that the inner product between two states $|1\rangle$ and $|2\rangle$ is $\langle 1| \eta |2\rangle$ and the expectation value of any operator $0$ is now $\langle 1| \eta 0 |2\rangle$, then $H$ will have only real expectation values, since $\eta^2 = 1, \eta^+ = \eta$, so that

$$\langle 1| \eta H |1\rangle^* = \langle 1| H^+\eta |1\rangle = \langle 1| \eta H \eta^2 |1\rangle = \langle 1| \eta H |1\rangle \tag{2.4}$$

(where $x^*$ denotes the complex conjugate of $x$). We can see from (2.2) that there are two conserved numbers:

$$N_1 = a^+a + b^+b$$

$$N_2 = c^+c + b^+b.$$

We may describe a complete set of states by $n_a, n_b, n_c)$, where the integers $n_a$, $n_b$, $n_c$ respectively denote the numbers of $\theta$, $V$ and $N$ particles in the state. Let us consider the states $|n - 1, 1, 0\rangle$ and $|n, 0, 1\rangle$; a linear combination of these states is an eigenstate of $H$ with eigenvalue $E$ if we can choose the complex numbers $A$ and $B$ so that

$$H (A |n - 1, 1, 0\rangle + B |n, 0, 1\rangle) = E (A |n - 1, 1, 0\rangle + B |n, 0, 1\rangle). \quad (2.5)$$

If we use (2) for $H$, (2.5) becomes

$$A [m |n - 1, 1, 0\rangle - g \sqrt{n} |n, 0, 1\rangle] + B [w |n, 0, 1\rangle + \sqrt{n} g |n - 1, 1, 0\rangle]$$

$$= E (A |n - 1, 1, 0\rangle + B |n, 0, 1\rangle). \quad (2.6)$$

There will be a non-trivial solution of (2.6) if $E$ satisfies the determinental equation

$$\det \begin{pmatrix} m - E & g \sqrt{n} \\ -g \sqrt{n} & w - E \end{pmatrix} = 0$$

or

$$(E - m) (E - w) + g^2 n = 0. \quad (2.7)$$

The roots of (2.7) are real provided

$$|g| \leqslant |m - w|/4n. \quad (2.8)$$

For a given value of $g$, $m$ and $w$ we see that the inequality (2.8) must be ultimately violated for large enough $n$. In other words there will always be a complex value for the energy. The solutions of (2.7) are

$$E = \tfrac{1}{2} (m + w) \pm \tfrac{1}{2} \sqrt{(m - w)^2 - 4ng} \quad (2.9)$$

so that these complex energies occur in conjugate pairs and are arbitrarily far from the real axis. However in a particular sector (with given value sof $N_1$ and $N_2$) the energies are always a bounded distance away from the real axis. It is these complex energies which we must now take account of when we try to set up a quantum field theory with an indefinite metric.

It is possible, of course, that there are theories with indefinite metric which never possess complex energy levels; one such is quantum electrodynamics in the Gupta–Bleuler formalism[13]. However the indefinite metric in the static Lee model is essentially dynamical, and is not needed when there is no inter-action, whilst that in quantum electrodynamics is even present for free fields and is not at all related to the interaction. So it is possible that theories with a "dynamical" indefinite metric do always have complex eigenvalues whilst theories with a non-dynamical (or kinematical) indefinite metric need not do so.

A general reason for complex energy values in the former case is that the energy level corresponding to the free particle in the above model has a negative length; as the interaction is switched on this state has to acquire complex energy, since all states with real energy are required to have positive length. Thus the requirement of unitarity appears to force the theory to acquire complex energies in those cases of a dynamical indefinite metric. This is quite different from the case of quantum electrodynamics, where the indefinite metric has at worst physical states of zero norm, which can in fact be removed rather simply from the theory.

## 3   QUANTUM MECHANICS WITH INDEFINITE METRIC

We will now consider how an indefinite metric may be used in ordinary quantum mechanics. We will do this by means of the "$\eta$-formalism", as it is called. This formalism treats the vector space of states of the system as a true Hilbert space $\mathscr{H}$, with a positive definite inner product between any two states $|A\rangle$, $|B\rangle$, denoted in the bra-ket notation of Dirac as $\langle A|B\rangle$, so that $\langle A|A\rangle \geqslant 0$, and $\langle A|A\rangle = 0$ requires $|A\rangle = 0$. The indefinite metric itself is introduced by a self-adjoint operator $\eta$ on $\mathscr{H}$ which satisfies

$$\eta^2 = \eta, \quad \eta^+ = \eta. \tag{3.1}$$

The *physical* overlap between two states $|A\rangle$, $|B\rangle$ is now $\langle A|\eta|B\rangle$, whilst the expectation value of an operator $O$ on $\mathscr{H}$ is given by $\langle A|\eta O|A\rangle$ when measured in the state $|A\rangle$. In order that this expectation value be real it is necessary that

$$\langle A|\,\eta O\,|A\rangle^* = \langle A|\,\eta O\,|A\rangle. \tag{3.2}$$

Since the left hand side of (3.2) is $\langle A|O^+\eta|A\rangle$ then we may satisfy (3.2) for general states $|A\rangle$ only if

$$O^+ = \eta O\eta. \tag{3.3}$$

Such an operator 0 is called pseudo-hermitian. It is possible to avoid the use of the $\eta$-formalism by considering $\mathscr{H}$ as an inner product space with inner product $\langle A\|B\rangle$ given by $\langle A|\eta|B\rangle$; the results obtained by avoiding explicit use of $\eta$ are identical to those obtained with its use. We choose the $\eta$-formalism as being slightly more transparent, though this is possibly a personal taste.

We will assume that there is no vector $|A\rangle$ which has zero overlap with all other states i.e. so that $\langle B|\eta|A\rangle = 0$ for all $|B\rangle$ in $\mathscr{H}$. The metric operator $\eta$

is then called non-degenerate; it is then possible to choose an orthonormal basis $|j\rangle$ $(j = 1, 2 \ldots)$ with

$$\eta\,|j\rangle = N_j\,|j\rangle, \quad N_j = \pm 1$$

and we have the completeness relation

$$1 = \sum_j |j\rangle\,N_j\,\langle j|. \tag{3.4}$$

Another property of importance in quantum mechanics is that of a symmetry transformation $U$. In order that such a transformation leave the overlap probability unchanged we need

$$|\langle A|\,\eta\,|B\rangle|^2 = |\langle A|\,U^{\mathsf{1}}\eta U\,|B\rangle|^2. \tag{3.5}$$

This can be satisfied if $U$ is either *pseudo-unitary*:

$$U^+\eta U = \eta \tag{3.6}$$

or *pseudo-anti-unitary:*

$$\langle A|\,U^+\eta U\,|B\rangle = \langle B|\,\eta\,|A\rangle \tag{3.7}$$

for any $|A\rangle$, $|B\rangle$ in $\mathscr{H}$.

Under a pseudo-unitary symmetry transformation $U$ an operator $O$ will transform as follows. We define the transformed operator $O'$ so that the matrix element of $O'$ (with $\eta$ included) between any two transformed states is equal to the matrix element of $O$ between the corresponding untransformed states:

$$\langle A|\,\eta O\,|B\rangle = \langle A'|\,\eta O'\,|B'\rangle \tag{3.8}$$

where $|A'\rangle = U|A\rangle$, $|B'\rangle = U|B\rangle$. Then (3.8) becomes

$$\langle A|\,\eta O\,|B\rangle = \langle A|\,U^+\eta O'U\,|B\rangle \tag{3.9}$$

so that

$$U^+\eta O'U = \eta O.$$

If we use the pseudo-unitarity of $U$ this latter equation becomes

$$\eta U^{-1}O'U = \eta O$$

so

$$O' = UOU^{-1} \tag{3.10}$$

which is the usual transformation of an operator when the indefinite metric is not present. Then $O$ is invariant under the symmetry if it commutes with the symmetry operator $U$.

In the case of a pseudo-anti-unitary symmetry operation the definition (3.8) is replaced by

$$\langle A|\,\eta O\,|B\rangle = \langle B'|\,\eta O'\,|A'\rangle. \tag{3.11}$$

But the right hand side of (3.11) is $\langle A|\,(U^{-1}O'U)^+\,\eta\,|B\rangle$, so that

$$\eta O = (U^{-1}O'U)^+\eta$$

or

$$O' = U\eta O^+\eta U^{-1}. \tag{3.12}$$

In the case of a pseudo-hermitian operator (3.12) reduces back to (3.10), as it should; in this case $O$ is invariant under the pseudo-unitary symmetry only if it commutes with $U$.

We may represent a pseudo-unitary operator $U$ as

$$U = e^{iH} \tag{3.13}$$

where $H$ is a pseudo-hermitian operator, since the right hand side can easily be shown to be pseudo-unitary:

$$(e^{iH})^+\,\eta e^{iH} = e^{-iH^+}\,\eta e^{iH} = \eta e^{-iH}\cdot e^{iH} = \eta.$$

as required. Similarly we may represent any pseudo-anti-unitary operator as the product $Ke^{iH}$, where $K$ is the complex conjugation operator and $H$ is pseudo-hermitian.

Let us now consider the properties of an $S$-matrix in this theory. We suppose that there is a complete set of states $|\alpha,\,\text{out}\rangle$ which correspond to outgoing states, where $\alpha$ is a suitable label; we also suppose that there is a complete set of ingoing states $|\beta,\,\text{in}\rangle$. The completeness here will take the form

$$\sum_\alpha |\alpha,\,\text{out}\rangle\,\langle\alpha,\,\text{out}|\,\eta = \sum_\beta |\beta,\,\text{in}\rangle\,\langle\beta,\,\text{in}|\,\eta = 1. \tag{3.14}$$

We define the $S$-matrix $S_{\beta\alpha}$ and $S$-operator by

$$|\alpha,\,\text{in}\rangle = S\,|\alpha,\,\text{out}\rangle \tag{3.15}$$

$$S_{\beta\alpha} = \langle\beta,\,\text{out}|\,\eta S\,|\alpha,\,\text{out}\rangle = \langle\beta,\,\text{out}|\,\eta\,|\alpha,\,\text{in}\rangle. \tag{3.16}$$

Then we prove the pseudo-unitarity of the $S$-operator, as follows[11].

$$\langle\alpha,\,\text{out}|\,S^+\eta S\,|\beta,\,\text{out}\rangle = \sum_\gamma \langle\alpha,\,\text{out}|\,S^+\eta\,|\gamma,\,\text{out}\rangle\,\langle\gamma,\,\text{out}|\,\eta S\,|\beta,\,\text{out}\rangle$$

$$= \sum_r \langle\gamma,\,\text{out}|\,\eta\,|\alpha,\,\text{in}\rangle^*\,\langle\gamma,\,\text{out}|\,\eta\,|\beta,\,\text{in}\rangle. \tag{3.17}$$

If we assume that we may choose the in and out states' labels so that $\eta$ has the same matrix elements between states with the same labels then (3.17) implies that

$$\langle \alpha, \text{out}| \, S^+ \eta S \, |\beta, \text{out}\rangle = \langle \alpha, \text{out}| \, \eta \, |\beta, \text{out}\rangle$$

and $S$ is pseudo-unitary. This causes trouble in the interpretation of $|S_{\beta\alpha}|^2$ as the probability of the transition from the state with labels $\alpha$ to that with labels $\beta$. For such an interpretation to be valid it is necessary that $\sum_{\alpha} |S_{\beta\alpha}|^2$ $= 1$. However from the second part of (3.17) we see that this sum has the value $\langle \beta \text{ in}| \, \eta \, |\beta \text{ in}\rangle$ which can be negative and indeed will be so for certain states if the indefinite metric is non-trivial.

It is possible to avoid this possibility by requiring that the physical states of the system always have positive length. If we denote these states as $|r\rangle$, with $r$ standing for the various labels of the state, then true unitarity, not just pseudo-unitarity, will hold if $S$ has zero matrix element between any physical state and any non-physical state (which we denote by $|c\rangle$). Thus if

$$\langle r| \, \eta S \, |c\rangle = 0 \tag{3.18}$$

then in the pseudo-unitarity condition (3.17) taken between two physical states we will have only contributions from physical intermediate states. Then

$$\sum_{\alpha} |S_{\gamma\alpha}|^2 = \sum_{r'} |S_{rr'}|^2 = \langle r| \, \eta \, |r'\rangle. \tag{3.19}$$

Since the right hand side of (3.19) is positive we may choose $|r\rangle$ to have unit length, so obtaining the true unitarity condition

$$\sum_{r'} |S_{rr'}|^2 = 1. \tag{3.20}$$

If the physical states form a complex vector subspace of the total space $\mathcal{H}$ then the Schwartz identity will be valid for $\langle A| \, \eta \, |B\rangle$:

$$|\langle A| \, \eta \, |B\rangle|^2 \leq |\langle A| \, \eta \, |A\rangle| \, |\langle B| \, \eta \, |B\rangle|. \tag{3.21}$$

Then we may interpret the quantity $|\langle A| \, \eta \, |B\rangle|^2/|\langle A| \, \eta \, |A\rangle|$ as the true probability that the physical state $|A\rangle$ is in the physical state $|B\rangle$; this means that the standard probability interpretation of quantum mechanics will be possible.

We may ensure that there are no transitions between physical and non-physical states if we consider the Hamiltonian $H$ of the system. Since it is an observable operator $H$ will be pseudo-hermitian; it will thus be able to have complex eigen-values $E_c$ with eigen-states which we again denote by $|c\rangle$. The real eigenvalues $E_r$ and corresponding eigenstates $|r\rangle$ will then be related to the complex ones as follows. Firstly

$$\langle r|\,\eta H\,|c\rangle = E_c\,\langle r|\,\eta\,|c\rangle = E_r\,\langle r|\,\eta\,|c\rangle$$

so

$$\langle r|\,\eta\,|c\rangle = 0. \tag{3.22}$$

Since we assume that $S$ commutes with $H$ (so that there is invariance under translations in time) then by similar arguments

$$\langle r|\,\eta S\,|c\rangle = 0. \tag{3.23}$$

If we impose the condition that *all* of the states $|r\rangle$ have positive length[11]

$$\langle r|\,\eta\,|r\rangle > 0, \quad \text{all } |r\rangle \tag{3.24}$$

then we have achieved a theory which has a consistent probability interpretation associated with it; the $S$-matrix will only relate the physical states $|r\rangle$, and will be unitary on the subspace of such states.

It is possible that the requirement that *all* of the states $|r\rangle$ have positive length be dropped and a unitary theory still be obtained. This appears difficult to achieve in general, unless the energies of the states $|r\rangle$ with negative length be disjoint from that of states with positive length. We will assume from now on that the conditions (3.23) and (3.24) are valid.

Let us analyse the complex eigenstates $|c\rangle$ further. These will satisfy

$$\langle c|\,\eta H\,|c'\rangle = E_{c'}\,\langle c|\,\eta\,|c'\rangle = E_c^*\,\langle c|\,\eta\,|c'\rangle \tag{3.25}$$

Then if $E_{c'} \neq E_c^*$ we have $\langle c|\,\eta\,|c'\rangle = 0$. When $E_{c'} = E_c^*$ it is possible that $\langle c|\,\eta\,|c'\rangle$ is not zero. We will assume[11] that each of the states $|c\rangle$ have their companion $|c'\rangle$ with $E_{c'} = E_c^*$ *and* $\langle c|\eta|c'\rangle \neq 0$. If $\mathrm{Im}\,E_c > 0$ we denote the state by $1 + c\rangle$, and its companion state $|c'\rangle$ with $E_{c'} = E^*$ will be denoted by $|-c\rangle$. Assuming that these states form a complete set we obtain the resolution of the identity

$$1 = \sum_r |r\rangle\,\langle r|\,\eta + \sum_c |+c\rangle\,\langle -c|\,\eta + \sum_c |-c\rangle\,\langle +c|\,\eta \tag{3.26}$$

(where the summation signs also denote integration over continuous ranges of parameters). The states $|r\rangle$ and $|\pm c\rangle$ may then be chosen so that

$$\langle r|\,\eta\,|r'\rangle = \delta_{rr'}, \quad \langle +c|\,\eta\,|-c'\rangle = \langle -c'|\,\eta\,|+c\rangle = \delta_{cc'}, \tag{3.27}$$

all other inner products zero.

We will find it useful to construct and discuss certain pseudo-projection operators $P_{\pm}$ and $P$. These are defined as follows:

$$P_+ = 1 - \sum_c |+c\rangle\langle -c|\,\eta \tag{3.28}$$

$$P_- = 1 - \sum_c |-c\rangle\langle +c|\,\eta \tag{3.29}$$

$$P = P_+ P_-. \tag{3.30}$$

From the definitions (3.28) and (3.29) we see that

$$P_+^+ = \eta P_- \eta, \quad P^+ = \eta P_+ \eta$$

whilst from (3.30) we have that $P$ is pseudo-hermitian

$$P^+ = \eta P \eta. \tag{3.31}$$

If we define

$$Q_+ = \sum_c |+c\rangle\langle -c|\,\eta, \quad Q_- = \sum_c |-c\rangle\langle +c|\,\eta \tag{3.32}$$

then

$$Q_+^2 = \sum_{c,c'} |+c\rangle\langle -c|\,\eta\,|+c'\rangle\langle -c'|\,\eta.$$

Using the inner products as given by (3.27) this becomes

$$Q_+^2 = \sum_{c,c'} |+c\rangle\,\delta_{cc'}\,\langle -c'|\,\eta = \sum_c |+c\rangle\langle -c|\,\eta = Q_+. \tag{3.33}$$

Also

$$Q_+ Q_- = \sum_{c,c'} |+c\rangle\langle -c|\,\eta\,|-c'\rangle\langle +c'|\,\eta = 0. \tag{3.34}$$

Similarly we may show that

$$Q_-^2 = Q_-, \quad Q_- Q_+ = 0. \tag{3.35}$$

Then we may use (3.33), (3.34) and (3.35) to obtain

$$P_+^2 = 1 - 2Q_+ + Q_+^2 = 1 - Q_+ = P_+ \tag{3.36}$$

$$P_-^2 = P_- \tag{3.37}$$

and

$$P^2 = (1 - Q_+ - Q_-)^2 = 1 - Q_+ - Q_- = P \tag{3.38}$$

An operator $R$ satisfying

$$R^2 = R, \quad R^+ = \eta R \eta$$

will be called a pseudo-projection operator; from (3.31) and (3.38) we see that $P$ is such an operator, though $P_+$ and $P_-$ are not, since they are not pseudo-hermitian.

We now see that $P_\pm$ and $P$ all commute with the Hamiltonian $H$. Consider, for example,

$$[P_+, H]_- = [\sum_c |+c\rangle \langle -c| \eta, H]_-$$

Now $\langle -c| \eta H = \langle -c| H^+ \eta = E^*_{-c} \langle -c| \eta$, so

$$[P_+, H]_- = \sum_c E^*_{-c} |+c\rangle \langle -c| \eta = \sum_c E_{+c} |+c\rangle \langle -c| \eta. \quad (3.39)$$

But $E^*_{-c} = E_{+c}$, so that the right hand side of (3.39) vanishes, as we wished to show. We may show that $H$ commutes with $P_-$ and $P$ in a similar fashion.

Let us now consider the time development of an operator $O$:

$$O(t) = e^{iHt} O\, e^{-iHt}. \quad (3.40)$$

Since $H$ has complex eigenvalues we know that certain matrix elements of $O(t)$ will increase indefinitely as $|t|$ increases. For example if we consider

$$\langle r| \eta O(t) |-c\rangle = \langle r| \eta\, e^{iHt} O\, e^{-iHt} |-c\rangle = e^{i(E_r - E_{-c})t} \langle r| \eta O |-c\rangle.$$

Then the matrix elements has exponential increase as $t \to -\infty$ unless $\langle r| \eta O |-c\rangle$ vanishes; this cannot be required for all operators $O(o)$. This exponential increase implies that we cannot apply Fourier transformations in $t$ in a simple fashion to any operator $O(t)$. More importantly there may be very few operators which have meaningful asymptotic limits as $|t|$ increases. In order to discuss such limits we may use the operators $P_\pm$ and $P$ to define regularisation procedures which remove such exponentially increasing factors.

For any operator $0(t)$ we define

$$O_+(t) = P_- O(t) P_+. \quad (3.41)$$

Then if we use (3.28), (3.29) and (3.40) this gives

$$O_+(t) = P_-\, e^{iHt} O\, e^{-iHt} P_+$$

$$= \sum_{c,c'} e^{i(E^*_{+c} - E_{+c'})t} |-c\rangle \langle +c| \eta O |+c'\rangle \langle -c'| \eta. \quad (3.42)$$

The exponential factor on the right of (3.42) has form

$$\exp [i\, \mathrm{Re}\, (E_{-c} - E_{+c'})\, t - \mathrm{Im}\, (E_{-c} - E_{+c'})\, t]. \quad (3.43)$$

We have defined $E_{-c}$ so as to have negative imaginary part whilst $E_{+c}$ has a positive imaginary part, so that $(E_{-c} - E_{+c'})$ has a negative imaginary part. This means that the exponent in (3.43) decreases exponentially as $t \to -\infty$, so that we expect $O_+(t)$ to have an asymptotic limit as $t \to -\infty$. Similarly we expect $O_-(b)$ to have an asymptotic limit as $t \to +\infty$, where

$$O_-(t) = P_+ O(t) P_- . \tag{3.44}$$

Finally we may project out the terms in $O(t)$ which increase exponentially as $t$ increases either to $+\infty$ or $-\infty$ by means of the pseudo-projection operator $P$ to give the regularised operator $O_{\text{reg}}(t)$:

$$O_{\text{reg}}(t) = P O(t) P \tag{3.45}$$

which has explicit time dependence

$$O_{\text{reg}}(t) = \sum_r e^{i(E_r - E_{r'})t} |r\rangle \langle r| \eta O |r'\rangle \langle r'| \eta . \tag{3.46}$$

This regularised operator evidently only has oscillatory contributions in $t$, so that we may apply Fourier transform methods to it.

## 4 QUANTUM FIELD THEORY WITH INDEFINITE METRIC

We now turn to the problem of setting up a quantum field theory with an indefinite metric. We wish to preserve as many of the axioms of quantum field theory as we can. We will not consider these axioms in too heavily a mathematical frame but more from a physical point of view; we will, however, use the numeration of the axioms given by Streater and Wightman[12]. These axioms are:

*O. Assumptions of a relativistic quantum theory*

The states of the theory are described by unit rays in an indefinite inner product space, which we will identify with a Hilbert space $\mathscr{H}$ with (indefinite) metric operator $\eta$ with $\eta^+ = \eta, \eta^2 = 1$. The relativistic transformation laws of the states is given by a continuous pseudo-unitary representation of the inhomogeneous $SL(2, C)$:

$$\{a, \Lambda\} \to U(a, \Lambda).$$

The pseudo-unitary operator $U(a, 1)$ defines the pseudo-hermitian operator $P_\mu$ by

$$U(a, 1) = e^{ia_\mu P^\mu}. \tag{4.1}$$

15  Aly, Particles and Fields

We interpret $P_\mu$ as the energy-momentum operator of the theory. There is a unique invariant vacuum state $|0\rangle$:

$$U(a, \Lambda)\,|0\rangle = |0\rangle.$$

### $O'$-Spectrum and indefinite metric assumptions

Here we need to add an extra set of conditions in order that the states of negative or zero length do not prevent a sensible physical interpretation to be given to the theory. We consider first the real eigenstates of $P_\mu$, which we denote by $|r\rangle$; we require that these form a subset of $\mathcal{H}$ which has *only* states of positive length

$$\langle r|\,\eta\,|r\rangle > 0. \tag{4.2}$$

It is this condition which we used in the previous section to ensure that the $S$ matrix between the physical states was a unitary matrix, and not just pseudo-unitary. The condition (4.2) will ensure unitary for the relativistic $S$-matrix in a similar fashion. We further assume that the *complete* spectrum of $P_\mu$ is invariant under any homogeneous Lorentz transformation.

### 4.1   Axioms for a field operator

We will assume that there is a field operator $\psi(x)$ which can be regarded as a distribution-valued operator on space-time ($x$ here denotes any real 4-vector $(x_0, \mathbf{x})$). This field operator can be used in the standard fashion to construct the Wightman functions

$$W(x_1, \ldots, x_n) = \langle 0|\,\eta\psi(x_1)\ldots\psi(x_n)\,|0\rangle. \tag{4.3}$$

### 4.2   Transformation law of the field

We assume that $\psi(x)$ is a neutral scalar field, so that

$$\psi^+(x) = \eta\psi(x)\,\eta \tag{4.4}$$

whilst

$$U(a, \Lambda)\,\psi(x)\,U(a, \Lambda)^{-1} = \psi(\Lambda x + a). \tag{4.5}$$

From (2.6), the invariance of the vacuum and the pseudo-unitarity of $U(a, \Lambda)$ follows the usual invariance of the Wightman functions. We may evidently consider other spin representations of $SL(2, C)$ but do not go into that here.

## 4.3  Local commutativity

We assume that $\psi(x)$ is microscopically local:

$$[\psi(x), \psi(y)]_- = 0 \quad \text{if} \quad (x - y)^2 < 0 \tag{4.6}$$

(where $x^2 = x_0^2 - \mathbf{x}^2$).

## 4.4  Asymptotic completeness

We have here to consider which part of the spectrum of $P_\mu$ will generate asymptotic states. We expect, indeed, that only the real part of this spectrum can be considered, since the complex part will always correspond to exponentially increasing states which have no asymptotic limit. Alternatively we may say that non-eigenstates cannot persist asymptotically in the past or the future, since we do not measure non-real values of energy or momentum with presently available experimental apparatus. This does not mean that there are not effects of the non-real eigenstates, but only that these effects cannot be measured directly.

We conclude that we may construct the asymptotic state spaces $\mathscr{H}_{\text{out}}$ and $\mathscr{H}_{\text{in}}$ which contain non-interacting states which are eigenstates of $P_\mu$ corresponding only to the real eigenvalues. For such spaces the asymptotic completeness condition cannot be the usual one: $\mathscr{H} = \mathscr{H}_{\text{out}} = \mathscr{H}_{\text{in}}$. But we may project out from the subspace spanned by the real eigenstates, so obtain a modified form of asymptotic completeness

$$\mathscr{H}_{\text{out}} = \mathscr{H}_{\text{in}} = P\mathscr{H}. \tag{4.7}$$

In terms of this asymptotic completeness we expect an $LSZ$-type[14] of asymptotic weak limit for the field operator

$$P\psi(x)\,P \to \psi_{\text{out}}(x) \quad \text{as} \quad x_0 \to +\infty \tag{4.8}$$

$$\to \psi_{\text{in}}(x) \quad \text{as} \quad x_0 \to -\infty.$$

In order for (4.8) to make sense as a weak limit in the whole of $\mathscr{H}$ we have to require that the real eigenvalues of $P_\mu$ correspond to the existence of real particles of mass $m$, whilst the non-real eigenvalues correspond to states which are annihilated by the asymptotic field operators. This annihilation corresponds to the fact that asymptotic particles cannot be annihilated or created in such non-real eigenstates at all times, since the resulting states would be non-real eigenstates so could not be asymptotic. Another way to see this annihilation is from the fact that $\psi_{\text{out}}$ or $\psi_{\text{in}}$ can themselves have weak asymptotic limits in the usual sense only if they annihilate $(I - P)\mathscr{H}$ in the sense of (4.8).

It would seem possible to extend the Haag–Ruelle–Hepp theory which essentially justified (4.8) for a positive-definite metric to this indefinite metric case; we will not bother about that here, but assume it can be done, and use the asymptotic condition (4.8) without further ado.

The use to which this asymptotic condition is put is to derive a reduction formula relating $S$-matrix elements to matrix elements of the complete interpolating field $\psi(x)$. Thus in terms of a set of in and out states

$$\left.\begin{aligned}
|\alpha, \text{in}\rangle &= \prod_i a^+_{\text{in},k_{\alpha_i}} |0\rangle \\
|\beta, \text{out}\rangle &= \prod_j a^+_{\text{out},k_{\beta_j}} |0\rangle
\end{aligned}\right\} \tag{4.9}$$

we define the $S$-matrix element $S_{\beta\alpha}$ as (see 3.16):

$$S_{\beta\alpha} = \langle\beta, \text{out}| \, \eta \, |\alpha, \text{in}\rangle.$$

We have still the orthogonality between the states $|r\rangle$ and $|c\rangle$

$$\langle c| \, \eta \, |r\rangle = 0, \tag{4.10}$$

so the argument for unitarity of $S$ goes through as before. The reduction formulae which determine $S_{\beta\alpha}$ in terms of the interpolating field $\psi$ must evidently be altered from the usual one[14] to take account of the altered form of asymptotic condition (4.8).

We should at this point remark that the asymptotic condition (4.8) is consistent with Lorentz covariance, since the projection operator $P$ commutes with the Lorentz transformation $U(a, \Lambda)$,

$$U(a, \Lambda)\, P U^{-1}(a, \Lambda) = P. \tag{4.11}$$

This follows since the action of $U(a, \Lambda)$ given by the left hand side of (4.11) is to replace $P$ by

$$P' = 1 - \sum_c |+\Lambda c\rangle \langle -\Lambda c| \, \eta - \sum_c |-\Lambda c\rangle \langle +\Lambda c| \, \eta \tag{4.12}$$

where $|\pm\Lambda c\rangle$ is the eigenstate with eigenvalue $\Lambda p^{(c)}_\mu$ of $P_\mu$ when $|\pm c\rangle$ has eigenvalue $p^{(c)}_\mu$ of $P_\mu$ (the $\pm$ sign corresponding to the sign of Im $p^{(c)}_0$). But we assumed in axiom $O'$ that the spectrum of $P_\mu$ is invariant under $\Lambda$ so that the right hand side of (4.12) is just equal to $P$ (noting that under $\Lambda$ the non-real eigenvalues $p^{(c)}_\mu$ remain non-real, the real ones remain real).

Returning to the reduction formula we see that following the usual arguments[14] we have that

$$S_{\beta\alpha} = (2\pi)^{-3/2(n+m)} \, i^{(n+m)} \int dy_1 \cdots dy_m \, dx_1 \cdots dx_n \, e^{i\left(\sum\limits_{j=1}^{m} p_j y_j - \sum\limits_{k=1}^{n} q_k x_k\right)} \times$$

$$\times \prod_{j=1}^{m} K_{y_j} \prod_{k=1}^{n} K_{y_k} \langle 0| \, \eta T \, [P\psi(y_1) \, P \cdots P\psi(y_m) \, P\psi(x_1) \, P \cdots P\psi(x_n) \, P] \, |0\rangle \tag{4.13}$$

where $p_j^2 = q_k^2 = m^2$ ($1 \leqslant j \leqslant m$, $1 \leqslant k \leqslant n$), $K_x = \square_x^2 - m^2$, and we have taken only one asymptotic free particle, of mass $m$. We have also assumed that $\eta$ commutes with the free fields $\psi_{\text{out}}$, $\psi_{\text{in}}$. This seems reasonable since $\eta$ is only effective when acting on the states with non-real energy. It is possible to generalise the reduction formula (4.13) when $\eta$ does not commute with $\psi_{\text{in}}$ or $\psi_{\text{out}}$, though we will not consider that further here. We see that the reduction formula expresses the $S$-matrix elements not as the on-mass-shell Fourier transforms of the usual Green's functions

$$\tau(x_1 \cdots x_n) = \langle 0| \, \eta T \, (\psi(x_1) \cdots \psi(x_n)) \, |0\rangle \tag{4.14}$$

but of the regularised Green's functions

$$\tau_{\text{reg}}(x_1 \cdots x_n) = \langle 0| \, \eta T \, (P\psi(x_1) \, P \cdots P\psi(x_n) \, P) \, |0\rangle. \tag{4.15}$$

Whilst the complete field operator $\psi(x)$ satisfies the condition of microscopic causality (4.6) this is not true for the correct interpolating field $P\psi(x) P$. Thus we have a non-locality introduced into the $S$-matrix elements. This non-locality would normally cause a break-down of Lorentz covariance due to the time ordered product of non-local fields not being a covariantly defined quantity. In the present case there is no breakdown of Lorentz invariance in the $S$-matrix elements, as we see from the definition (4.9) (and the fact that the in or out fields $\psi_{\text{in}}$, $\psi_{\text{out}}$ transform under the Lorentz group according to the pseudo-unitary representation $U|a, \Lambda)$ of $SL(2, c)$). Thus the indefinite metric may be regarded as a means of introducing non-locality in a covariant fashion; this non-locality arises from the non-real energy levels, so these play an important, though indirect, role in the theory.

We remark that the reduction formula (4.13) involves the Fourier transform of the regularised Greens' function $\tau_{\text{reg}}(x_1 \cdots x_n)$; this Fourier transform is expected to be well defined, since each regularised field $P\psi(x) P$ entering in $\tau_{\text{reg}}$ should only have oscillatory dependence on the space-time variable $x$. Thus the asymptotic condition (4.8) appears not only natural but necessary. We will turn in the next section to discussing the validity of this condition for the Lee model with indefinite metric and the solution of this model in the $N$–$\theta$ sector.

## 5   LEE MODEL WITH INDEFINITE METRIC

Qe take the same Hamiltonian as discussed by Lee and Wick[11], being the extension of the model we discussed in Section 2 to the case of a moving $\theta$ particle. The Hamiltonian is

$$H = m_0 V^+ V + \sum_k \omega_k a_k^+ a_k + g\left[ V^+ N \sum_k \frac{u(\omega)}{(2\omega)^{1/2}} a_k - VN^+ \sum_k \frac{u(\omega)}{(2\omega)^{1/2}} a_k^+ \right]$$

$$(5.1)$$

where $\omega = (k^2 + \mu^2)^{1/2}$, $\mu$ is the mass of the $\theta$-particle, $m_\sigma^0$ the bare mass of the $V$-particle, $u(\omega)$ a suitable cut-off function and $g$ the bare coupling constant. The non-zero equal time commutation relations are

$$[a_k, a_{k'}^+]_- = \delta_{kk'}, \quad [V, V^+]_- = [N, N^+]_- = 1. \qquad (5.2)$$

We see that as for the model in § 2, $\eta = (-1)^{V^+ V}$, and the $H$ is a pseudo-hermitian operator.

In order to derive the asymptotic condition we follow the arguments of Curtis and Maxon[16]. They show by means of a generalised Lippman–Schwinger equation that the matrix elements of $N(t)$ and $a_k(t)$ converge to the free in- or outfield matrix elements as $t \to \mp\infty$. These matrix elements are between asymptotic states, so have positive length. In the case of the indefinite metric Lee model the argument goes through precisely as in reference (16), though now only for matrix elements between states with real energy levels (where we assume that it is possible to choose the value of $g$ in (5.1) so that such states always have positive length). In order to obtain an asymptotic condition as an operator condition we have to consider the case when the states have non-real energy. To do this we realise that the in or out states containing various non-interaction $N$ and $\theta$ particles have real energy, so must be orthogonal to the states with non-real energy. Thus for a 2-particle out state and any state $|c\rangle$ with non-real energy,

$$\langle c| \, \eta \, |N\theta_k \text{ out}\rangle = 0. \qquad (5.3)$$

Then

$$\langle c| \, \eta a_{k,\text{out}}^+ |N\rangle = 0$$

so that the asymptotic limit of $a(t)$ as $t \to +\infty$ must contain the projection operator $P$. Since we also have[16]

$$\langle r| \eta a_{k,\text{out}}^{+}|N\rangle = \lim_{t \to +\infty} \langle r| e^{-i\omega_k t} \eta a_k^{+}(t)|N\rangle \tag{5.4}$$

for any state $|r\rangle$ with real eigenvalue, we may solve (5.3) and (5.4) by

$$a_{k,\text{out}}^{+} = \lim_{t \to +\infty} e^{-i\omega_k t} P a_k^{+}(t). \tag{5.5}$$

We have shown that (5.5) is a solution of the asymptotic limits (5.3) and (5.4) in the $N$–$\theta$ sector only; the arguments of reference (16) allow us simply to extend this to show that (5.5) satisfies the asymptotic condition

$$\langle c| \eta a_{k,\text{out}}^{+} |N\theta_{k_1} \cdots \theta_{k_n} \text{ out}\rangle = 0 \tag{5.6}$$

$$\langle r| \eta a_{k,\text{out}}^{+} |N\theta_{k_1} \cdots \theta_{k_n} \text{ out}\rangle = \lim_{t \to +\infty} \langle r| e^{-i\omega_k t} \eta a_k^{+}(t) |N\theta_{k_1} \cdots \theta_{k_n} \text{ out}\rangle$$

for any state $|N\theta_{k_1} \cdots \theta_{k_n} \text{ out}\rangle$ composed of one $N$ and $n\theta$ particles.

If we also consider the complex conjugate conditions which arise from (5.6) we obtain the adjoint to (5.5):

$$a_{k,\text{out}} = \lim_{t \to \infty} e^{i\omega_k t} a_k(t) P^{+}. \tag{5.7}$$

We notice that (5.6) has arisen when $a_{k,\text{out}}^{+}$ acts to the left on states with real energy, so we can replace it by the equivalent equation:

$$a_{k,\text{out}}^{+} = \lim_{t \to \infty} e^{-i\omega_k t} P a_k^{+}(t) P \tag{5.8}$$

and (5.7) by the adjoint of (5.8):

$$a_{k\ \text{out}} = \lim_{t \to \infty} e^{i\omega_k t} P^{+} a_k(t) P^{+}. \tag{5.9}$$

We notice that (5.8) and (5.9) are just what we expected from the general arguments of the previous section, since $P a_k^{+}(t) P$ and its adjoint have purely oscillatory time development. This is more than is needed for the discussion of the limit as $t \to +\infty$; we could have worked with the half-regularised operator $P_{+} a_k^{+}(t) P_{-}$, which has no exponentially increasing terms as $t \to +\infty$, or with $P_{-} a_k^{+}(t) P_{+}$, which has the same behaviour as $t \to -\infty$. The first of these is used by Lee and Wick[11], but we will work with the fully regularised fields $P a_k^{+}(t) P$, $P^{+} a_k(t) P^{+}$ which allows the simplest forms of reduction formulae to be given.

Finally we remark that the use of (5.8) in place of (5.7) is justified by the orthogonality which exists between the states with real and non-real energy; since $a_{\text{out}}$ and $a_{\text{out}}^{+}$ are initially defined as acting on states of real energy and create states with real energy it is necessary that when they act on states with non-real energy they give zero. This also implies that $\langle c|\, \eta a_{\text{out}}^{+}\, |c'\rangle = 0$ for any non-real energies $E_c$, $E_{c'}$; this is correct, since such non-real energy states give no contributions to asymptotic states.

In a similar fashion we have the asymptotic condition

$$\psi_{N,\text{out}}^{+} = \lim_{t \to +\infty} P\psi_N^{+}(t)\, P \tag{5.10}$$

whilst the in-operator asymptotic limits are

$$a_{N,\text{in}}^{+} = \lim_{t \to -\infty} e^{-i\omega_k t}\, Pa_k^{+}(t)\, P \tag{5.11}$$

$$\psi_{N,\text{in}}^{+} = \lim_{t \to -\infty} P\psi_N^{+}(t)\, P \tag{5.12}$$

and their adjoints. We do not consider asymptotic limits of the $V$ operator, since these are not needed to build up asymptotic states, since these cannot contain $V$ particles, they being of negative length so unphysical.

We may now use the asymptotic limits (5.8) to (5.12) to obtain the reduction formulae generalising those of reference (16); they are exactly as in that reference except that we have the regularised Green's function in place of the complete Green's function. Thus the scattering of $n$ $\theta$-particles and one $N$-particle to $m$ $\theta$-particles and one $N$-particle is described by

$$S_{\alpha\beta} = \delta_{\alpha\beta} - (m!\, n!)^{-1/2} \int_{-\infty}^{+\infty} dt' \int_{-\infty}^{+\infty} dt \cdot \exp\left[ it'\left(\sum_{\mu=1}^{m}\omega_{\mu}'\right) - it\left(\sum_{\mu=1}^{m}\omega_{\nu}\right)\right] \times$$

$$\times \left(i\frac{d}{dt'} - \sum_{\mu=1}^{m}{}'\,\omega_{\mu}'\right)\left(i\frac{d}{dt} - \sum_{\nu=1}^{n}\omega_{\nu}\right)\tau_{\text{reg}}(t', b) \tag{5.13}$$

and $\tau_{\text{reg}}(t', t)$ is defined as

$$\tau_{\text{reg}}(t', t) = \langle 0|\, \eta T\left[PN(t')\, P\prod_{\mu=1}^{m}(a_{k_{\mu}'}(t')\, P)\cdot PN^{+}(t)\, P\prod_{\nu=1}^{n}(a_{k_{\nu}}^{+}(t)\, P)\right]|0\rangle. \tag{5.14}$$

Since we have invariance under time translation we may consider simply the function $\tau_{\text{reg}}(s, 0)$. We will conclude this section by obtaining this regularised Green's function for $N$–$\theta$ scattering.

The method we use to obtain this function, and will use in the next section to discuss the modified Lee model, is by means of the coupled Green's functions equations which can be derived from the Heisenberg equations of motion following from the Hamiltonian (5.1).

These Heisenberg equations are

$$i \frac{d}{dt} N(t) = -g \sum_k \left( u(\omega)/\sqrt{2\omega} \right) V(t) a_k^+(t) \tag{5.15}$$

$$\left( i \frac{d}{dt} - m_0 \right) V(t) = g \sum_k \left( u(\omega)/\sqrt{2\omega} \right) N(t) a_k(t) \tag{5.16}$$

$$\left( i \frac{d}{dt} - \omega \right) a_k(t) = -g \left( u(\omega)/\sqrt{2\omega} \right) N^+(t) V(t). \tag{5.17}$$

As in reference (16) we may derive from (5.15), (5.16), (5.17) the following coupled differential equations for the complete Green's functions

$$\tau_1(s) = \langle 0| \, \eta T \, [V(s) \, V^+] \, |0\rangle \tag{5.18}$$

$$\tau_2 (s, \omega) = \langle 0| \, \eta T \, [N(s) \, a_k(s) \, V^+] \, |0\rangle \left( \sqrt{2\omega}/u(\omega) \right) \tag{5.19}$$

$$\tau_3 (s, \omega) = \langle 0| \, \eta T \, [V(s) \, N^+ a_k^+] \, |0\rangle \left( \sqrt{2\omega}/u(\omega) \right) \tag{5.20}$$

$$\tau_4 (s, \omega, \omega') = \langle 0| \, \eta T \, [N(s) \, a_k (s) \, N^+ a_{k'}^+] \, |0\rangle \left( \sqrt{4\omega\omega'}/u(\omega) \, u(\omega') \right) \tag{5.21}$$

which are

$$\left( i \frac{d}{ds} - m_0 \right) \tau_1(s) = i\delta (s) + g \sum_k [u^2(\omega)/2\omega] \tau_2 (s, \omega) \tag{5.22}$$

$$\left( i \frac{d}{ds} - \omega \right) \tau_2 (s, \omega) = -g\tau_1 (s) \tag{5.23}$$

$$\left( i \frac{d}{ds} - \omega \right) \tau_3 (s, \omega) = -g\tau_1 (s) \tag{5.24}$$

$$\left( i \frac{d}{ds} - \omega \right) \tau_4 (s, \omega, \omega') = i [2\omega/u^2(\omega)] \delta_{k,k'} - g\tau_3 (s, \omega'). \tag{5.25}$$

We have denoted by $N$, $V$, $a_k$, etc., the corresponding fields at time $t = 0$. We now attempt to solve the coupled differential equations together with the boundary conditions $\tau_2 (0, \omega) = \tau_3 (0, \omega) = 0$ (so that $\tau_2 (s, \omega) = \tau_3 (s, \omega)$ from (5.23) and (5.24). It is very simple to achieve this by using Fourier transforms. But we have already seen that the complete Green's functions $\tau_1, \tau_2,$ $\tau_3$, etc., will increase exponentially in $s$ for large $|s|$, due to possible non-real energy levels. Whilst we cannot use direct Fourier transform methods we can still attempt to use an integral representation which will do for us as much as the Fourier transformation will in solving (5.22), (5.23), (5.24) and (5.25). Thus we take, for $i = 1, 2, 3$ and $4$,

$$\tau_i(s) = \frac{1}{i} \int_c e^{-iWs} \hat{\tau}_i(W)\, dW \tag{5.26}$$

(where we leave out explicit description of the variables $\omega$ and $\omega'$ in the functions $\tau_i(s)$). We hope to be able to choose the contour $C$ so as to be parallel to the real axis, though suitable displaced from it; we will see that such a choice will be possible so that (5.26) is invertible in straightforward fashion. If we insert (5.26) into (5.22) to (5.25) we obtain a set of equations of form

$$\int_c F(W)\, e^{-iWs}\, dW = 0. \tag{5.27}$$

We solve this equation by taking the integrand $F(W)$ to vanish identically on $W$; this gives the set of algebraic equations

$$(W - m_0)\, \hat{\tau}_1(W) = \hat{\delta}(W) + g \sum_k [u^2(\omega)/2\omega]\, \hat{\tau}_2\,(W, \omega) \tag{5.28}$$

$$(W - \omega)\, \hat{\tau}_2\,(W, \omega) = -g\hat{\tau}_1\,(W) \tag{5.29}$$

$$(W - \omega)\, \hat{\tau}_4\,(W, \omega, \omega') = [2\omega/u^2(\omega)]\, \delta_{k,k'} - g\hat{\tau}_2\,(W, \omega') \tag{5.30}$$

In (5.28) we have defined $\hat{\delta}(W)$ to be the inverse of $\delta(s)$ under the transformation (5.26), so that

$$\delta(s) = \frac{1}{i} \int_c e^{iWs}\, \hat{\delta}(W)\, dW. \tag{5.31}$$

We will investigate $\hat{\delta}(W)$ and the nature of the inverse transformation in more detail in a minute; we will assume that $\hat{\delta}(W)$ is an analytic function of $W$ and show later that this is justified.

We now solve (5.28), (5.29) and (5.30) algebraically to give

$$\hat{\tau}_2\,(W, \omega) = -g\hat{\tau}_1\,(W)/(W - \omega) \tag{5.32}$$

$$\hat{\tau}_1(W) = \hat{\delta}(W)/h(W) \tag{5.33}$$

where

$$h(W) = (W - m_0) + g^2 \sum_k{}' \frac{u^2(\omega)}{2\omega} \cdot \frac{1}{(W - \omega)} \qquad (5.34)$$

Finally

$$\hat{\tau}_4(W, \omega, \omega') = \frac{2\omega\delta_{kk'}}{u^2(\omega)(W - \omega)} + \frac{g^2\hat{\delta}(W)}{(W - \omega)(W - \omega')h(W)}. \qquad (5.35)$$

The analytic structure of $\hat{\tau}_1(W)$ is basically given by that of $h(W)$; in particular the poles and branch points of $\hat{\tau}_1(W)$ are given by the zeros and branch points of $h(W)$. As carefully described by Lee and Wick[11] this structure is as follows. There is a branch cut along the real $W$-axis from $\mu$ to $+\infty$, together with 2 complex zeros, at $E_{\pm c}$ (with $E_{+c} = E_{-c}^*$). These complex poles are ensured to be non-real provided $h(W) < 0$ for all $W < \mu$; the latter condition can easily be seen to be satisfied for any positive value of $g^2$ if $m_0^0 > \mu$, since both terms on the right hand side of (5.35) will be negative for $W < \mu$. For $m_0^0 < \mu$ it is necessary that $g^2$ is large then some positive quantity $g_{\text{crit}}^2$ in order that $h(W) < 0$ for all $W < \mu$, as can be seen by a slightly more detailed analysis[11].

From the above analysis we see that we can take the contour $C$ in (5.26) to be parallel to the real axis but above the pole $E_{+c}$ in the upper half plane. Thus if $E_{\pm c} = m_V \pm i\Gamma$, we can take $C = [W : \operatorname{Im} W = \gamma, \gamma > \Gamma]$. With such a definition we see that for $s < 0$ we may complete the contour $C$ by a large semi-circle in the upper half plane, and since there are no poles or other singularities of $\hat{\tau}_i$ in the region enclosed by $C$ and this semi-circle (which also gives a vanishing contribution for very large radius) then $\tau_i(s) = 0$ for $s < 0$. This is correct as we see by inspection of (5.18)–(5.21). For $s > 0$ we can complete $C$ by a contour into the lower half plane. This contour encloses the poles at $E_{\pm c}$ and the branch cut from $W = \mu$ to $+\infty$, so the right hand side of (5.26) has contributions from these singularities; these contributions correspond exactly to those of intermediate states which arise when the complete Green's functions (5.18)–(5.21) are expanded in intermediate states following (3.26). Indeed such an integral representation is precisely that extension of the energy-analytic representation (EAR) to take account of non-real energies and levels.

We now see how we may invert (5.26) using the form of $C$ given in the previous paragraph, since it now becomes a Laplace transform:

$$\hat{\tau}(W) = \int_{-\infty}^{+\infty} \tau(s) \, e^{iWs} ds \qquad (5.36)$$

for any $W$ on $C$. Then in particular $\hat{\delta}(W) = 1$; this is evidently an analytic function of $W$, so that our solutions (5.32), (5.33) and (5.35) are correct.

We now wish to use these solutions to obtain the quantities of interest as far as scattering theory goes, that is the *regularised* Green's functions. We can obtain these very simply from (5.36); we wish to obtain a function which has purely oscillatory behaviour for large positive $s$ and which may be written as the projection onto real energy states of the complete Green's function. Each of the Green's functions $\tau_1, \tau_2, \tau_3. \tau_4$ have the representation

$$\tau_i(s) = \theta(s) \int_c e^{-iWs}\hat{\tau}_i(W)\, dW \tag{5.37}$$

where we have included the $\theta$-function explicitly in (5.3). Now if we consider any of the functions $\tau_i(s)$ explicitly in (5.18)–(5.21) we see that they have the form

$$\theta(s) \langle 0|\, \eta A_i\,(s)\, B_i\, |0\rangle \tag{5.38}$$

where $A_i$ and $B_i$ are suitable operators, $B$ being time-independent. If we use (3.26) we obtain

$$\begin{aligned}
\tau_i(s) = \theta(s) \Big\{ &\sum_r \langle 0|\, \eta A_i\,(0)\, |r\rangle \langle r|\, \eta B_i\, |0\rangle\, e^{-iE_r s} \\
&+ \sum_{+c} \langle 0|\, \eta A_i\,(0)\, |+c\rangle \langle -c|\, \eta B_i\, |0\rangle\, e^{-iE_{+c} s} \\
&+ \sum_{-c} \langle 0|\, \eta A_i\,(0)\, |-c\rangle \langle +c|\, \eta B_i\, |0\rangle\, e^{-iE_{-c} s} \Big\}.
\end{aligned} \tag{5.39}$$

We can write the right hand side of (5.39) in the form of (5.37), though now with contour $C'$ along the positive real axis, together with two small circles, one round $E_{+c}$ and the other round $E_{-c}$. We can also distort the contour $C$ and get exactly $C'$; the integrands of the right hand side of (5.39) and the continued form of (5.37) must then be identical. We can obtain $\tau_{\text{reg}}$ from the expression (5.39) by removing the contributions from the intermediate states $|\pm c\rangle$ i.e. from the small circles round $E_{\pm c}$ in $C'$. This corresponds in (5.37) to replacing the contour $C$ by a contour encircling the real axis for $W > \mu$ in a clockwise fashion. On opening this contour out we obtain that

$$\tau_{\text{reg},i}\,(s) = \theta(s) \left\{ \int_{-\infty+i\varepsilon}^{+\infty+i\varepsilon} e^{-iWs}\, \hat{\tau}_i(W)\, dW - 2\pi i\, e^{-iE_{-c} s}\, \tau_i'(E_{-c}) \right\}. \tag{5.40}$$

The last term on the right hand side of (5.40) is not present in the propagator defined by Lee and Wick[11], but since it is exponentially damped as $s \to +\infty$

it does not contribute on-the-mass-shell; the Lee–Wick propagator can be obtained if we use the purely outgoing asymptotic limit of (3.42), so the regularised Green's function is obtained by moving $C$ down to just above the real axis in (5.31).

## 6  EXTENDED LEE MODEL WITH INDEFINITE METRIC

In this section we will consider the extended Lee model with indefinite metric[11]. We do this for two reasons; firstly, it was pointed out by Lee and Wick that ordinary perturbation theory gives the wrong answers in this case (though not in the $N$–$\theta$ sector for the indefinite metric Lee model of the previous section). We want to show that the use of the coupled Green's function equations for this extended model also give the right answers. The second reason for our study of this model is that it gives an example of the greater simplicity of the Green's function equations as compared to methods using eigenvalue and eigenfunction methods.

The model itself is obtained from the Lee model with indefinite metric by addition of two new particles $V'$ and $\theta'$ allowing the further transition

$$V' \rightleftharpoons V + \theta'. \tag{6.1}$$

The extra terms to be added to the Hamiltonian (5.1) are

$$m_0' V'^+ V' + {\sum_k}' \omega_k' b_k^+ b_k + g' \left[ V'^+ V \sum_k \left( u'(\omega')/\sqrt{2\omega'} \right) b_k \right.$$
$$\left. - V'V^+ \sum_k \left( (u'(\omega')/\sqrt{2\omega'}) b_k^+ \right) \right] \tag{6.2}$$

where $\omega' = \sqrt{\mu'^2 + k^2}$ and $\mu'$ is the $\theta'$-meson's mass. The operators $V'$ and $b_k$ satisfy the non-zero commutation relations

$$[V', V'^+]_- = 1, \quad [b_k, b_{k'}^+]_- = \delta_{kk'}. \tag{6.3}$$

We have that the total Hamiltonian, equal to the sum of the terms in (5.1) and (6.2) is still a pseudo-hermitian operator with the same metric operator as there. We wish to solve the equations of motion which arise from this model to obtain the $N\theta\theta'$ scattering $S$-matrix element. We will do that by starting with the Heisenberg equations of motion. These are:

$$\left( i \frac{d}{ds} - m_0 \right) V(s) = \sum_k \left[ g\left( u(\omega)/\sqrt{2\omega} \right) N(s)\, a_k(s) \right.$$
$$\left. - g' \left( u'(\omega')/\sqrt{2\omega'} \right) V'(s)\, b_k^+(s) \right] \tag{6.4}$$

$$i \frac{d}{ds} N(s) = g \sum_k \left[ u(\omega)/\sqrt{2\omega} \right] V(s) \, a_k^+(s) \tag{6.5}$$

$$\left( i \frac{d}{ds} - \omega \right) a_k(s) = -g \left[ u(\omega)/\sqrt{2\omega} \right] V(s) \, N^+(s) \tag{6.6}$$

$$\left( i \frac{d}{ds} - \omega' \right) b_k(s) = -g' \left[ u'(\omega')/\sqrt{2\omega'} \right] V'(s) \, V^+(s) \tag{6.7}$$

$$\left( i \frac{d}{ds} - m_0' \right) V'(s) = g' \sum_k \left[ u'(\omega')/\sqrt{2\omega'} \right] V(s) \, b_k(s). \tag{6.8}$$

The relevant Green's functions are:

$$\tau_1'(s) = \langle 0| \, \eta T \, [V'(s) \, V'^+] \, |0\rangle$$

$$\tau_2' (s, \omega_1') = \langle 0| \, \eta T \, [V(s) \, b_{k_1}(s) \, V'^+] \, |0\rangle$$

$$\tau_3' (s, \omega_1, \omega_2') = \langle 0| \, \eta T \, [N(s) \, a_{k_1}(s) \, b_{k_2}(s) \, V'^+] \, |0\rangle$$

$$\tau_4' (s, \omega') = \langle 0| \, \eta T \, [V'(s) \, V^+ b_k^+] \, |0\rangle$$

$$\tau_5' (s, \omega_1, \omega_2') = \langle 0| \, \eta T \, [V(s) \, b_{k_1}(s) \, V^+ b_{k_2}^+] \, |0\rangle \tag{6.9}$$

$$\tau_6' (s, \omega_1, \omega_2', \omega_3') = \langle 0| \, \eta T \, [N(s) \, a_{k_1}(s) b_{k_2}(s) \, V^+ b_{k_3}^+] \, |0\rangle$$

$$\tau_7' (s, \omega_1, \omega_1') = \langle 0| \, \eta T \, [V'(s) \, N^+ a_{k_1}^+ b_{k_2}^+] \, |0\rangle$$

$$\tau_8' (s, \omega_1, \omega_2', \omega_3') = \langle 0| \, \eta T \, [V(s) \, b_{k_2}(s) \, N^+ a_{k_1}^+ b_{k_3}^+] \, |0\rangle$$

$$\tau_9' (s, \omega_1, \omega_2, \omega_3', \omega_4') = \langle 0| \, \eta T \, (N(s) \, a_{k_1}(s) \, b_{k_3}(s) \, N^+ a_{k_2}^+ b_{k_4}^+] \, |0\rangle$$

all being in the same sector with the same values of the conserved quantities $N_{V'} + N_V + N_N$, $N_N - N_\theta$, $N_{V'} + N_\theta$, where $N_A$ is the number operator for the particle of type $A$. From (6.4)–(6.8) and the definitions (6.9) we may derive the coupled equations for the Green's functions (6.9).

$$\left( i \frac{d}{ds} - m_0' \right) \tau_1'(s) = i\delta (s) + \sum_k g' \frac{u'(\omega')}{\sqrt{2\omega'}} \, \tau_2' (s, \omega') \tag{6.10}$$

$$\left( i \frac{d}{ds} - m_0 - \omega' \right) \tau_2' (s, \omega') = g \sum_{k_1} \frac{u(\omega_1)}{\sqrt{2\omega_1}} \, \tau_3' (s, \omega_1, \omega') - g' \frac{u'(\omega')}{\sqrt{2\omega'}} \, \tau_1'(s) \tag{6.11}$$

$$\left( i \frac{d}{ds} - \omega_1 - \omega_2' \right) \tau_3' (s, \omega_1, \omega_2') = - g \frac{u(\omega_1)}{\sqrt{2\omega_1}} \, \tau_2' (s, \omega_2'). \tag{6.12}$$

The differential equations for $\tau'_4$ to $\tau'_9$ involve $\tau'_1, \tau'_2, \tau'_3$; we will not consider these further here, since we are concerned most immediately with the $V'$ propagator. We will solve (6.10), (6.11) and (6.12) in a similar fashion to (5.22)–(5.25) of the preceding section, that is, by means of the transform (5.26), though with a different contour $C$. The equations for the transforms $\hat{\tau}'_1, \hat{\tau}'_2, \hat{\tau}'_3$ become

$$(W - m'_0)\, \hat{\tau}'_1(W) = i\hat{\delta}\,(W) + \sum_k g' \frac{u'(\omega')}{\sqrt{2\omega'}}\, \hat{\tau}'_2\,(W, \omega') \qquad (6.13)$$

$$(W - m_0 - \omega')\, \hat{\tau}'_2\,(W, \omega') = g \sum_{k_1} \frac{u(\omega_1)}{\sqrt{2\omega_1}}\, \hat{\tau}'_3\,(W, \omega_1, \omega') - g' \frac{u'(\omega')}{\sqrt{2\omega'}}\, \hat{\tau}'_1(W) \qquad (6.14)$$

$$(W - \omega_1 - \omega'_2)\, \hat{\tau}'_3\,(W, \omega_1, \omega'_2) = -g\, \frac{u(\omega_1)}{\sqrt{2\omega_1}}\, \hat{\tau}'_2\,(W, \omega'_2). \qquad (6.15)$$

We may solve (6.13)–(6.15) algebraically to obtain

$$\hat{\tau}'_1(W) = (W - m'_0)^{-1} \left\{ \hat{\delta}(W) + \sum_k g'\left(u'(\omega')/\sqrt{2\omega'}\right) \hat{\tau}'_2\,(W, \omega') \right\} \qquad (6.16)$$

$$\hat{\tau}'_3\,(W, \omega_1, \omega'_2) = -gu\,(\omega_1)\, \hat{\tau}'_2\,(W, \omega'_2)/\left[\sqrt{2\omega_1}\,(W - \omega_1 - \omega'_2)\right] \qquad (6.17)$$

where

$$(W - m_0 - \omega')\, \hat{\tau}'_2\,(W, \omega') = -g^2 \sum_k u^2(\omega_1)\, \hat{\tau}'_2\,(W, \omega')/[2\omega_1\,(W - \omega_1 - \omega')]$$

$$- \left\{g'u'\,(\omega')/\left[\sqrt{2\omega'}\,(W - m'_0)\right]\right\}$$

$$\left\{\hat{\delta}(W) + \sum_{k_1} g'\left(u'(\omega'_1)/\sqrt{2\omega'_1}\right) \hat{\tau}'_2\,(W, \omega'_1)\right\}.$$

$$(6.18)$$

We may rewrite (6.18) as

$$h\,(W - \omega')\, \hat{\tau}'_2\,(W, \omega') = - \frac{g'u'\,(\omega')}{\sqrt{2\omega'}\,(W - m'_0)} \times$$

$$\times \left[\hat{\delta}(W) + \sum_{k_1} \frac{g'u'\,(\omega'_1)}{\sqrt{2\omega'_1}}\, \hat{\tau}'_2\,(W, \omega'_1)\right] \qquad (6.19)$$

where $h(z)$ is defined by (5.34). Whereas (6.19) is an integral equation for $\hat{\tau}'_2\,(W, \omega)$ it still has a very simple form, essentially like

$$f(x)\, F(x) = g(x) + h(x) \int K(y)\, F(y)\, dy \qquad (6.20)$$

where we have replaced the variable $\omega'$ by $x$ and

$$
F(x) = \hat{\tau}'_2(W, x), \quad f(x) = h(W - x), \quad g(x) = -\frac{g'u'(x)\,i\hat{\delta}(W)}{\sqrt{2x}(W - m'_0)},
$$

$$
h(x) = -\frac{g'u'(x)}{\sqrt{2x}(W - m'_0)}, \quad K(x) = \frac{g'u'(x)}{\sqrt{2x}}.
$$

$$\tag{6.21}$$

We may trivially solve (6.20) to obtain

$$
F(x) = \frac{g(x)}{f(x)} + \frac{h(x)}{f(x)} \cdot \frac{\int [K(y)\,g(y)/f(y)]\,dy}{1 - \int [h(y)\,K(y)/f(y)]\,dy}. \tag{6.22}
$$

The factor

$$
1 - \int dy \cdot h(y)\,K(y)/f(y) = 1 + g'^2 \sum_k \frac{u'^2(\omega')}{2\omega'(W - m'_0)}\,h^{-1}(W - \omega')
$$

will be denoted by $(W - m'_0)^{-1}\,\mathscr{H}(W)$, so that $\mathscr{H}(W)$ is given by

$$
\mathscr{H}(W) = (W - m'_0) + g'^2 \sum_k \frac{u'^2(\omega')}{2\omega'}\,h^{-1}(W - \omega') \tag{6.23}
$$

and is the same as the function $\mathscr{H}_\infty(W)$ defined in equation (3.18) of Lee and Wick[11]. As shown in that reference, $\mathscr{H}$ has a cut along the real axis for $W \geqslant (\mu + \mu')$ and a pole on the real axis if $V'$ is stable (which otherwise shows up as an unstable particle pole on an unphysical sheet in the standard fashion). It also has two cuts parallel to the real axis, one form $(m_0 + \mu' + i\Gamma)$ to $\infty + i\Gamma$ and the other from $m_0 + \mu' - i\Gamma$ to $\infty - i\Gamma$. We remark that these branch cuts correspond to the presence of an extra $\theta'$ particle in the complex eigenstates $|\pm c\rangle$ of energy $m_0 \pm i\Gamma$ which we discussed in the previous section.

If we now explicitly calculate $\hat{\tau}'_2(W, \omega')$ using (6.22) and (6.23) we find

$$
\hat{\tau}'_2(W, \omega') = -\frac{g'u'(\omega')\,\hat{\delta}(W)}{\sqrt{2\omega'}(W - m'_0)\,h(W - \omega')} \times
$$

$$
\times \left[ 1 - \frac{g'^2}{\mathscr{H}(W)} \sum_{k_1} \frac{u'^2(\omega'_1)}{2\omega'_1 h(W - \omega'_1)} \right] \tag{6.24}
$$

whilst

$$\hat{t}'_1(W) = \frac{\hat{\delta}(W)}{(W - m'_0)} - \sum_k \frac{g'^2 u'^2(\omega')}{2\omega' h(W - \omega')} \frac{\hat{\delta}(W)}{(W - m'_0)^2} \times$$

$$\times \left[1 - \frac{g'^2}{\mathscr{H}(W)} \sum_{k_1} \frac{u'^2(\omega'_1)}{2\omega'_1 h(W - \omega'_1)}\right]$$

$$= \frac{\hat{\delta}(W)}{(W - m'_0)} \left\{1 - \frac{[\mathscr{H}(W) - W + m'_0]}{(W - m'_0)} \times \right.$$

$$\left. \times \left[1 - \frac{1}{\mathscr{H}(W)}(\mathscr{H}(W) - W + m'_0)\right]\right\} = \frac{\hat{\delta}(W)}{\mathscr{H}(W)}. \qquad (6.25)$$

We may determine the correct contour $C$ by a similar method as in the preceding section; as there $C$ may be chosen to be parallel to the real axis and above all the complex singularities of $\mathscr{H}(W)^{-1}$, so above $i\Gamma$. Again as before we can obtain the regularised Green's function by replacing $C$ by an integral just above the real axis and one encircling the lower cut in an anti-clockwise manner. This latter contribution gives no contribution for large positive $s$, so does not contribute to the $S$-matrix elements. Hence we obtain the same effective $V'$ propagator as in reference (11) (equation (6.31)).


## 7   IMPLICATIONS

We have shown in the preceding two sections that our field theory axioms in the case of an indefinite metric which we presented in Section 4 are valid in certain models. We wish to consider in this section the implications of these axioms for elementary particle processes. Let us first consider the affect of the indefinite metric on symmetries.

We showed in Section 4 that the indefinite metric is consistent with the proper inhomogeneous Lorentz group. It will evidently be consistent with any internal symmetries and with charge conjugation invariance. Let us consider the other symmetries of interest, $P'$, $T$, and $P'CT$ where $P'$ is the parity operator. Again $P'$ can easily be a symmetry (as it is for the Lee model or its extended form), thought now represented by a pseudo-unitary symmetry when acting on the fields (being unitary on the asymptotic states).

Let us, then, consider the possible $T$-invariance of a theory with indefinite metric. We define the pseudo-unitary operator $T$ as acting on the field

$\phi\,(\mathbf{x},\,t)$ by

$$T\phi\,(\mathbf{x},\,t)\,T^{-1} = \phi\,(\mathbf{x},\,-t). \tag{7.1}$$

We will consider only a single field $\phi$ and take it to be pseudo-hermitian. Then the transformation (7.1) will ensure that $T$ acts in a standard fashion on the particles of the field $\phi$. It is immediate that if $\phi\,(\mathbf{x},\,t)$ is the correct interpolating field then (7.1) implies time reversal invariance of $S$-matrix elements. However the indefinite metric implies that the field operator $P\phi\,(x)\,P$ is the correct interpolating field, where $P$ is the projection operator (3.30). Then we will obtain time reversal invariance of the $S$-matrix elements if $T$ and $P$ commute. To show that this can be obtained let us write

$$T = \eta K V \tag{7.2}$$

where $K$ is the operator of complex conjugation whilst $V$ is a unitary operator, and $V\phi\,(\mathbf{x},\,t)\,V^{-1} = \phi\,(\mathbf{x},\,-t)$.

We assume that $\eta K V$ commutes with $H$:

$$[\eta K V,\,H]_{-} = 0. \tag{7.3}$$

This condition is quite natural, for instance for the Lee model with indefinite metric; commutation of $\eta$ with $H$ produced $H^{+}$ which reverts to $H$ on commutation with $K$. Then the state $\eta K V\,|\pm c\rangle$ is also an eigenstate of $H$ belonging to the eigenvalue $E_{\pm c}$, so assuming non-degeneracy of energy levels

$$\eta K V\,|\pm c\rangle = \alpha_{\pm}\,|\pm c\rangle \tag{7.4}$$

where $\alpha_{\pm}$ are complex numbers. Also

$$\langle +c|\,\eta\eta K V\,|-c\rangle = \alpha_{-}\,\langle +c|\,\eta\,|-c\rangle = \alpha_{+}^{-1}\,\langle -c|\,\eta\,|+c\rangle \tag{7.5}$$

so that, using the normalisation $\langle +c|\,\eta\,|-c\rangle = 1$

$$\alpha_{-} = \alpha_{+}^{-1}. \tag{7.6}$$

We now explicitly evaluate the commutator bracket $[\eta K V,\,P]_{-}$:

$$[\eta K V,\,P]_{-} = \sum_{c} [\eta K V\,|-c\rangle\,\langle +c|\,\eta - |-c\rangle\,\langle +c|\eta^{2}K V$$

$$+\; \eta K V\,|+c\rangle\,\langle -c|\,\eta - |+c\rangle\,\langle -c|\eta^{2}K V]$$

$$= \sum_{c} [\alpha_{-}\,|-c\rangle\,\langle +c|\,\eta - \alpha_{+}^{-1}\,|-c\rangle\,\langle +c|\,\eta + \alpha_{+}\,|+c\rangle\,\langle -c|\,\eta$$

$$-\; \alpha_{-}^{-1}\,|-c\rangle\,\langle +c|\,\eta = 0.$$

Thus we can deduce immediately from the resulting relation

$$TP\phi\,(\mathbf{x}, t)\,(TP)^{-1} = \phi\,(\mathbf{x}, -t)$$

the condition of time-reversal invariance for the regularised Green's functions:

$$\tau_{\mathrm{reg}}^{T}\,(x_1 \cdots x_n) = \tau_{\mathrm{reg}}\,(\mathbf{x}_n, -x_{n0};\,\ldots;\,\mathbf{x}_1, -x_{10}) \qquad (7.7)$$

where $\tau_{\mathrm{reg}}^{T}$ is the time-transformed regularised Green's function

$$\langle 0_T|\,T\,(\phi_{\mathrm{reg}}^{T}(x_1) \cdots \phi_{\mathrm{reg}}^{T}(x_n))\,|0_T\rangle \qquad (7.8)$$

with $|0_T\rangle = T\,|0\rangle$, $\phi_{\mathrm{reg}}^{T}(x) = TP\phi\,(x)\,(TP)^{-1}$.

From parity and $C$ invariance we also obtain $PCT$ invariance, and so the equality of particle and antiparticle masses and lifetimes. We note that very general proof of the $TCP$ theorem given by Jost[12] cannot be extended to the case of an indefinite metric due to the breakdown of local commutativity for the correct interpolating field $P\phi\,(x)\,P$. We remark parenthetically that the $T$ operator of (7.2) has an extra $\eta$ beyond the usual antiunitary form $KV$; this extra $\eta$ will behave as the identity operator on the asymptotic in or out states, commuting with all the in and out fields. Thus in terms of the physical states it can be neglected; in terms of the dynamics of the system as expressed by the total Hamiltonian, this is certainly not true, $\eta$ being unequal to the identity in a very non-trivial fashion. Still, usual time reversal operations on particle observables are unchanged from a theory with positive metric.

We turn finally to the discussion of the possible breakdown of causality and dispersion relations. We see that the correct interpolating field $P\phi\,(x)\,P$ is not microscopically local, so that we can no longer obtain dispersion relations for the scattering amplitudes. Indeed from the results of our discussion of the Lee model we expect there to be complex poles of the scattering amplitude *on the physical sheet*, together with possible branch cuts. These singularities evidently cause a breakdown of dispersion relations in their usual form involving an integral purely along the real axis. What will be the effect of these complex poles? There will be two types of effect, one microscopic, the other macroscopic.

Let us consider the microscopic effect of complex poles on the physical sheet. By microscopic we mean concerned with small number of elementary particles. Then in the elastic scattering amplitude we expect a non-real pole on the physical sheet, together with its complex conjugate, to give a contribution.

$$\lambda/[S - (M + i\Gamma)^2] + \lambda^{+}/[S - (M - i\Gamma)^2]. \qquad (7.9)$$

The order of magnitude of $\lambda$ and $\Gamma$ will be $O(g^2)$ for small $g$, as in the case of weak interactions, when we take the indefinite metric as arising from the intermediate $W$ boson (as from the $V$ particle in the Lee model discussed in Section 5). These poles evidently destroy the dispersion relations, though only by a small amount if $M$ is very large. Thus if we consider the forward $\pi$–$N$ dispersion relations[17], where again the discontinuity across the real axis is given by the total cross-section, by the optical theorem, the dispersion relations now are:

$$\frac{1}{2}\,\omega\,\mathrm{Re}\,[F_{\pi^-}(\omega) - F_{\pi^+}(\omega)] - \frac{\omega^4}{4\pi^2}\,P\int_m^\infty \frac{q'd\omega'}{\omega'^2\,(\omega'^2 - \omega^2)}\,(\sigma_{\pi^-}(\omega') - \sigma_{\pi^+}(\omega'))$$

$$= 2f^2 + \frac{\omega^2}{2\pi^2}\int_m^\infty \frac{q'd\omega'}{\omega'^2}\cdot\frac{\sigma_{\pi^-}(\omega') - \sigma_{\pi^+}(\omega')}{2}$$

$$+ \frac{1}{2}\,\omega\left[\frac{\lambda}{\omega - \omega_0} + \frac{\lambda^+}{\omega - \omega_0^+}\right] \tag{7.10}$$

where $\omega_0 = [(M + i\Gamma)^2 - M_N^2 - \mu^2]/2M_N$, $M_N$ and $\mu$ being the mass of the nuclear and pion respectively. We have assumed only a contribution arising from a non-real pole with zero charge, so from a neutral intermediate boson (taken as scalar for simplicity). We see that if $M$ is large, say $\gtrsim 10$ BeV, then the extra term on the right of (7.10) behaves as a term linear in $\omega$; for small $\Gamma$ we have that this term will be of order $\omega g^2/M$. In the case of weak neutral intermediate vector boson this term is equal to $MGw$, where $G$ is the weak coupling constant, so has slope (in units of the nucleon mass) of $(M/M_N)\,M_N^2 G$ $= (M/M_N)\,10^{-5}$. This would seem to put an upper limit on the mass of the intermediate vector boson, which we consider in more detail elsewhere. It is evident that a more careful analysis may show up the presence of such complex poles in certain scattering channels. There are also other effects, such as the variation of the phase shifts[11] which we hope to return to elsewhere.

Finally let us consider the macroscopic effects of the breakdown of local commutativity for the correct interpolating field $P\phi\,(x)\,P^{-1}$. These effects will be of the form of the scattering of a beam of particles before it seems to reach a scattering centre. As was shown by Lee and Wick[11] it is not possible to observe any violation of causality by means of observing only the motion of the average position of a wave packet; it is necessary to measure the change in shape of the wave packet. The effects of a complex pole for a single particle are difficult to observe due to the fact that the "accausality" distance $\Gamma^{-1}$

is exceedingly small, being less than $10^{-11}$ cms. or $10^{-13}$ cms. for weak and electromagnetic interactions respectively[11]. However it is possible that coherence of this accausality over macroscopic distance may arise from many-body effects, so that an effective transmission of signals at a speed faster than light can be achieved. This has been suggested[18] to occur in quasars or pulsars, which are thought to be composed of matter in a highly condensed state. It would seem of interest to investigate this possibility further.

## Acknowledgement

I would like to thank J. M. Charap for a very helpful discussion.

## References


1. P. A. M. Dirac, *Proc. Roy. Soc. (London)*, **A114**, 243 (1927).
2. J. R. Oppenheimer, *Phys. Rev.*, **35**, 461 (1930).
3. P. A. M. Dirac, *Proc. Roy. Soc. (London)*, **A167**, 148 (1938).
4. See, for example, the preface and papers therein in the collection of selected papers by J. Schwinger, *Quantum Electrodynamics*, Dover Publ., New York (1958).
5. K. Symanzik, lectures at the Summer School for High Energy Physics, Hercegnovi, Yugoslavia (1961); J. G. Taylor, *Il Nuovo Cimento. Suppl.*, *I*, **1**, 857 (1963).
6. See, for example, paper IV of J. G. Taylor, Ref. 5.
7. R. Delbourgo, A. Salam and J. Strathdee, "Infinities in the theory of vector mesons", International Centre for Theoretical Physics (unpublished) (1969).
8. P. A. M. Dirac, *Proc. Roy. Soc. (London)*, **A180**, 1 (1942).
9. See, for example, E. C. G. Sudarshan, *Indefinite Metric and Non Local Field Theories*, *Proc. 14th Solvay Conference, (1967)* Wiley & Son, N.Y., and references quoted there.
10. G. Källen and W. Pauli, *Det Kongelige Danske Videnskabernes Selskab, Mat.-fys. Meddelsler*, **30**, No. 7 (1955).
11. T. D. Lee and G. C. Wick, *Nucl. Phys.*, **B9**, 209 (1969).
12. See, for example, R. F. Streater and A. S. Wightman, *PCT Spin and Statistics and All That*, Benjamin, New York (1964), or R. Jost, "The general theory of quantized fields", *Lectures in Applied Mathematics*, Vol. IV, American Math. Soc. (1965).
13. S. N. Gupta, *Proc. Phys. Soc.*, **63**, 681 (1950); K. Bleuler, *Helv. Phys. Acta*, **23**, 567 (1950).
14. H. Lehmann, K. Symanzik and W. Zimmermann, *Il Nuovo Cim.*, **1**, 205 (1955).
15. R. Haag, *Phys. Rev.*, **112**, 669 (1958); D. Ruelle, *Helv. Phys. Acta*, **35**, 147 (1962); K. Hepp, *Acta Phys. Austreatica*, **17**, 85 (1963).
16. R. B. Curtis and M. S. Maxon, *Phys. Rev.*, **137**, B, 996 (1965).
17. See, for example, S. Gasiorowicz, "The application of dispersion relations in quantum field theory", *Fortschritte der Physik*, **8**, 669 (1960) or S. Gasiorowicz, *Elementary Particle Physics*, Wiley and Sons, New York (1966), Ch. 23.
18. S. Bludman and M. Ruderman, *Phys. Rev.*, **170**, 117 (1968).

# Singular Interactions in Nonrelativistic Quantum Theory

H. H. ALY

*Southern Illinioes University*

W. GUTTINGER and H. J. W. MULLER

*Universität München, Germany*

## Contents

# 1   INTRODUCTION

The investigation of singular potentials in nonrelativistic quantum theory has been motivated by the desire to get a better understanding of the singular nature of unrenormalizable and marginally singular interactions in relativistic quantum theory. The physical analogy of field theoretic singular interactions with singular potential scattering, however, breaks down for short distances because no probabilistic interpretation is available for the field theoretic matrix elements (in the sense of an $L^2$ integrability boundary condition) in virtue of creation and annihilation processes during the interaction, except for the case of asymptotically in-or-out-going states (i.e., for large spacelike separations of field excitations). Nevertheless a certain formal analogy between the field theoretic and the potential case survives if the former is supplied with a certain definition via Euclidean space-time concepts at the expense of dispensing with the conventional interpretation of interaction in terms of particle exchange. The principal merit of the studies on singular potentials may therefore be found in that they reinitiated attempts to extract information from unrenormalizable and marginally singular field and $S$-matrix theories which are supposed to govern weak and vector boson interactions but seemed to be untractable up to quite recently.

The quantum theory of repulsive singular potentials supplies a good ground to test various programs developed for relativistic calculations. We shall consider singular potentials which fall off fast enough (faster than $r^{-3}$) at infinity so that a phase shift is defined for $k^2 = 0$. This property is ensured by the conditions imposed on the potentials and given below. Together with these properties, the potentials are chosen of the types which have either a pole, a branch point or an essential singularity.

In this review article we shall consider various properties of singular potentials and their applications.


## 2  PRELIMINARY CONSIDERATIONS

For the discussion to follow it is useful to have the following theorem:

Let $x_1, x_2$ ($x_1 < x_2$) be two arbitrary points of any interval $(a, b)$ of the real axis, along which the function $F(z)$ is analytic. Then if $w$ is any solution of the equation

$$\frac{d^2 w}{dz^2} + F(z)\, w(z) = 0 \tag{2.1}$$

and if throughout $(a, b)$

$$\text{either} \quad \text{Re}\,(F(z)) \leqq 0$$

$$\text{or} \quad \text{Im}\,(F(z)) \text{ does not change sign,}$$

then $w$ can have at most one zero in $(a, b)$.

*Proof* (Method of Green's transforms):

Set $w = w_1$, $dw/dz = w_2$, $F(z) = g_1(z) + ig_2(z)$. Then we have the simultaneous equations

$$\left.\begin{aligned} dw_1/dz &= w_2 \\ dw_2/dz &= -F(z)\, w_1. \end{aligned}\right\} \tag{2.2}$$

From these we obtain

$$w_2\, d\bar{w}_1 + \bar{w}_1 dw_2 = |w_2|^2\, d\bar{z} - F(z)\, |w_1|^2\, dz \tag{2.3}$$

(the bar meaning complex conjugation), and by integrating between limits $x_1, x_2$ along the real axis and separating real and imaginary parts, we obtain

$$w_1 \bar{w}_2 \Big|_{x_1}^{x_2} + \int_{x_1}^{x_2} \bar{F}(z)|w_1|^2\, dx = \int_{x_1}^{x_2} |w_2|^2\, dx,$$

i.e.

$$\left.\begin{aligned} \text{Re}\,[w_1 \bar{w}_2]_{x_1}^{x_2} + \int_{x_1}^{x_2} g_1(x)\,|w_1|^2\, dx &= \int_{x_1}^{x_2} |w_2|^2\, dx \\ \text{Im}\,[w_1 \bar{w}_2]_{x_1}^{x_2} &= \int_{x_1}^{x_2} g_2(x)\,|w_1|^2\, dx. \end{aligned}\right\} \tag{2.4}$$

Suppose now, $w_1$ has more than one zero in $(a, b)$, and let two consecutive zeros be $x_1, x_2$. Then $w_1(x_1) = 0 = w_1(x_2)$ and

$$\left.\begin{array}{c}
\displaystyle\int_{x_1}^{x_2} g_1(x)\,|w_1|^2\,dx = \int_{x_1}^{x_2} |w_2|^2\,dx \\[4mm]
\displaystyle 0 = \int_{x_1}^{x_2} g_2(x)\,|w_1|^2\,dx.
\end{array}\right\} \tag{2.5}$$

Thus, if $g_1(x)$ is negative, or if $g_2(x)$ does not change sign, our assumption must be wrong; hence $w_1$ has at most one zero in $(a, b)$.

From this theorem we immediately obtain as a corollary: The solution $w = w_1$ oscillates within $(a, b)$ if

$$\text{i)} \quad \operatorname{Re} F(z) \geqq 0$$

and

$$\text{ii)} \quad \operatorname{Im} F(z) \text{ changes sign or vanishes identically.}$$

## 3   DEFINITION OF THE POTENTIAL

In the present context we shall consider spherically symmetric potentials satisfying the following conditions:

a) $V(z)$ is analytic in the half-plane $\operatorname{Re} z > 0$,

b) along any ray $z = r \exp[i\sigma]$, $|\sigma| < \pi/2$,

$$\int_{r_0}^{\infty} |\exp(\nu z) \cdot v(z)|\,dr$$

converges for each $r_0 > 0$ and $\nu < \mu$ (range of potential), and

$$\text{R):} \quad \int_0^{\infty} x\,|V(x)|\,dx < \infty,$$

i.e. the first absolute moment of $V$ is finite, or

$$\text{C):} \quad \lim_{x \to 0} x^2 V(x) = V_0 > 0$$

in such a way that

$$\int_0^{\infty} x\left|V(x) - \frac{V_0}{x^2}\right| dx < \infty,$$

or

$$\text{S):} \quad \int_0^c \frac{1}{x^2 V^{1/2}(x)}\, dx < \infty, \quad \text{or} \quad \int_c^\infty x^2\, |V(x)|\, dx < \infty,$$

for any $c > 0$.

Potentials satisfying conditions a), b) and R) are called regular potentials, those satisfying conditions a), b) and C) are called potentials of centrifugal type, and those satisfying a), b) and S) are called singular potentials. Further, if the potential is real and positive (or negative) in some interval $[0, x_0]$, then it is called repulsive (or attractive). Moreover, if the potential is repulsive, then there are always two types of solutions near $r = 0$:

$$\text{those with} \qquad \lim_{r \to 0} - 0$$

$$\text{and those with} \quad \lim_{r \to 0} \neq 0,$$

whereas, if the potential is attractive, this subdivision is no longer possible, since

$$\lim_{r \to 0} (\text{any solution}) \neq 0.$$

Potentials of centrifugal type will not be discussed at great length in the present context; for a detailed discussion see (for instance) Meetz (1).


## 4   ANALYTIC PROPERTIES OF THE SOLUTIONS AND THE SCATTERING AMPLITUDE

### 4.1   Regular potentials

Using only conditions a) and b) it can be shown that:

The Jost solutions $f(\pm k, l, x)$ defined by the asymptotic boundary conditions

$$\lim_{x \to \infty} \exp(\pm ikx) f(\pm k, l, x) = 1 \tag{4.1}$$

are entire functions of $\lambda$ and are analytic in the $k$-plane cut along the positive and negative imaginary axes. For physical (integral) $\lambda$ the region of analyticity consists of the $k$-plane cut along the imaginary axis from $i\mu/2$ to $i\infty$ and from $-i\mu/2$ to $-i\infty$ with a pole of the $l$th order at the origin.

Furthermore, it can be shown that (for regular potentials):

A solution $\varphi(k, l, x)$ exists with the boundary condition

$$\lim_{x \to 0} x^{-l-1}\varphi(k, l, x) = 1, \quad \operatorname{Re} l > -\tfrac{1}{2}.$$

$\varphi(k, l, x)$ is an entire function of $k$, analytic in $\mathrm{Re}\, l > -\frac{1}{2}$, and is called the regular solution.

The solution $\varphi$ may be reexpressed as a linear combination of the (linearly independent) Jost solutions:

$$\varphi(k, l, x) = \frac{1}{2ik}\,[f(k, l)f(-k, l, x) - f(-k, l)f(k, l, x)], \quad (4.2)$$

where the coefficients $f(\pm k, l)$ are the Jost functions. The $S$-matrix is then defined by the quotient

$$S(k, l) = e^{i\pi l}\,\frac{f(k, l)}{f(-k, l)}. \qquad (4.3)$$

Of course, the existence of the equivalent expression

$$S(k, l) = e^{i\pi l}\,\lim_{x \to 0}\left\{\frac{f(k, l, x)}{f(-k, l, x)}\right\} \qquad (4.4)$$

follows from the existence of the regular solution. Expression (4.4) is equal to

$$e^{i\pi l}\,\frac{f(k, l, 0)}{f(-k, l, 0)} \qquad (4.5)$$

provided each of the limits $\lim f(\pm k, l, 0)$ exists separately, i.e. (4.5) follows from (4.4) if $f(\pm k, l, 0) = f(\pm k, l)$.

Alternatively, from the existence of the limit (4.4) one may deduce the existence of the regular solution. This is the method adopted by Limić (2) for potentials satisfying conditions a), b) and C), or a), b) and S).

Of course, the relations (4.1), (4.2) are valid in general and do not depend on the type of potential considered.

## 4.2  Repulsive singular potentials

### 4.2.1  Existence of S

*Theorem:* $S(k, l)$ as given in (4.4) exists for each $k$ outside the cuts on the imaginary axis if $f(k, l, 0) \neq 0$ (zeros of $S$).

*Proof:* Consider the differential coefficient of $f(k, x)/f(-k, x)$ in the half-open interval $(0, x_1]$:

$$\frac{d}{dx}\left[\frac{f(k, x)}{f(-k, x)}\right] = \frac{f'(k, x)f(-k, x) - f'(-k, x)f(k, x)}{f^2(-k, x)}$$

$$= \frac{W[f(-k, x)f(k, x)]}{f(k, x)f(-k, x)} \cdot \left\{\frac{f(k, x)}{f(-k, x)}\right\}, \qquad (4.6)$$

where $W(f(-k, x), f(k, x)) = -2ik$. Hence

$$\frac{f(k, x)}{f(-k, x)} = \frac{f(k, x)}{f(-k, x)} \exp\left[2ik \int_x^{x_1} \frac{dy}{f(k, y)f(-k, y)}\right]. \qquad (4.7)$$

The theorem follows, if we can show that this integral exists for $x \to 0$. For this reason consider equation (4.1), where now

$$F(x) = -\left[V(x) + \frac{l(l+1)}{x^2} - k^2\right]. \qquad (4.8)$$

For $\mathrm{Re}\, V(z) > 0$ (i.e. $V$ repulsive) and any finite $k^2$, we can always find an interval $[x_1, x_2]$ within an interval $(0, x_0)$ of the real axis, where $\mathrm{Re}\, F(x) \leqslant 0$. Then by Section 1.1 the solution $f$ of (4.1) can vanish at most once within $[x_1, x_2]$ or $(0, x_0)$. Thus we can always choose an interval $(0, x_1]$ within which $f(\pm k, x)$ has no zeroes (remembering that $f(k, 0) \neq 0$). Hence the integral in (4.7) exists in the limit $x \to 0$; q.e.d.

### 4.2.2  Existence of the regular solution

The function

$$\varphi(k, x) = f(k, x) - S(k)f(-k, x) \qquad (4.9)$$

$(f(\pm k, x) \neq 0)$ is a regular solution at the origin.

*Proof:* For $x \in (0, x_1]$ we have by (4.7)

$$\lim_{x \to 0} \varphi(k, x) = \lim_{x \to 0}\left[f(k, x) - f(k, x)\exp\left(2ik \int_0^x \frac{dy}{f(k, y)f(-k, y)}\right)\right]$$

$$= -2ik \lim_{x \to 0} f(k, x) \int_0^x \frac{dy}{f(k, y)f(-k, y)} = 0.$$

q.e.d.

### 4.2.3  Analytic properties of the regular solution in $k$ and $l$

An easy way of studying the analytic properties of the solutions of the radial wave equation is to find an appropriate integral equation.

Consider the equation

$$\varphi_i''(x) = F(x)\,\varphi_i(x) \qquad (4.10)$$

and let $\varphi_1(x)$, $\varphi_2(x)$ be a pair of linearly independent solutions of (4.10), i.e. set

$$W[\varphi_1, \varphi_2] = -1. \qquad (4.11)$$

Let also

$$\varphi_1(x) = Q(x) \exp\left[\int_x^{x_0} \sqrt{W(y)}\, dy\right] \qquad (4.12)$$

for some finite $x_0$. Then $F(x)$ is related to $Q$ and $W$ by

$$F(x) = W(x) + \frac{Q''(x)}{Q(x)} - 2\sqrt{W(x)}\,\frac{Q'(x)}{Q(x)} - \frac{W'(x)}{2\sqrt{W(x)}}\,, \qquad (4.13)$$

and $\varphi_2$ is found to be

$$\varphi_2(x) = Q(x) \exp\left[\int_x^{x_0} \sqrt{W(y)}\, dy\right]\int_0^x \frac{dy}{Q^2(y)}\, \exp\left[-2\int_y^{x_0} \sqrt{W(z)}\, dz\right]$$

$$(4.14)$$

(by (4.11), (4.12)).

From now on we shall consider only potentials of classes C) and S).
Potentials of class C):
Write

$$V(x) = V_0/x^2$$

and

$$W(x) = \frac{V_0 + l(l+1)}{x^2}\,, \qquad (4.15)$$

$$D_x = \frac{d^2}{dx^2} - W(x),$$

and choose $Q(x)$ to be the regular solution of

$$Q''(x) - 2\sqrt{W(x)}\,Q' - \frac{W'}{2\sqrt{W}}\,Q = 0, \qquad (4.16)$$

so that $Q(x)$ is the regular solution at the origin. Then

$$\varphi_i''(x) = W(x)\,\varphi_i(x), \quad \lambda = 1, 2, \qquad (4.17)$$

and the radial Schrödinger equation is

$$\varphi''(x) = [W(x) - k^2]\,\varphi(x). \qquad (4.18)$$

Clearly

$$F(x, y) = \varphi_1(x)\,\varphi_2(y) - \varphi_2(x)\,\varphi_1(y) \qquad (4.19)$$

is also a solution of (4.17). Consider now

$$\int_0^y F(x, y)(-k^2)\, \varphi(x)\, dx = \int_0^y F(x, y)\, [\varphi''(x) - W(x)\, \varphi(x)]\, dx$$

$$= \int_0^y [F(x, y)\, \varphi''(x) - F''(x, y)\, \varphi(x)]\, dx$$

$$= [F(x, y)\, \varphi'(x) - F'(x, y)\, \varphi(x)]_0^y. \qquad (4.20)$$

Evaluating (4.20) with the proper boundary conditions, we obtain

$$\varphi(k, x) = \varphi_1(x) + \int_0^x F(x, y)(-k^2)\, \varphi(k, y)\, dy. \qquad (4.21)$$

Now $F(x, y)$ does not contain $k$; hence the kernel is an entire function of $k$ (cf. Poincaré's theorem (3)) and the equation may be iterated and each term is analytic in $k$. The convergence of the series can also be proved (Limic) (2). Thus:

For potentials behaving like $1/x^2$ near $x = 0$, there exists a unique function $\varphi_1(x)$ such that the regular solution $\varphi(k, l, x)$ with the boundary condition

$$\lim_{x \to 0} \frac{\varphi(k, l, x)}{\varphi_1(x)} = 1 \qquad (4.22)$$

is an entire function of $k$ (i.e. analytic everywhere except at $\infty$, where it has an essential singularity).

Potentials of Class S):

We now write

and choose

$$\left. \begin{aligned} W(x) &= V(x) \\ Q(x) &= W^{-1/4}(x) \end{aligned} \right\}. \qquad (4.23)$$

Then

$$\left. \begin{aligned} D_x \varphi_i &= 0, \quad i = 1, 2; \quad D_x = \frac{d^2}{dx^2} - V(x) - \frac{Q''(x)}{Q(x)}, \\[2mm] \frac{Q''(x)}{Q(x)} &= \frac{5}{16}\, (V'/V)^2 - \frac{1}{4}\, V''/V), \end{aligned} \right\} \qquad (4.24)$$

$$\varphi_2(x) = \frac{1}{2}\, Q(x) \exp\left[ -\int_x^{x_0} \sqrt{W(y)}\, dy \right], \qquad (4.25)$$

and

$$\varphi_1(x)\,\varphi_2(x) = \tfrac{1}{2}W^{1/2}(x). \tag{4.26}$$

The radial wave equation may be written

$$D_x\varphi\,(x) = \left[\frac{l\,(l+1)}{x^2} - k^2 - \frac{Q''(x)}{Q(x)}\right]\varphi(x). \tag{4.27}$$

Then by the same procedure as above we find that

$$\varphi\,(k,x) = \varphi_2(x) + \int_0^x F\,(x,y)\left[k^2 - \frac{l\,(l+1)}{y^2} + \frac{Q''(y)}{Q(y)}\right]\varphi(y)\,dy. \tag{4.28}$$

By (4.24), (4.26) we see that the kernel of this integral equation behaves like $[y^2 V\,(y)]^{-1}$ near $y = 0$. For the class of potentials considered here (i.e. $S$), this is an integrable function at the origin, and hence the function $\varphi(x)$ defined above exists. In the same way as above it follows that: $\varphi\,(k,l,x)$ is an entire analytic function of $k$ and analogously (in this case) also of $l$ for $0 \leqslant x < \infty$, and satisfies the boundary condition

$$\lim_{x\to 0} \frac{\varphi\,(k,l,x)}{\varphi_2(x)} = 1. \tag{4.29}$$

### 4.2.4   *Analytic properties of the solutions in $G = g^2$*
The scattering amplitude $A_1(k)$ is defined by

$$A_l(k) = \frac{1}{2ik}\,[S(k) - 1], \tag{4.30}$$

so that a solution $\psi(r)$ may be written $(\lambda = 1 + \tfrac{1}{2})$

$$\left.\begin{aligned} \psi\,(k,r) &= g\,(k,r) + A_l(k)f\,(-k,r), \\[2mm] \text{where}\qquad g\,(k,r) &= -\frac{1}{2ik}\,[f\,(k,r) - f\,(-k,r)] \end{aligned}\right\}. \tag{4.31}$$

Of course, these relations are true in general and are not restricted to regular potentials.

Following Pais and Wu (4) we now examine the analytic properties of the solutions of the wave equation as a function of the coupling parameter $G$, and hence show that for regular as well as singular potentials, the scattering amplitude $A_1(k)$ may be defined as

$$A_l(k) = -\lim_{r \to 0} \left\{ \frac{g(\lambda, k, r)}{f(\lambda, -k, r)} \right\}, \tag{4.32}$$

where this expression is more general than (4.4) combined with (4.30).

Now $A_1(k)$ may also be found by using power series expansions in $G$ without the introduction of a cut-off (cf. Tiktopoulos and Treiman (5)). This follows from the fact, justified below, that $g(k, r)$ and $f(-k, r)$ are for $r \neq 0$ entire functions in the coupling constant $G$, i.e. their power expansions are everywhere uniquely defined.

Considering first of all only the case of $S$-waves, we have

$$D_r \psi(r) = GV(r) \psi(r), \quad D_r \equiv \frac{d^2}{dr^2} + k^2, \quad k \geqslant 0 \quad (0 \leqslant r \leqslant \infty) \tag{4.33a}$$

For $r \to \infty$ we set

$$\psi(k, r) \sim \frac{1}{k} \sin kr + A(k) e^{ikr}, \tag{4.33b}$$

where $A(k)$ is to be determined. Then, since we are considering a repulsive potential, the set of all solutions $\psi$ may be divided into two classes (cf. Section 3) denoted by $S$ and $L$, such that for any $\psi_1 \in \{\psi^S\}$, $\psi_2 \in \{\psi^L\}$;

$$\{\psi\} = \{\{\psi^S\}, \{\psi^L\}\} \quad \text{and} \quad \lim_{r \to 0} \left( \frac{\psi_1(r)}{\psi_2(r)} \right) = 0. \tag{4.34}$$

Clearly, the solution linearly independent of $\psi_1$ belongs to $\{\psi^L\}$ and vice versa. Then there exists always a solution $\psi_S \in \{\psi^S\}$ which is real. Set

$$\psi(k, r) - \text{const } \psi_s(k, r), \tag{4.35}$$

and define a function $F$ by

$$F(r', r) \equiv \frac{1}{k} \sin k (r' - r). \tag{4.36}$$

Clearly

$$D_r F(r', r) = 0$$

and (writing $V \equiv GV(r)$)

$$G \int_{r>0}^{\infty} F(r', r) V(r') \psi(r') \, dr' = \left[ F(r', r) \frac{d}{dr'} \psi(r') - \frac{d}{dr'} F(r', r) \cdot \psi(r') \right]_r^{\infty}$$

$$= \psi(k, r) - \psi_0(k, r), \tag{4.37}$$

where

$$\psi_0\,(k,r) = \frac{1}{k}\,\sin kr + A(k)\,e^{ikr}. \qquad (4.38)$$

Thus, defining two functions $g$ and $f$ by

$$g\,(k,r) = \frac{1}{k}\,\sin kr + G\int_r^\infty dr'F\,(r',r)\,V(r')\,g\,(k,r'), \qquad (4.39)$$

$$f(-k,r) = e^{ikr} + G\int_r^\infty dr'F\,(r',r)\,V(r')\,f\,(-k,r') \qquad (4.40)$$

we see that

$$g\,(k,r) + A(k)\,f\,(-k,r) = \psi\,(k,r).$$

Clearly $f\,(\pm k,r)$ are the Jost functions defined by the limit (4.1).

Equations (4.39), (4.40) are Volterra integral equations defined for any $r > 0$. By Section 3 we see that all iteration integrals exist. Hence: $f, g$ are entire functions of $G$. This does not imply that $\psi\,(k,r)$ is an entire function of $G$, since nothing has so far been said about the $G$-dependence of $A(k)$.

We now want to show that $A(k)$ is determined by the limit (4.32).

Let $\psi_S,\,\psi^*$ be linearly independent solutions of the radial wave equation. Then $\psi_S \in \{\psi^S\}$, $\psi^* \in \{\psi^L\}$, and

$$\left.\begin{aligned} f\,(-k,r) &= \alpha\psi_S + \alpha'\psi^* \\ g\,(k,r) &= \beta\psi_S + \beta'\psi^* \end{aligned}\right\}, \qquad (4.41)$$

so that

$$\text{const.}\,\psi_S = \psi = g\,(k,r) + A(k)\,f\,(-k,r)$$

$$= ((\beta + A\alpha)\,\psi_S + (\beta' + A\alpha')\,\psi^*. \qquad (4.42)$$

Hence

$$\beta' + A\alpha' = 0.$$

If i) $\alpha' = \beta' = 0$, then $f$ is proportional to $g$, which contradicts (4.39), (4.40).

Thus ii) assume $\alpha' = 0$, $\beta' \neq 0$ or vice versa. Then by (4.42) $\psi\,(k,r)$ does not exist (see, however Pais and Wu (6)). Finally iii) if $\alpha' \neq 0$, $\beta' \neq 0$, we have

$$A = -\beta'/\alpha'. \qquad (4.43)$$

Now

$$\lim_{r\to 0} -\left(\frac{g\,(k,r)}{f\,(-k,r)}\right) = -\lim_{r\to 0}\left(\frac{\beta\psi_S + \beta'\psi^*}{\alpha\psi_S + \alpha'\psi^*}\right) = -\lim_{r\to 0}\left(\frac{\beta\psi_S/\psi^* + \beta'}{\alpha\psi_S/\psi^* + \alpha'}\right)$$

$$= -\beta'/\alpha'.$$

by (4.34), Hence

$$A(k) = \lim_{r \to 0} \left( - \left( \frac{g(k, r)}{f(-k, r)} \right) \right). \qquad (4.44)$$

In the general case of $l \neq 0$ we have the following modifications:
Equation (4.43b) is replaced by

$$\psi_l(k, r) \sim \frac{1}{k} \sin\left( kr - \frac{l\pi}{2} \right) + A_l(k)\, e^{ikr} \qquad (4.45)$$

and the integral equations (4.39), (4.40) become

$$g_l(k, r) - e^{-(i\pi l)/2} \left( \frac{\pi r}{2k} \right)^{1/2} J_{l+1/2}(kr) - \frac{\pi}{2}\, G \int_r^\infty dr' F_l(r', r)\, V(r')\, g_l(k, r'),$$

$$(4.46)$$

$$f_l(k, r) = e^{(-i\pi/2)(l+1)} \left( \frac{\pi kr}{2} \right)^{1/2} H^{(2)}_{l+1/2}(kr) - \frac{\pi}{2}\, G \int_r^\infty dr' F_l(r', r) \times$$

$$\times\ V(r') f_l(k, r'), \qquad (4.47)$$

where

$$F_l(r', r) = \sqrt{rr'}\, [J_{l+1/2}(kr')\, Y_{l+1/2}(kr) - J_{l+1/2}(kr)\, Y_{l+1/2}(kr'). \qquad (4.48)$$

The general arguments, of course, remain unchanged, and (4.44) becomes

$$A_l(k) = -\lim_{r \to 0} \left( \frac{g_l(k, r)}{f_l(-k, r)} \right). \qquad (4.49)$$

In the case of regular potentials (again taking $l = 0$ for simplicity) we can
define the counterpart-Volterra equations to (4.39), (4.40), i.e. for $\psi_S$, $\psi^*$,
say, we have

$$\psi_S(k, r) = \frac{1}{k} \sin kr + G \int_0^r dr' F(r', r)\, V(r')\, \psi_S(k, r')$$

$$\psi^*(k, r) = \left\{ \begin{matrix} e^{ikr} \\ \cos kr \end{matrix} \right\} + G \int_0^r dr' F(r', r)\, V(r')\, \psi^*(k, r'). \qquad (4.50)$$

In each case the iterated series converge uniformly in $r$, and

$$\lim_{r \to 0} \psi_S = 0, \quad \lim_{t \to 0} \psi^* = 1.$$

Thus, by (4.42)

$$\lim_{r \to 0} f(-k, r) = \alpha', \quad \lim_{r \to 0} g(k, r) = \beta'$$

and hence

$$A(k) = - \left\{ \frac{\lim\limits_{r \to 0} g\,(k,\,r)}{\lim\limits_{r \to 0} f\,(-k,\,r)} \right\},$$

as claimed. (For specific examples see Pais and Wu (6)).

Consider again the integral equation for the solution $g$ of the radial wave equation. We observe that if we let $r$ approach zero, then the $n$th iteration behaves near the origin like

$$g^{(n)} \underset{r \to 0}{\sim} G^n r^{l+1-n(\beta-2)}, \tag{4.51}$$

if $GV \sim Gr^{-\beta}$. Thus for regular potentials ($\beta < 2$) the series solutions will exist down to $r = 0$ and will converge for sufficiently small values of the coupling constant $G$. But for singular potentials ($\beta > 2$) all iteration integrals would exist down to $r = 0$ only if, for all $n$,

$$\frac{2l + 1}{\beta - 2} > n. \tag{4.52}$$

Thus the series does *not* exist for any value of $G$ other than zero. The inequality (4.52) does, however, show that for a given value of $l$, the first few terms of the series may exist, and thereafter all other terms yield divergent results. This behaviour near the origin indicates that $g\,(k,\,r)$ is highly non-analytic in the coupling constant $G$ at $r = 0$ and that there is a branch point at $G = 0$. This discussion applies equally to the regular solution, for which the integral equation is frequently written

$$\psi_R\,(k,\,r) = \left(\frac{\pi k r}{2}\right)^{1/2} J_{l+1/2}\,(kr) + \frac{\pi}{2}\,G \int_0^\infty \sqrt{rr'}\ \times$$

$$\times\,\{J_{l+1/2}\,(kr')\,Y_{l+1/2}\,(kr)\,\theta\,(r-r')$$

$$+\,J_{l+1/2}\,(kr)\,Y_{l+1/2}\,(kr')\,\theta\,(r'-r)\}\,V(r')\,\psi\,(kr')\,dr'. \tag{4.53}$$

(Note: $g\,(k,\,r)$ is not the regular solution; $g$ and $\psi_R$ satisfy different boundary conditions. Thus $\lim\limits_{r \to 0} \psi_R\,(k,\,r) \sim r^{l+1}$,

$$\lim_{r \to \infty} \psi_R\,(k,\,r) \sim \sin\left(kr - \frac{l\pi}{2}\right) + \tan \delta_l \cos\left(kr - \frac{l\pi}{2}\right).$$

   The possibility of avoiding divergent results by the introduction of a cutoff parameter $\varepsilon > 0$ at the lower limit in integral equations valid for regular potentials has been widely discussed. However, we have seen from the above that the theory of scattering contains a built-in limiting operation (i.e. the limits (4.4), (4.32)), which avoids the introduction of a cutoff as an extraneous computational device. Nevertheless it is instructive to understand the mechanism of this procedure since in most cases it yields the correct result. So consider the simple example of $S$-waves and the potential

$$GV_\varepsilon (r) = \theta (r - \varepsilon) Gr^{-4}. \tag{4.54}$$

The phase-shift is obtained from the relation

$$\tan \delta_l = - \frac{G}{k} \int_0^\infty \left( \frac{\pi kr}{2} \right)^{1/2} J_{l+1/2} (kr) V(r) \psi_R (k, r) dr, \tag{4.55}$$

where $\psi_R (k, r)$ is given by (4.53). In the evaluation of (4.55) we obtain in general integrals over products of Bessel functions. For later convenience we quote the general formula

$$\int_\alpha^\infty dt\, J_\mu(t)\, J_{\pm\mu}(t)\, \frac{1}{t^{1+2\mu}}$$

$$= \frac{\pi \Gamma (2\gamma + 1)}{2^{2\gamma+1} \Gamma^2 (\gamma + 1) \Gamma (\gamma + \mu + 1) \Gamma (\gamma - \mu + 1)} \times \frac{1}{\sin\left\{ \left( \frac{\mu \pm \mu}{2} - \gamma \right) \pi \right\}}$$

$$- \frac{1}{2^{2\gamma+1} \left( \gamma - \frac{\mu \pm \mu}{2} \right) \Gamma (\mu \pm \mu + 1)} \left( \frac{\alpha}{2} \right)^{\mu \pm \mu - 2\gamma} \times$$

$$\times 2F_3 \left( \frac{\mu \pm \mu + 1}{2}, -\gamma + \frac{\mu \pm \mu}{2}, \pm\mu + 1, \mu + \frac{\mu \pm \mu}{2} + 1, \right.$$

$$\left. -\gamma + \frac{\mu \pm \mu}{2} + 1 \,\middle|\, -\alpha^2 \right). \tag{4.56}$$

Integrating (4.55) from $\varepsilon$ to $\infty$, we find that for the potential (4.54)

$$\tan \delta_0 = -k \left[ \frac{G}{\varepsilon} + \frac{1}{3} \frac{G^2}{\varepsilon_3} + \frac{1}{5} \frac{G^3}{\varepsilon^5} + \cdots \right]$$

$$= -k \sqrt{G} \tanh \left[ \sqrt{G}/\varepsilon \right] \tag{4.57}$$

provided in every order of $G$ only the leading terms for $\varepsilon \to 0$ are retained. For finite $\varepsilon$ this expression is analytic in $G$. Passing to the limit $\varepsilon \to 0$, we obtain

$$\tan \delta_0 = -k \sqrt{G}. \tag{4.58}$$

Thus there is now a branch point at the origin of the $G$-plane. The cutoff procedure therefore leads to finite results. Tiktopoulos and Treiman (5) have also known that the result (4.58) is in fact the correct one.

### 4.2.5   *The phase-shift for the potential $G/r^m$*

The integral (4.56) may be used for the calculation of $g_1(k, r), f_1(k, r)$ and hence $A_1(k)$ in the case of a potential

$$GV(r) = G/r^m. \tag{4.59}$$

We find

$$g_l(k, r) = (-i)^l \left(\frac{\pi r}{2k}\right)^{1/2} J_{l+1/2}(kr) - \frac{G\pi}{2}(-i)^l \left(\frac{\pi r}{2k}\right)^{1/2} k^{m-2} \times$$

$$\times \frac{1}{\sin\{(l + \tfrac{1}{2})\pi\}} \cdot \left[ J_{l+1/2}(kr) \left\{ \frac{A}{\sin(1 - m/2)\pi} \right.\right.$$

$$- \frac{F(2)}{2^{m-1}\left(\dfrac{m}{2} - 1\right)}\left(\frac{2}{kr}\right)^{m-2} \right\} - J_{-l-1/2}(kr) \left\{ \frac{A}{\sin\left(l + \dfrac{3}{2} - \dfrac{m}{2}\right)\pi} \right.$$

$$\left.\left. - \frac{F(1)}{2^{m-1}\left(\dfrac{m}{2} - l - \dfrac{3}{2}\right)\Gamma(2l + 2)}\left(\frac{kr}{2}\right)^{2l-m+3} \right\} \right] + O(G^2), \tag{4.60}$$

$$f_l(k, r) = (-i)^{l+1}\left(\frac{\pi kr}{2}\right)^{1/2} \frac{i}{\sin\{(l + \tfrac{1}{2})\pi\}} \left[ J_{-l-1/2}(kr) - e^{i(l+1/2)\pi} \times\right.$$

$$\times J_{l+1/2}(kr) + \frac{G\dfrac{\pi}{2}(-i)^{l+2}k^{m-2}}{[\sin\{(l + \tfrac{1}{2})\pi\}]^2}\left(\frac{\pi kr}{2}\right)^{1/2} \times$$

$$\times\, J_{l+1/2}\,(kr)\,\left\{\frac{A}{\sin\left(-l-\dfrac{m-1}{2}\right)\pi} - \left(\frac{kr}{2}\right)^{-2l-m+1}\times\right.$$

$$\left.\times\, \frac{F(3)}{2^{m-1}\left(l+\dfrac{m-1}{2}\right)\Gamma(-2l)}\right\} + e^{i(l+1/2)\pi}\, J_{-l-1/2}\,(kr)\,\times$$

$$\times\,\left\{\frac{A}{\sin\left(l-\dfrac{m}{2}+\dfrac{3}{2}\right)\pi} - \left(\frac{kr}{2}\right)^{2l-m+3}\times\right.$$

$$\left.\times\, \frac{F(1)}{2^{m-1}\left(-l+\dfrac{m-3}{2}\right)\Gamma(2l+2)}\right\}$$

$$-\,\{J_{-l-1/2}(kr) + e^{i(l+1/2)\pi}\, J_{l+1/2}(kr)\}\times$$

$$\times\,\left\{\frac{A}{\sin\left(1-\dfrac{m}{2}\right)\pi} - \left(\frac{kr}{2}\right)^{-(m-2)}\frac{F(2)}{2^{m-1}\left(\dfrac{m-2}{2}\right)}\right\}\Bigg]\Bigg] + O(G^2),$$

$$(4.61)$$

where

$$A = \frac{\pi\Gamma(m-1)}{2^{m-1}\,\Gamma^2\left(\dfrac{m}{2}\right)\Gamma\left(l+\dfrac{m+1}{2}\right)\Gamma\left(-l+\dfrac{m-1}{2}\right)},$$

$$F(1) = {}_2F_3\left(l+1, l-\frac{m}{2}+\frac{3}{2}, l+\frac{3}{2}, 2l+2, l-\frac{m}{2}+\frac{5}{2}\,\bigg|\, -(kr)^2\right),$$

$$F(2) = {}_2F_3\left(\frac{1}{2}, 1-\frac{m}{2}, -l+\frac{1}{2}, l+\frac{3}{2}, -\frac{m}{2}+2\,\bigg|\, -(kr)^2\right),$$

$$F(3) = {}_2F_3\left(-l, -l-\frac{m}{2}+\frac{1}{2}, -l+\frac{1}{2}, -2l, -l-\frac{m}{2}+\frac{3}{2}\,\bigg|\, -(kr)^2\right)$$

$$(4.62)$$

Hence

$$A_l(k) = -\lim_{r \to 0} \left( \frac{g_l(k, r)}{f_l(k, r)} \right)$$

$$= (-1)^{l+1} \frac{\pi^2 G}{2^m} k^{m-3} \frac{\Gamma(m-1)}{\Gamma^2\left(\dfrac{m}{2}\right) \Gamma\left(l + \dfrac{m+1}{2}\right) \Gamma\left(-l + \dfrac{m-1}{2}\right)} \times$$

$$\times \frac{1}{\sin\left\{\left(l + \dfrac{3}{2} - \dfrac{m}{2}\right)\pi\right\}} + O(G^2), \tag{4.63}$$

provided

$$2 < m < 2l + 3. \tag{4.64}$$

For $\tan \delta_l$ we obtain

$$\tan \delta_l \simeq$$

$$\frac{(-1)^{l+1} \pi^2 G k^{m-2} \Gamma(m-1)}{\left[2^m \Gamma^2\left(\dfrac{m}{2}\right) \Gamma\left(l + \dfrac{m+1}{2}\right) \Gamma\left(-l + \dfrac{m-1}{2}\right) \sin\left\{\left(l + \dfrac{3}{2} - \dfrac{m}{2}\right)\pi\right\} + i(-1)^{l+1} \pi^2 G k^{m-2} \Gamma(m-1)\right]}$$
$$\tag{4.65}$$

For a more detailed discussion of this low-energy expansion see Section 4.6. We remark at this stage, that (4.65) is to be treated with care, since the non-analyticity of the scattering amplitude at $G = 0$ implies that a power expansion in $G$ does not converge. (See also Section (4.47).

### 4.2.6  *Jost functions and the S-matrix*
We conclude from Section 1.31 that the Jost functions $f(\pm k, l)$ are given by

$$f(\pm k, l) = W[f(\pm k, l, r), \varphi(k, l, r)]. \tag{4.66}$$

Their region of analyticity is the common region of analyticity of the functions $f, f', \varphi, \varphi'$. Now, it can be shown (Bottino *et al.* (1)) that the derivatives of the Jost solutions $f$ have the same analytic properties as the $f$'s themselves; also the regions of analyticity of $\varphi'$ are readily seen to be the same as those of $\varphi$, since $\varphi'$ satisfies a similar integral equation. Thus the Jost function $f(\pm k, l)$ is analytic in the topological product of the $k$-plane cut along the $(\pm)$-imaginary axis and the half-plane $\mathrm{Re}\, l > -\frac{1}{2}$; the latter may be extended to the entire $l$-plane for potentials more singular than those of centrifugal type. For physical $l$ and regular potentials the cuts in the $k$-plane proceed from $\pm i\mu/2$ to $\pm\infty$ with a pole of the $l$th order at the origin. The $S$-matrix, of course, is a meromorphic function of $l$ (for each $k$) in the domain of intersection of the regions of analyticity of the Jost functions $f(\pm k, l)$. Limic (2) has shown that the $S$-matrix approaches a constant with increasing $k$ for potentials of class C) (as for regular potentials), whereas for singular potentials it has an essential singularity at infinity.

## 4.3 Canonical form of the radial wave equation

We now consider the potential

$$V(r) = g^2/r^\eta, \quad \eta > 2. \tag{4.67}$$

In order to avoid infrared difficulties (logarithmic divergences) we must have $\eta > 1$. The coupling parameter $g^2$ has dimensions (length)$^{-2}$; hence, as we have seen before, the case $\eta = 2$, where $g^2$ is dimensionless, plays a very special role: In fact, it divides the potentials into regular and singular ones. Setting $\lambda = l + \frac{1}{2} \cdot r = zr_0$, the radial wave equation becomes

$$\frac{d^2\psi}{dz^2} + \left[ k^2 r_0^2 - \frac{l(l+1)}{z^2} - \frac{g}{r_0^{\eta-2} z^\eta} \right] \psi = 0. \tag{4.68}$$

We choose the free parameter $r_0$ such that the kinetic and potential terms have the same coefficients, i.e.

$$k^2 r_0^2 = \frac{g^2}{r_0^{\eta-2}} = f^2, \quad r_0 = \left(\frac{g}{k}\right)^{2/\eta}, \quad f = k\left(\frac{g}{k}\right)^{2/\eta}. \tag{4.69}$$

The radial Schrodinger equation now assumes the canonical form

$$\frac{d^2\psi}{dz^2} + \left[ f^2 - \frac{f^2}{z^\eta} - \frac{\lambda^2 - \frac{1}{4}}{z^2} \right] \psi = 0. \tag{4.70}$$

Here $f$ is a dimensionless parameter. Again we note the difference between regular and singular potentials: In the first case $(1 - 2/\eta < 0)$ $f$ decreases with increasing energy, and so we understand, why the Born approximation gives the right high-energy limit; in the case of singular potentials, $f$ increases with increasing energy, so that a high-energy expansion implies an expansion in inverse powers of $f$. Thus in the singular case, low energy implies weak coupling, whereas high energy implies strong coupling.

## 4.4 Symmetry properties of the solutions

If we now set

$$\psi(z) = z^{1/2}\Phi(z), \quad \alpha = \frac{2}{\eta - 2} \tag{4.71}$$

we obtain

$$\frac{d^2\Phi}{dz^2} + \frac{1}{z}\frac{d\Phi}{dz} + \left(f^2 - \frac{f^2}{z^{2+2/\alpha}} - \frac{\lambda^2}{z^2}\right)\Phi = 0. \tag{4.72}$$

Further, setting

$$\frac{1}{z} = y^{\alpha}, \tag{4.73}$$

eq. (4.72) becomes

$$\frac{d^2\Phi}{dy^2} + \frac{1}{y}\frac{d\Phi}{dy} + \left(\frac{f^2\alpha^2}{y^{2\alpha+2}} - \alpha^2 f^2 - \frac{\alpha^2\lambda^2}{y^2}\right)\Phi = 0. \tag{4.74}$$

Thus if $\Phi(z, f, \lambda, \alpha)$ is a solution of (4.72), then so is $\Phi\left(z^{-1/\alpha}, c^{-i\pi/2}f\alpha, \lambda\alpha, \frac{1}{\alpha}\right)$. In particular the Jost solutions, defined by the limit

$$\left.\begin{aligned}\lim_{z\to\infty}\psi^{(\pm)}(z) &\simeq e^{\mp ifz} = e^{\mp ikr}, \\ \psi^{(\pm)}(z) &= z^{1/2}\Phi^{(\pm)}(z),\end{aligned}\right\} \tag{4.75}$$

may be written

where

$$\left.\begin{aligned}\lim_{z\to\infty}\Phi^{(\pm)}(z, f, \lambda, \alpha) &\simeq z^{-1/2}e^{\mp ifz} \\ \lim_{z\to 0}\Phi^{(\pm)}\left(z^{-1/\alpha}, e^{-i\pi/2}\alpha f, \alpha\lambda, \frac{1}{\alpha}\right) &\simeq z^{1/2\alpha}e^{\mp\alpha fz - 1/\alpha}\end{aligned}\right\} \tag{4.76}$$

Thus the regular solution may be written

$$\psi_R(z, f, \lambda, \alpha) = \psi^{(+)}\left(z^{-1/\alpha}, e^{-i\pi/2}\alpha f, \alpha\lambda, \frac{1}{\alpha}\right) \underset{z\to 0}{\simeq} z^{\eta/4}\exp\left[-\alpha g/r^{1/\alpha}\right]. \tag{4.77}$$

The corresponding irregular solution is obviously given by

$$\Psi_{IR}(z, f, \lambda, \alpha) = \psi_R((-1)^{2/\eta}z, (-1)^{2/\eta}f, \lambda, \alpha) \underset{z\to 0}{\simeq} z^{\eta/4}\exp\left[+\alpha g/r^{1/\alpha}\right]. \tag{4.78}$$

For many purposes it is convenient to define these solutions more specifically in terms of integral equations. Thus, since the equation

$$\frac{d^2\psi}{dr^2} - \frac{g^2}{r^\eta}\,\psi = 0 \tag{4.79}$$

is exactly soluble in terms of Besel functions, we could define the regular solution by the equation

$$\psi(r) = -\sqrt{r}\,H_n^{(2)}(\theta) + \frac{\pi}{2i\,(2-\eta)}\int_0^r F(r',r)\,\psi(r')\left[\frac{l(l+1)}{r'^2} - k^2\right]dr',$$

where

$$F(r',r) = \sqrt{rr'}\,[H_n^{(1)}(v')\,H_n^{(2)}(v) - H_n^{(1)}(v)\,H_n^{(2)}(v')],$$

$$v(r) = \frac{ig}{2\eta}\,r^{2n},\quad v' = v(v'),\quad n = \frac{1}{4}\,(2-\eta). \tag{4.80}$$

Here $\psi$ is normalized such that

$$\psi(r) \underset{t\to 0}{\sim} \sqrt{\frac{\eta-2}{\pi g}}\,e^{i\pi/4\,(3-\eta/2)}\,r^{\eta/4}\exp\left[-\frac{2g}{\eta-2}\,\frac{1}{r^{\eta-2/2}}\right]. \tag{4.81}$$

Similarly we have

$$\psi_R(\omega) = \sqrt{\frac{\pi\alpha r}{2}}\,g\,e^{-i\pi/2\,(\alpha\lambda+1)/2}\,H_{\alpha\lambda}^{(2)}(\omega) + \frac{\pi}{4}\,k^2 g^{4/\eta-2}$$

$$\int_\omega^\infty F(\omega,\omega')\left(\frac{\omega'}{\omega}\right)^{\alpha/2}\left(e^{-i\pi/2}\,\frac{\alpha}{\omega'}\right)^{2(1+\alpha)}\psi_R(\omega')\,d\omega',$$

$$F(\omega,\omega') = \sqrt{\omega\omega'}\,[H_{\alpha\lambda}^{(2)}(\omega)\,H_{\alpha\lambda}^{(1)}(\omega)' - H_{\alpha\lambda}^{(2)}(\omega')\,H_{\alpha\lambda}^{(1)}(\omega)]$$

$$\omega(r) = e^{-i\pi/2}\,\alpha f z^{-1/2} = e^{-i\pi/2}\,g\alpha/r^{1/\alpha},\quad \omega' = \omega(r'),$$

$$\alpha\lambda = (2l+1)/(\eta-2). \tag{4.82}$$

Here $\psi(r)$ is normalized such that

$$\psi(r) \underset{r\to 0}{\sim} r^{\eta/4}\exp\left[-\frac{2g}{\eta-2}\,\frac{1}{r^{\eta-2/2}}\right]. \tag{4.83}$$

The derivation of these equations is quite simple. We infer from them immediately that $\psi_R$ satisfies the following symmetry properties:

$$\left. \begin{aligned} \psi_R(\lambda, k, g, r) &= \psi_R(\lambda, -k, g, r) \\ \psi_R(\lambda, k, g, r) &= \psi_R(-\lambda, k, g, r) \end{aligned} \right\} \tag{4.84}$$

The second of these relations follows from the property of Hankel functions:

$$H^{(1,2)}_{-\nu}(z) = e^{\pm i\pi\nu}\, H^{(1,2)}_{\nu}(z).$$

Furthermore, we can choose the irregular solution $\psi_{IR}$ such that

$$\psi_{IR}(\lambda, k, g, r) = \psi_R(\lambda, k, -g, r). \tag{4.85}$$

Hence

$$W[\psi_R, \psi_{IR}] = -2g. \tag{4.86}$$

For $\omega \neq 0$ (i.e. $r \neq \infty$) the iteration integrals in (4.82) all exist.

For the Jost solution we obtain the following integral equation

$$\psi^{(+)}(r) = \sqrt{\frac{\pi k r}{2}}\, e^{-i\pi/2\,(\lambda+1/2)}\, H^{(2)}_{\lambda}(kr) + g^2\, \frac{\pi}{4i} \int_r^{\infty} F(r, r')\, \frac{1}{r'^{\eta}}\, \psi(r')\, dr',$$

$$F(r, r') = \sqrt{rr'}\, [H^{(2)}_{\lambda}(kr)\, H^{(1)}_{\lambda}(kr') - H^{(2)}_{\lambda}(kr')\, H^{(1)}_{\lambda}(kr)].$$

For

$$r \to \infty: \quad \psi^{(+)}(r) \sim e^{-ikr}. \tag{4.87}$$

Again we observe that $\psi^{(+)}$ satisfies certain symmetry relations:

$$\left. \begin{aligned} \psi^{(+)}(\lambda, k, g, r) &= \psi^{(+)}(\lambda, k, -g, r) \\ \psi^{(+)}(\lambda, k, g, r) &= \psi^{(+)}(-\lambda, k, g, r) \end{aligned} \right\} \tag{4.88}$$

Furthermore, the Jost solution $\psi^{(-)}$ may be chosen such that

$$\psi^{(-)}(\lambda, k, g, r) = \psi^{(+)}(\lambda, -k, g, r). \tag{4.89}$$

Hence

$$W[\psi^{(+)}, \psi^{(-)}] = 2ik \tag{4.90}$$

For $r \neq 0$ the iteration integrals in (4.87) all exist.

We note finally that the integral equations (4.82), (4.87) are related to each other by the inversion formula (4.77), so that either may be obtained from the other. Also, in our earlier notation we have

$$f(\pm k, r) = \psi^{(\pm)}(k, r). \tag{4.91}$$

### 4.5  The *S*-matrix

We are now interested in the analytic continuation of the solutions $\psi_R$. $\psi_{IR}$ and $\psi^{(\pm)}$ into one another. So we write

$$\left.\begin{aligned}
\psi_R (z, f, \lambda, \alpha) &= A (f, \lambda, \alpha)\, \psi^{(+)} (z, f, \lambda, \alpha) + B (f, \lambda, \alpha)\, \psi^{(-)} (z, f, \lambda, \alpha) \\
\psi_{IR} (z, f, \lambda, \alpha) &= \bar A (f, \lambda, \alpha)\, \psi^{(+)} (z, f, \lambda, \alpha) + \overline B (f, \lambda, \alpha)\, \psi^{(-)} (z, f, \lambda, \alpha)
\end{aligned}\right\}$$

$$(4.92)$$

We also note that the interchanges

$$k \to -k, \quad g \to -g$$

are equivalent to the replacements

$$\left.\begin{aligned}
z &\to (-1)^{2/n}\, z \\
f &\to -(-1)^{2/n} f
\end{aligned}\right\}, \qquad
\left.\begin{aligned}
z &\to (-1)^{2/n}\, z \\
f &\to (-1)^{2/n} f
\end{aligned}\right\}
\tag{4.93}$$

respectively. The symmetry relations of Section 4.4 now read

$$\left.\begin{aligned}
\psi_{R,IR} (z, f, \lambda, \alpha) &= \psi_{R,IR} ((-1)^{2/n}\, z, -(-1)^{2/n} f, \lambda, \alpha) \\
\psi_{R,IR} (z, f, \lambda, \alpha) &= \psi_{IR,R} ((-1)^{2/n}\, z, (-1)^{2/n} f, \lambda, \alpha) \\
\psi^{(\pm)} (z, f, \lambda, \alpha) &= \psi^{(\pm)} ((-1)^{2/n}\, z, (-1)^{2/n} f, \lambda, \alpha) \\
\psi^{(\pm)} (z, f, \lambda, \alpha) &= \psi^{(\mp)} ((-1)^{2/n}\, z, -(-1)^{2/n} f, \lambda, \alpha)
\end{aligned}\right\}
\tag{4.94}$$

Hence

$$\left.\begin{aligned}
\bar A (f, \lambda, \alpha) &= A ((-1)^{2/n} f, \lambda, \alpha),\ \overline B (f, \lambda, \alpha) = B ((-1)^{2/n} f, \lambda, \alpha), \\
A (f, \lambda, \alpha) &- B ((-1)^{2/n} f, \lambda, \alpha),\ B (f, \lambda, \alpha), = A (-(-1)^{2/n} f, \lambda, \alpha)
\end{aligned}\right\}
\tag{4.95}$$

and

$$\left.\begin{aligned}
\psi_R (z, f, \lambda, \alpha) &= A (f, \lambda, \alpha)\, \psi^{(+)} (z, f, \lambda, \alpha) \\
&\quad + A (-(-1)^{2/n} f, \lambda, \alpha)\, \psi^{(-)} (z, f, \lambda, \alpha) \\
\psi_{IR} (z, f, \lambda, \alpha) &= A ((-1)^{2/n} f, \lambda, \alpha)\, \psi^{(+)} (z, f, \lambda, \alpha) \\
&\quad + A (-f, \lambda, \alpha)\, \psi^{(-)} (z, f, \lambda, \alpha)
\end{aligned}\right\}
\tag{4.96}$$

Then

$$S_l^{(n)}(k) = -\,\frac{A (-(-1)^{2/n} f, \lambda, \alpha)}{A (f, \lambda, \alpha)}.
\tag{4.97}$$

In particular for $\eta = 4$

$$S_l^{(4)}(k) = -\frac{A\,(\pm if, \lambda, \alpha)}{A\,(f, \lambda, \alpha)}.$$

(4.98)

On the other hand, if we reexpress $\psi^{(-)}$ as a linear combination of regular and irregular solutions, we obtain

$$\psi^{(-)}(z, f, \lambda, \alpha)$$

$$= \frac{A\,((-1)^{2/\eta} f, \lambda, \alpha)\,\psi_R\,(z, f, \lambda, \alpha) - A\,(f, \lambda, \alpha)\,\psi_{IR}\,(z, f, \lambda, \alpha)}{A\,((-1)^{2/\eta} f, \lambda, \alpha)\,A\,(-(-1)^{2/\eta} f, \lambda, \alpha) - A\,(f, \lambda, \alpha)\,A\,(-f, \lambda, \alpha)},$$

(4.99)

and the condition for poles (e.g. Regge poles) is

$$A\,(f, \lambda, \alpha) = 0.$$

(4.100)

The Jost functions $f(\pm k, l) \equiv F^{(\pm)}(f, \lambda, \alpha)$ are given by

$$F^{(\pm)}(f, \lambda, \alpha) = W\,[\psi^{\pm}(z, f, \lambda, \alpha), \psi_R\,(z, f, \lambda, \alpha)].$$

(4.101)

They obey the relationship

$$F^{(\pm)}(-(-1)^{2/\eta} f, \lambda, \alpha) = F^{(\mp)}(f, \lambda, \alpha).$$

(4.102)

The $S$-matrix may be written

$$S = \frac{F^{(+)}(f, \lambda, \alpha)}{F^{(-)}(f, \lambda, \alpha)}$$

(4.103)

Alternatively, inverting (4.76), (4.77) we obtain

and
$$\left.\begin{aligned}
\lim_{z \to 0} \Phi_R\,(z, f, \lambda, \alpha) &\simeq z^{1/2\alpha} e^{-\alpha f z - 1/\alpha} \\[2mm]
\Phi_R\,(z, f, \lambda, \alpha) &= \Phi^{(+)}\left(z^{-1/\alpha}, e^{-i\pi/2}\, \alpha f, \alpha\lambda, \frac{1}{\alpha}\right)
\end{aligned}\right\}$$

(4.104)

so that

and
$$\left.\begin{aligned}
\Phi^{(\pm)}(z, f, \lambda, \alpha) &= \Phi_R\,(z^{-1/\alpha} e^{\pm i\pi/2} \alpha f, \alpha\lambda, \frac{1}{\alpha} \\[2mm]
\lim_{z \to \infty} \Phi^{(\pm)}(z, f, \lambda, \alpha) &\simeq z^{-1/2} e^{\mp i f z}
\end{aligned}\right\}.$$

(4.105)

The solutions $\Phi_R, \Phi^{(\pm)}$ may be matched at the invariant point $Z = Z^{-1/\alpha}$, i.e. $z = +1$. The physical wave function at infinity is again

$$\psi\,(z, f, \lambda, \alpha) = A\,[\psi^{(+)}(z, f, \lambda, \alpha) - S\psi^{(-)}(z, f, \lambda, \alpha)]$$

(4.106)

and at the origin

$$\psi(z, f, \lambda, \alpha) = B\psi_R(z, f, \lambda, \alpha). \tag{4.107}$$

Continuity at the point of connection requires that

$$\left.\frac{\dfrac{d}{dz}\Phi^{(+)}(z, f, \lambda, \alpha) - S\dfrac{d}{dz}\Phi^{(-)}(z, f, \lambda, \alpha)}{\Phi^{(+)}(z, f, \lambda, \alpha) - S\Phi^{(-)}(z, f, \lambda, \alpha)}\right]_{z=1} = \left.\frac{\dfrac{d}{dz}\Phi_R(z, f, \lambda, \alpha)}{\Phi_R(z, f, \lambda, \alpha)]}\right]_{z=1}$$

Hence

$$S = \left.\frac{\Phi_R(z, f, \lambda, \alpha)\dfrac{d}{dz}\Phi^{(+)}(z, f, \lambda, \alpha) - \Phi^{(+)}(z, f, \lambda, \alpha)\dfrac{d}{dz}\Phi_R(z, f, \lambda, \alpha)}{\Phi_R(z, f, \lambda, \alpha)\dfrac{d}{dz}\Phi^{(-)}(z, f, \lambda, \alpha) - \Phi^{(-)}(z, f, \lambda, \alpha)\dfrac{d}{dz}\Phi_R(z, f, \lambda, \alpha)}\right]_{z=1}$$

$$= \left.\frac{\Phi_R(1, f, \lambda, \alpha)\dfrac{1}{\alpha}\dfrac{d}{dy}\Phi_R\left(y, i\alpha f, \alpha\lambda, \dfrac{1}{\alpha}\right) + \Phi_R\left(1, i\alpha f, \alpha\lambda, \dfrac{1}{\alpha}\right)\dfrac{d}{dy}\Phi_R(y, f, \lambda, \alpha)}{\Phi_R(1, f, \lambda, \alpha)\dfrac{1}{\alpha}\dfrac{d}{dy}\Phi_R\left(y, -i\alpha f, \alpha\lambda, \dfrac{1}{\alpha}\right) + \Phi_R\left(1, -i\alpha f, \alpha\lambda, \dfrac{1}{\alpha}\right)\dfrac{d}{dy}\Phi_R(y, f, \lambda, \alpha)}\right]_{y=1}$$

$$\tag{4.108}$$

Thus, once the solution $\Phi_R$ (or one of the other solutions) is known, the $S$-matrix may be evaluated.

## 4.6   The low-energy (weak-coupling) limit

We now want to discuss in more detail the low-energy behaviour of the phase shift for scattering by the potential $V(r) = g^2/r^n, n > 2$ (cf. also Section 4.2.5). So it is desirable to have integral equations for $\psi$ in the regions $z \gtrless 1$, i.e. (4.82) and (4.87), or (4.47). But for clarity in the present context we rewrite these in a modified notation.

For $z < 1$ it is convenient to use the variable $1/z = y^\alpha$, and to write

$$\left.\begin{aligned}\psi_R(z) &= z^{1/2}\Phi(z), \quad \Phi = y^{-1/2}\varphi, \\ v &= \alpha f y, \nu = \alpha\lambda.\end{aligned}\right\} \tag{4.109}$$

Then $\varphi$ satisfies the equation

$$\frac{d^2\varphi}{dv^2} - \left[1 + \frac{v^2 - \frac{1}{4}}{v^2}\right]\varphi = -\left(\frac{\alpha f}{v}\right)^{2\alpha+2}\varphi, \qquad (4.110)$$

and the corresponding integral equation becomes

$$\varphi(v) = \sqrt{\frac{2v}{\pi}}\, K_v(v) + \int_v^\infty V(v')\, \varphi_0\,(v',v)\, \varphi(v')\, dv'$$

$$\varphi_0\,(v',v) = \sqrt{vv'}\,[I_v(v')\, K_v(v) - I_v(v)\, K_v(v')], \qquad (4.111)$$

$$V(v) = -\left(\frac{\alpha f}{v}\right)^{2\alpha+2},$$

$I$ and $K$ being modified Bessel and Hankel functions respectively (it is easily verified that $\varphi(v)$ satisfies the boundary conditions of the regular solution). On the other hand, for $z > 1$ we have (setting $z = w/f$ in equation (4.72))

$$\frac{d^2\psi}{d\omega^2} + \left[1 - \frac{\lambda^2 - \frac{1}{4}}{\omega^2}\right]\psi = \left(\frac{f}{\omega}\right)^{2+2/\alpha}\psi, \qquad (4.112)$$

$$\psi = \sqrt{\frac{\pi\omega}{2}}\, e^{-i\pi/2(\lambda+1/2)}\, H_\lambda^{(2)}(\omega) + \int_\omega^\infty V(\omega')\, \psi_0\,(\omega',\omega)\, \psi(\omega')\, d\omega',$$

$$\varphi_0\,(\omega',\omega) = \frac{1}{i}\,\frac{\pi}{4}\,\sqrt{\omega\omega'}\,[H_\lambda^{(2)}(\omega)\, H_\lambda^{(1)}(\omega') - H_\lambda^{(2)}(\omega')\, H_\lambda^{(1)}(\omega)],$$

$$V(\omega) = \left(\frac{f}{\omega}\right)^{2+2/\alpha}. \qquad (4.113)$$

Of course, because of the symmetry of the solutions, it is sufficient to know and to solve just one of the integral equations (4.111), (4.113). But in view of the complexity of the transformation of one into the other, it is useful to have both equations to refer to. The iterative solution of (4.113) in powers of $f$ has recently been discussed at great length by del Guidice and Galzenati (8). The complete analytic expression of the first iteration has already been given in Section 4.2.5 (cf. eq. (4.61)). We observe that there are several critical values of $\eta = \eta_c$, for which this iterative solution breaks down. In particular, when $\lambda\alpha = 1$, $\sin \pi (\lambda - 1/\alpha) = 0$. This condition separates the values of $\eta$ into two sets: those for which $\lambda\alpha = \dfrac{2l + 1}{\eta - 2} \gtrless 1$ or $\eta \gtrless 2l + 3$.

In the general theory of scattering by singular potentials this condition is of considerable significance. We have already seen (cf. (4.52)) that for $\lambda\alpha > 1$ the first term of the Born approximation gives a finite result, whereas for $\lambda\alpha < 1$ all terms of the Born series diverge. For $\lambda\alpha = 1$ a logarithmic term appears. Similarly a scattering length (in the customary sense) cannot be defined unless $\eta > 2l + 3$ (cf. O'Malley *et al.* (9), Levy and Keller (10)). (Thus, for $\eta = 4$ and $S$-wave scattering length can be defined, whereas a $p$-wave scattering length requires modification of the usual method).

Del Guidice and Galzenati (8) have derived expansions of $\psi_R$ and $\psi^{(+)}$ in rising powers of $f$ by expanding the various Bessel functions and collecting terms of the same power of $f$. For $S$-waves (and $\eta \neq \eta_c$, e.g. 3, 4, 5) they obtain

$$
\begin{aligned}
\psi_R(z) = {} & \frac{\Gamma\left(\dfrac{1}{\eta - 2}\right) z}{\sqrt{\pi}\,(\eta - 2)^{1/2 - 1/(\eta - 2)}} \left[ f^{1/2 - 1/(\eta - 2)} \right. \\[2ex]
& + \left( \frac{z^{2-\eta}}{\eta - 3} - \frac{\eta - 2}{6} z^2 \right) f^{5/2 - 1/(\eta - 2)} + 0 \left( f^{5/2 + 3/(\eta - 2)} \right) \Bigg] \\[2ex]
& + \frac{\Gamma\left(\dfrac{1}{\eta - 2}\right)}{\sqrt{\pi}\,(\eta - 2)^{1/2 + 1/(\eta - 2)}} \left[ f^{1/2 + 1/(\eta - 2)} \right. \\[2ex]
& + \left( \frac{z^{2-\eta}}{\eta - 1} - \frac{\eta - 2}{2} z^2 \right) f^{5/2 + 1/(\eta - 2)} + 0 \left( f^{5/2 + 5/(\eta - 2)} \right) \Bigg], \quad (4.114)
\end{aligned}
$$

$$
\begin{aligned}
\psi^{(-)}(z) = {} & 1 - izf - \left[ \frac{1}{2} z^2 - \frac{z^{2-\eta}}{(\eta - 1)(\eta - 2)} \right] f^2 \\[2ex]
& + i \left[ \frac{z^3}{6} - \frac{z^{3-\eta}}{(\eta - 2)(\eta - 3)} \right] f^3 + 2^{\eta - 2}\, \Gamma(1 - \eta)\, e^{i\eta\pi/2}\, f^\eta \\[2ex]
& + ie^{\eta - 2}\Gamma(1 - \eta)\, e^{i\eta\pi/2}\, zf^{\eta + 1} + 0(f^4). \quad (4.115)
\end{aligned}
$$

The $S$-matrix element or phase-shift may be calculated by using (4.108) or via Jost functions. In this way it is found that (for $\eta \neq \eta_c$)

$$
\tan \delta_0 = \frac{\Gamma\left(-\dfrac{1}{\eta-2}\right)}{\Gamma\left(\dfrac{1}{\eta-2}\right)} \left[ (\eta-2)^{-2/(\eta-2)} f^{1+2/(\eta-2)} \right.
$$

$$
\left. + \frac{\Gamma^2\left(-\dfrac{2}{\eta-2}\right)\Gamma\left(-\dfrac{3}{\eta-2}\right)}{\Gamma\left(\dfrac{1}{\eta-2}\right)\Gamma\left(-\dfrac{4}{\eta-2}\right)} \frac{f^{3+6/\eta-2}}{(\eta-2)^{1+6/\eta-2}} \right]
$$

$$
\div \frac{\sqrt{\pi}}{4} \cdot \frac{\Gamma\left(\dfrac{1}{2}-\dfrac{\eta}{2}\right)}{\Gamma\left(\dfrac{\eta}{2}\right)} f^\eta + 0\left(f^{\eta+1+2/(\eta-2)}\right). \quad (4.116)
$$

Del Guidice and Galzenati (8) have also calculated corresponding expressions for the critical cases $\eta_c = 3, 4, 5$. We note that for $\eta \gtrless 3$:

$$
f^{1+2/(\eta-2)} \lessgtr f^\eta,
$$

so that in (4.116) these terms change their positions when $\eta$ crosses the value $\eta = 3$.

For all rational values of $\eta$ the low-energy expansion of $\tan \delta$ contains logarithmic terms; consequently $\tan \delta$ is not analytic at $f = 0$, where it has a branch point of the logarithmic type.

For irrational values of $\eta$ the expansion contains terms of the type

$$
f^{l\eta+m/(\eta-2)+\eta/(2)} \qquad\qquad (l,\, m,\, n \text{ integers}),
$$

and thus leads to a nonalgebraic branch-point at $f = 0$.

Now since the expansion in rising powers of $f$ is equivalent to a small-coupling-constant-expansion or a low-energy-expansion, the branch-point at $f = 0$ represents singularities in both the $k$- and $g$-planes. This branch-point is, of course, due to both the long range of the potential and to its singular behaviour at the origin; in other words, it is both an infra-red and an ultraviolet divergence. In fact:

$$\ln f = \left(1 - \frac{2}{\eta}\right) \ln k + \frac{2}{\eta} \ln g;$$

thus in the complex $k$-plane the cut due to the potential comes up to the origin, a characteristic feature of the long range potential, which could be removed by introducing an exponential fall-off, e.g.

$$V(r) = g^2 e^{-\mu t}/r^\eta.$$

The other term, $\ln g$, is due to the highly singular nature of the potential at the origin.

### 4.7  The high-energy (strong-coupling) limit

We have seen in Section 4.6 that the low-energy expansion (generally also obtainable in the case of singular potentials) does not differ considerably in its general form from that for regular potentials. We now show that it is really the high-energy region where the physical difference between both types of potentials is most pronounced.

In this case it is easiest, however, to calculate the phase-shift directly from the defining limit (cf. Bertocchi *et al.* (11))

$$\delta = \lim_{R \to \infty} (\eta\pi + \tfrac{1}{2}l\pi - k'_n R), \tag{4.117}$$

where $R$ (assumed large) is the radius of a sphere enclosing the interacting system, and $k'_n$ is the momentum of the scattered wave. By requiring the wave function to be single-valued on the surface of the sphere, one obtains for the repulsive potential the expression

$$\delta = \delta_0 + \delta'$$

where

$$\delta_0 = -\frac{1}{2} \int_{\gamma(\infty)} r' \, dr' \, \frac{d}{dr'} \left[ k^2 - V(r') - \frac{\left(l + \frac{1}{2}\right)^2}{r'^2} \right]^{1/2} + \frac{\pi}{2}\left(l + \frac{1}{2}\right),$$

$$\delta' = \frac{1}{2i} \int_{\gamma(\infty)} g'(x) \, dx, \tag{4.118}$$

$\gamma(\infty)$ being a contour surrounding the right-hand cut of the function $\sqrt{-V(x)}$, $x = \ln r$ in the plane of complex $x$. To a first approximation in $f$ the other integral may be neglected.

Evaluating $\delta_0$ in the present case, one finds in general

$$\delta \sim -k \left(\frac{g}{k}\right)^{2/\eta}. \tag{4.119}$$

More specifically, for $l = 0$ Bertocchi *et al.* (11) obtain

$$\frac{1}{k}\left(\frac{k}{g}\right)^{2/\eta}\left[\delta - \frac{\pi}{4}\right] = -\frac{1}{2\eta} B\left(1 - \frac{1}{\eta}, \frac{1}{2}\right)$$
$$- \frac{\eta + 1}{24\eta} B\left(\frac{1}{\eta}, \frac{1}{2}\right) \cdot \frac{1}{k^2}\left(\frac{k}{g}\right)^{4/\eta},$$
$$+ O\left(\frac{1}{k^4}\left(\frac{k}{g}\right)^{8/\eta}\right). \tag{4.120}$$

where $B(\mu, \nu)$ is the Euler-function defined by

$$B(\mu, \nu) = \frac{\Gamma(\mu)\,\Gamma(\nu)}{\Gamma(\mu + \nu)}.$$

A similar result was recently derived by Limic (2); in fact, he obtained

$$S(l, k) \sim \exp\left[i\pi\left(l + \tfrac{1}{2}\right) - 2ikz_0\left(1 + \tau\right)\right]$$

where

$$V(z_0) = k^2, \quad V \propto \frac{1}{z^\alpha}, \quad \tau = 0\left(\frac{1}{\alpha}\right).$$

We see from (4.119) that for the repulsive potential and $\eta \to \infty$, we have a repulsive core with the phase-shift proportional to the momentum. (We recall that for a perfectly reflecting sphere of radius $a$, i.e. $V(r) = \infty$ for $r < a$ and 0 for $r > a$, $\delta_l = -ka$). This behaviour is completely different from that for regular potentials, where the phase-shift goes to zero for $k \to \infty$. Of course, it is only at high energies that the particle can penetrate so deep as to feel the effect of the singular part of the potential. It has been conjectured (cf. Bertocchi *et al.* (11)) that at high energies the phase-shift might continue to yield resonances in each state of angular momentum. But, as Dombey (12) has shown, the high-energy limit of scattering by a repulsive singular potential is very similar to that of scattering by a partially reflecting sphere whose transparency increases slowly with energy. Thus for $k^2 \to \infty$ no resonances are expected.

## 4.8  Effective-range theory for singular potentials

### 4.8.1  General considerations

The low-energy expansion for the phase-shift

$$k^{2l+1} \cot \delta_l = -\frac{1}{A(l)} + \frac{1}{2} r_0(l)\, k^2 + O(k^4) \qquad (4.121)$$

has been of considerable value for investigating $N\text{–}N$ scattering phenomena in nuclear physics. There, of course, the interactions involve only short-range regular potentials. We now want to discuss the validity of this approach in the case of singular potentials. In particular we shall see, that only for certain values of $l$ and (cf. (4.52)) can a scattering length $A(l)$ and an effective-range $r_0$ be defined. It is an assumption inherent in customary effective-range theory that the potentials have only a short range. Thus, in the case of long range potentials, i.e. potentials which fall off no faster than any power of $1/r$, one would expect some alterations of the method to become necessary; in fact, it is due to this slow fall-off at infinity that the range is ordinarily defined may become infinite.

Again we consider potentials which are more singular than the centrifugal term. (For a repulsive potential proportional to $1/r^2$, the phase-shift does exist, but is energy-independent; hence an expansion of the type (4.121) cannot be defined).

The origin of the breakdown of the effective-range expansion for long-range singular potentials may be understood as follows (cf. O'Malley *et. al* (9) and Levy and Keller (10)). For short-range potentials the asymptotic behaviour of the solutions of the radial wave equation is obtained by neglecting the potential, so that the regular solution ($\psi(0) = 0$) approaches asymptotically linear combination of the free solutions, the coefficients being determined by the phase $\delta_l$. Thus for $r \to \infty$:

$$\psi(r) = \psi_\infty(r) = \text{const}\,[krj_l\,(kr) - \tan \delta \cdot \eta_l\,(kr)],$$

$$\psi_\infty(r) \xrightarrow[r \to \infty]{} \text{const}\left[\sin\left(kr - \frac{l\pi}{2}\right) + \tan\delta \cos\left(kr - \frac{l\pi}{2}\right)\right] \qquad (4.122)$$

and

$$\psi_\infty(r) \xrightarrow[r \to 0]{} \text{const}\left[\frac{(kr)^{l+1}}{(2l+1)!!} + \tan\delta\, \frac{(2l-1)!!}{(kr)^l}\right], \qquad (4.123)$$

where $a!! = a\,(a-2)\,(a-4)\ldots$

We note that for $l = 0$

$$\psi_\infty(r) \sim \text{const} \, [kr + \tan \delta], \tag{4.124}$$

so that

$$A(0) = -\frac{\tan \delta}{k} \tag{4.125}$$

is the $s$-wave scattering-length, as is well-known. In general, of course, the scattering-length is defined by

$$\frac{1}{A(l)} = -\lim_{k \to 0} k^{2l+1} \cot \delta_l. \tag{4.126}$$

Now in the case of regular potentials the terms left out in (4.123) fall-off more rapidly than any power of $1/r$. We note also, that the asymptotic form (4.123) is simply a lincar combination of the linearly independent solutions of the radial wave equation for $k = 0 = V(r)$. For long-range potentials, $\psi_\infty(r)$ will no longer have this form in the limit $k \to 0$, because now there are terms which may dominate the $1/r^{+l}$-term. To see this in more detail, consider the solutions of the wave equation for the potential $V(r) = g^2/r^n$ in the limits $k = 0$ and $r \to \infty$. The solutions are easily found to be

$$\left. \begin{array}{c} r^{1/2} J_m(x), \quad r^{1/2} N_m(x) \\[2mm] m = \dfrac{2l + 1}{n - 2}, \quad x = \dfrac{2g}{n - 2} \, r^{-1/2(n-2)} \end{array} \right\}. \tag{4.127}$$

where

From the asymptotic forms of these solutions we see that

a) for $m \neq$ integer,

$$r^{1/2} J_m(x) \sim r^{-l}, \; r^{1/2} N_m(x) \sim r^{l+1}; \tag{4.128a}$$

hence a scattering length in the customary sense can always be defined, and

b) for $m = $ integer,

$$r^{1/2} J_m(x) \sim r^{-l} \quad \text{but} \quad r^{1/2} N_m(x) \sim \left\langle \frac{1}{r^l} \ln r, \frac{1}{r^l} \right\rangle \quad \text{for} \quad n \leqslant 2l + 3,$$

or

$$r^{1/2} N_m(x) \sim \left\langle \frac{1}{r^l} \ln r, \frac{1}{r^l}, r^{n-3l-3} \right\rangle \quad \text{for} \quad n > 2l + 3, \tag{4.128b}$$

where $\langle \cdots \rangle$ means "dominant term of ..."; hence for $n < 2l + 3$ the scattering-length cannot be defined. The new terms in (4.128) originate directly from the potential. Compared with the behaviour of corresponding solutions for regular potentials, they show that one cannot entirely neglect the long-range potential at infinity, no matter how large one chooses.

Now, if $A(l)$ does not exist, then the question of the existence of $r_0(l)$ does not arise. But if $A(l)$ does exist, then the existence and meaning of $r_0$ has to be reexamined.

For simplicity consider $s$-waves and $n \geqq 4$. Then we have solutions $\psi$ (for $V = 0$) with the boundary conditions

$$\left.\begin{array}{l} \psi(0) = 0, \quad \psi(r) \to \cos kr + \cot \delta \cdot \sin kr, \\[2ex] \psi_0(0) = 0, \quad \psi_0(r) \to r - \dfrac{r}{A(0)}, \end{array}\right\} \tag{4.129}$$

the subscripts implying $k = 0$.

Similarly we define corresponding solutions $\phi$, $\phi_0$ of the free Schrödinger equation, so that for all $r$

$$\left.\begin{array}{l} \phi(r) = \cos kr + \cot \delta \cdot \sin kr, \\[2ex] \phi_0(r) = 1 - r/A(0). \end{array}\right\} \tag{4.130}$$

Here we have chosen the normalization such that for $r \to \infty$, $\psi \to \phi$. Then, directly from the respective differential equations, we obtain

$$k \cot \delta = -\frac{1}{A(0)} + k^2 \int_0^\infty (\phi\phi_0 - \psi\psi_0)\, dr. \tag{4.131}$$

This exact expression is not restricted to any particular type of potential. But in the case of regular potentials it is now argued that $V \sim 0$ for $r$ larger than some $R$, so that the main contribution to the integral comes from the region $0 < r < R$. Then for very small $k^2$: $\phi \sim \phi_0$, $\psi \sim \psi_0$, and the effective range is defined by

$$\frac{1}{2} r_0 = \int_0^\infty (\phi_0^2 - \psi_0^2)\, dr. \tag{4.132}$$

Does this integral exist for $n \geqq 4$?

Take, for instance, $n = 4$. Then $m = \frac{1}{2}$ and $\psi_0(r) \underset{r \to 0}{\sim} r^0, r'$; thus the integral does exist, and $r_0$ is finite. But consider now a value $n_0$ of $n$ such that $(2l + 1)/(n - 2)$ is an integer for $l \neq 0$ (e.g. $n = 7$, $l = 2$), then

$$\psi_0(r) \sim r^{n-3\,l-3}\,(r^{-2})$$

and the (corresponding) integral does not exist, i.e. $r_0$ is infinite.

We conclude therefore: In the case of singular potentials $A(1)$ may be defined if $n > 2l + 3$. But the next term of the expansion exists only for even more restricted values of $n$. In general we can say: for any long-range, highly singular potential the effective-range expansion will break down after a certain number of terms, whereas for short-range, nonsingular potentials these difficulties do not arise.

### 4.8.2    *The Calogero-Wu approximation for the scattering length*

We shall now discuss briefly another method which has recently been suggested for approximating the scattering amplitude in the case of repulsive potentials strongly singular at the origin. These methods have been developed and tested by Calogero (13) using an idea of Wu (14).

For simplicity we restrict ourselves again to $s$-wave scattering. Then, on integrating equation (4.6),

i.e.

$$\frac{d}{dx}\left[\frac{f(k, x)}{f(-k, x)}\right] = \frac{W[f(-k, x), f(k, x)]}{f^2(-k, x)},$$

and using the boundary conditions of the Jost solutions $f(\pm k, x)$, we obtain

$$\frac{f(k, x)}{f(-k, x)} = e^{-2ikx} + 2ik \int_x^\infty dx\left[\frac{1}{f^2(-k, x)} - e^{-2ikx}\right] \quad (4.133)$$

Hence, on using (4.4), we have for the $s$-wave amplitude $A_0(k)$

$$A_0(k) = \frac{1}{2ik}\left[\lim_{x\to 0}\left(\frac{f(k, x)}{f(-k, x)}\right) - 1\right]$$

$$= \int_0^\infty dx\left[\frac{1}{f^2(-k, x)} - e^{-2ikx}\right]. \quad (4.134)$$

Similarly (by differentiating $f(-k, x)/f(k, x)$ we obtain

$$A_0(k) = \frac{\displaystyle\int_0^\infty dx\left[\frac{1}{f^2(k, x)} - e^{2ikx}\right]}{1 - 2ik\displaystyle\int_0^\infty dx\left[\frac{1}{f^2(k, x)} - e^{2ikx}\right]}. \quad (4.135)$$

The $s$-wave zero-energy scattering amplitude is thus seen to be

$$A_0(0) = \int_0^\infty dx\left[\frac{1}{f(x)^2} - 1\right]. \quad (4.136)$$

The validity of (4.134), (4.135) is not restricted to any particular type of potential. They may be evaluated by a systematic iterative procedure using the integral equations

$$f(\pm k, r) = e^{\mp ikr} + \frac{G}{k} \int_r^\infty dr' \sin k\,(r' - r)\, V(r') f(\pm k, r'). \quad (4.137)$$

Although $f$ is represented as a power series in $G$, $A$ is not—as required, since the nonanalyticity of $A$ in $G$ at $G = 0$ implies that a power expansion in $G$ does not converge.

Without delving into rigorous proofs of convergence, we just state that in general this procedure does not give rise to infinities, no matter what type of potential is considered. The divergence of $f(-k, r)$ at the origin does not cause any trouble in this case. Furthermore, as we have seen before, $f(-k, r)$ is an entire function of $G$ for all $r \neq 0$. Hence iteration of its integral equation (4.137) yields a convergent expansion. Thus the only possible divergences could arise in successive iteration-integrals in $A$. This question is again closely related to the analytic properties of $A$. Finally we note that $A$ is rather insensitive to the detailed behaviour of $f(-k, r)$ near $r = 0$, where $f$ diverges: this again is a reflection of the physical fact, that the scattering particle never really comes very close to the origin.

Apart from these considerations, Calogero (13) has also tested the usefulness of his approximations for some particular examples. In his case the agreement with known exact results is remarkably good.

### 4.8.3 *Another approximation for the phase-shift*

Finally we wish to discuss briefly another approximation-procedure which has the advantage that it avoids the explicit calculation of the wave-functions; instead it yields directly an expression for the phase-shift. This method was first formulated by Levy and Keller (10), who developed it specifically for long-range potentials. At the same time it was discussed independently by Calogero (13) for short-range regular potentials. Subsequently Calogero (13) examined and tested the validity of further approximations which may be derived from the fundamental first-order differential equation for the phase-shift; however, the validity of the method for singular potentials appears to have been shown only by Levy and Keller (10).

The radial wave equation

$$\psi''(r) + \left[ k^2 - \frac{l(l + 1)}{r^2} - V(r) \right] \psi(r) = 0 \qquad (4.138)$$

has two linearly independent solutions $\hat{j}_l\,(kr)$, $\hat{n}_l\,(kr)$ when $V(r) = 0$:

$$\hat{j}_l\,(kr) = \left(\frac{\pi kr}{2}\right)^{1/2} J_{l+1/2}\,(kr), \quad \hat{n}_l\,(kr) = \left(\frac{\pi kr}{2}\right)^{1/2} N_{l+1/2}(kr); \quad (4.139)$$

their Wronskian is given by

$$W\,[\hat{j}_l\,(kr),\,\hat{n}_l\,(kr)] = k. \tag{4.140}$$

The regular solution $\psi(r)$ may therefore be written

$$\psi(r) = A(r)\,[\hat{j}_l\,(kr) - \tan\delta(r)\cdot\hat{n}_l\,(kr)], \tag{4.141}$$

$$\psi(0) = 0,$$

where $A(r)$ and $\delta(r)$ are, as yet, unknown functions of $r$. $\delta(r)$ called the "phase-function", satisfies the boundary condition

$$\delta(0) = 0.$$

From the asymptotic behaviour of $\hat{j}_l$, $\hat{n}_l$, i.e.

$$\left.\begin{aligned} \hat{n}_l(z) \xrightarrow[z\to\infty]{} &-\cos\left(z - \frac{l\pi}{2}\right), \\[2ex] \hat{j}_l(z) \xrightarrow[z\to\infty]{} &\sin\left(z - \frac{l\pi}{2}\right), \end{aligned}\right\} \tag{4.142}$$

we see that for $r \to \infty$

$$\psi \sim \sin\left(kr - \frac{l\pi}{2} + \delta(\infty)\right),$$

thus $\delta(\infty)$ is the phase-shift we wish to calculate.

Differentiating (4.141), we obtain

$$\psi'(r) = A(r)\,[\hat{j}'\,(kr) - \tan\delta(r)\cdot\hat{n}'_l\,(kr)], \tag{4.143}$$

where we have set

$$A'(r)\,[\hat{j}_l\,(kr) - \tan\delta(r)\,\hat{n}_l\,(kr)] = A(r)\,\hat{n}_l\,(kr)\cdot\delta'(r)\cdot\sec^2\delta(r). \tag{4.144}$$

Differentiating (4.143), substituting the result in (4.138) and eliminating $A'(r)$ with the help of (4.144), we obtain

$$\frac{d}{dr}\,[\tan\delta(r)] = -\frac{1}{k}\,V(r)\,[\hat{j}_l\,(kr) - \tan\delta(r)\cdot\hat{n}_l\,(kr)]^2,$$

or

$$\delta'(r) = -\frac{1}{k} V(r) \left[\cos \delta(r) \cdot \hat{\jmath}_l(kr) - \sin \delta(r) \cdot \hat{n}_l(kr)\right]^2. \quad (4.145)$$

$\delta(r)$ is the phase produced by the potential up to the distance $r$.

For simplicity we now restrict ourselves to $s$-waves (for the general case cf. Calogero (13)). Then

$$\delta'(r) = -\frac{1}{k} V(r) \cdot \sin^2 [kr + \delta(r)]. \quad (4.146)$$

Integrating this expression, we have

$$\delta = -\frac{1}{k} \int_0^\infty dr\, V(r) \cdot \sin^2 (kr + \delta(r)). \quad (4.147)$$

The main contribution to the phase is expected to come from the region near $r = 0$, where the potential is a maximum. Thus from (4.146) and (4.147)

$$\delta(r) = -kr + \sin^{-1} \left[\sqrt{\frac{-\delta'(r)k}{V(r)}}\right]$$

$$\sim -kr + \frac{k}{\sqrt{V(r)}} + \frac{k}{4} \cdot \frac{V^1(r)}{V^2(r)} + \cdots. \quad (4.148)$$

The first term in this expansion corresponds to the phase-shift produced by a hard sphere of radius $r$. In order to obtain an expression reasonably valid at all energies, we need an approximation which reproduces not only the behaviour (4.148) for small $r$, but also that for $r$ large. Calogero (13) suggested the following approximation

$$\delta(r) \sim -kr + \frac{kr}{1 + r\sqrt{V(r)}} \quad (4.149)$$

for $r \to 0$ it reproduces the first two terms of (4.148), whereas for $r \to \infty$ it approches zero for a potential proportional to $1/r^m$, $m > 2$ (this may be assumed to be reasonable for $k \to 0$). Inserting (4.149) into (4.147) we obtain

$$\delta \simeq -\frac{1}{k} \int_0^\infty dr\, V(r) \sin^2 \left[kr \left\{r\sqrt{V(r)} + 1\right\}^{-1}\right]. \quad (4.150)$$

This approximation suppresses the behaviour of the wave function in the interior region (i.e. between $r = 0$ and $r = \infty$), as it was constructed from small-$r$ and large-$r$ approximations. We note that it is real only for repulsive potentials. Furthermore, the nonanalyticity in the coupling constant is an inherent feature of the approximation.

We note that in the limit $k \to 0$, the expression (4.150) yields an approximation for the scattering-length $A(0)$:

$$A(0) = \lim_{k \to 0} \left( \frac{\delta}{k} \right) = \int_0^\infty dr \, r^2 V(r) \left[ 1 + r \sqrt{V(r)} \right]^{-2}. \qquad (4.151)$$

The higher the energy, the closer does the particle approach the scattering centre; so we expect the behaviour of $\delta(r)$ near $r \sim 0$ to yield the dominant contribution at high energies. Substituting (4.148) into (4.147), we have

$$\delta = -\frac{1}{k} \int_0^\infty dr \, V_0(r) \sin^2 \left[ k \left\{ \frac{1}{\sqrt{V(r)}} + \frac{1}{4} \frac{V'(r)}{V^2(r)} \right\} \right]. \qquad (4.152)$$

Calogero (13) has evaluated this integral for several potentials; e.g. for a potential proportional to $1/r^m$, $m > 4$, he obtains

$$\delta \sim k^{1 - (2/m)}.$$

This is consistent with the expression (4.119) obtained above.

For further discussions of these approximations we refer to Calogero (13); the relationship between (4.147) and the effective-range expansion has been studied by Levy and Keller (10).

## 4.9 Attractive singular potentials

We now want to consider potentials which are highly singular but attractive. For the general arguments that follow (due to Case (15)) the exact functional form of the potential (other than its singular behaviour at the origin) and the value of $l$ are unimportant; so we consider only the case of $s$-waves and a potential

$$V(r) = -g^2/r^n, \quad n > 2.$$

Furthermore, we are interested solely in the resulting eigenvalues and eigenstates; hence we set

$$-E = K^2$$

We then have the Schrodinger-equation

$$\frac{d^2\phi}{dr^2} + [-k^2 + g/r^n]\phi = 0. \qquad (4.153)$$

The results of Section 2 tell us immediately that the solutions $\phi$ oscillate near $r = 0$. The two linearly independent solutions have the behaviour

$$\begin{aligned}
\phi &\sim r^{n/4} \exp\left[ \pm ig \int^r \frac{dr}{r^{n/2}} \right] \\
&= r^{n/4} \exp\left[ \mp ig \, \frac{2}{n-2} \cdot \frac{1}{r^{(n-2)/2}} \right] \\
\text{or} \qquad \phi &\simeq Ar^{n/4} \cos\left[ \frac{2}{n-2} \cdot \frac{g}{r^{(n-2)/2}} + B \right]
\end{aligned} \qquad (4.154)$$

(WKB-approximation).

Consider now two eigenfunctions $\phi_1, \phi_2$ belonging to eigenvalues $-K_1^2$, $-K_2^2$. Then

$$\left. \begin{aligned}
\phi_1'' + [-k_1^2 + g^2/r^n]\,\phi_1 &= 0, \\
\phi_2'' + [-k_2^2 + g^2/r^n]\,\phi_2 &= 0,
\end{aligned} \right\} \qquad (4.155)$$

and, as usual, we require $\phi_1, \phi_2$ to vanish at infinity. Hence

$$\begin{aligned}
(K_1^2 - K_2^2) \int_0^\infty \phi_1 \phi_2 \, dr &= \left[ \phi_2 \frac{d\phi_1}{dr} - \phi_1 \frac{d\phi_2}{dr} \right]_0^\infty \\
&= \left( \phi_2 \frac{d\phi_1}{dr} - \phi_1 \frac{d\phi_2}{dr} \right)_0. \qquad (4.156)
\end{aligned}$$

Writing

$$\phi_i \cong A_i r^{n/4} \cos\left( \frac{2}{n-2} \cdot \frac{g}{r^{(n-2)/2}} + B_i \right) \qquad (4.157)$$

near $r = 0$, we find

$$(K_1^2 - K_2^2) \int_0^\infty \phi_1 \phi_2 \, dr = A_1 A_2 \, g \sin (B_1 - B_2). \qquad (4.158)$$

Thus, if all eigenfunctions $\phi_i$ are required to behave as

$$\phi_i \sim A_i r^{n/4} \cos\left( \frac{2}{n-2} \cdot \frac{g}{r^{(n-2)/2}} + B \right), \qquad (4.159)$$

where $B$ is a fixed phase common to all eigenfunctions, these will be orthogonal. Conversely, if we require the eigenfunctions to vanish at infinity and to have a fixed phase $B$ at the origin, then the eigenvalue-problem is uniquely defined. In the case of regular potentials the condition of quadratic integrability of the eigenfunctions suffices to guarantee the completeness of the ortho-normal set. In the singular case there are many ways of defining a solution regular at the origin when the potential is attractive; so it is necessary to impose a further condition. This condition, of course, defines the new meaning of regularity. It also shows that only a single parameter is required to describe uniquely the scattering at small distances for singular, attractive potentials. Case (15) has pointed out that this parameter may be directly related to a physically plausible cutoff, as one would not expect a singular potential to reproduce an actual physical interaction in the neighbourhood of the origin. But the exact physical significance of Case's parameter is by no means clear.

Finally we add for the sake of completeness: Even for potentials singular and attractive, the spectrum of eigenvalues is discrete (this follows from the quadratic integrability of the wave functions), it extends, however, to minus infinity, since the Hamilton operator is now not bounded from below. Hence, unless one introduces a cutoff or an infinitely high repulsive wall close to the origin (with hard core boundary conditions), these attractive potentials are of little physical importance, since no real system has an infinite binding-energy or ground-state.

## 5   CLASSES OF SINGULAR POTENTIALS

In this section we shall survey various classes of singular potentials defined below, in connection with inapplicability of the peratization technique first introduced by Feinberg and Pais (16) for calculations in the field theory of the intermediate vector boson. This technique deals with arranging the series of the perturbation expansion in terms of the coupling constants of the potentials, where the most singular parts in terms of a cutoff are isolated in each order of perturbation. One then tries to sum up these series hoping that meaningful finite results can be obtained, when the cutoff (an invariant quantity) tends to zero. This procedure of isolating singular terms according to their degree of singularity and their summation, continued until one is left with only finite parts of the perturbation series, is referred to as peratization.

It may be pointed out that in the case of vector-meson electrodynamics it is possible in some cases to isolate the leading terms by this method of expansion since these terms come exclusively from a logarithmically divergent lowest order radiative correction. This means, it is adequate to assume the existence of the limit of certain infinite series to some series we encounter in dealing with singular potentials. If this process of isolating singular terms in the series is successful then the limits need not to be calculated. In contrast to this situation, we notice that in the case of weak interactions such limits must be evaluated from the start. This has been done for an infinite subset of graphs. Since it is evident that one cannot test these procedures in a field theoretic context, where the field theory itself might not even exist, various authors utilized the close analogy with the theory of repulsive singular potentials, where each term in the Born series diverges.

We define a singular potential $V(r)$ by

$$\text{(i)} \int_a^\infty r\,|V(r)|\,dr \to \infty \qquad \text{for any fixed value of } b > 0,$$

$$\text{and (ii)} \int_c^\infty r^2\,|V(r)|\,dr < \infty \quad \text{for any } c > 0.$$

These two conditions ensure that the phase shift $\delta$ is defined for $k^2 = 0$.

Various classes of potentials were considered by different authors to investigate the behaviour of the wave function and the phase shifts for scattering by repulsive singular potentials.

Khuri and Pais (17) as well as Tiktopoulos and Treiman (5) discussed scattering by pure inverse power potentials

$$V(r) = gr^{-n}; \quad g > 0, \quad n > 3 \tag{5.1}$$

while Aly *et al.* (18) considered potentials having a branch point, i.e.

$$V(r) = g\,(\ln r)^2/r^4 \tag{5.2}$$

$$V(r) = g\,(\ln r)^2/r^4 - g^{1/2}r^{-3} \tag{5.3}$$

and others having an essential singularity, i.e.

$$V(r) = g\,\frac{(e^{2/r} + v_1^2)}{r^4}, \quad g > 0, \quad v_1 > 0 \tag{5.4}$$

Wu (14) discussed scattering by a potential with a branch point type singularity

$$V(r) = -g\ln r/r^4, \quad g > 0. \tag{5.5}$$

In considering the class of potentials given by Eq. (5.1), it has been shown (14, 17, 18) that the correct zero energy scattering amplitude can be recovered by the technique of peratization.

### 5.1  Power potentials

Let us then start with the repulsive power potentials of Eq. (5.1) for which the radial Schrodinger equation

$$\frac{d^2\psi}{dr^2} + \left[ k^2 - \frac{l(l+1)}{r^2} - gr^{-n} \right] \psi(k, l, r) = 0 \tag{5.6}$$

has the solution

$$\psi(\mathbf{x}) = \frac{1}{r} \sum_l \psi(k, l, r)\, y_{10}(\theta) \tag{5.7}$$

Throughout this discussion we shall deal with the calculation of the zero energy scattering amplitude. This facilitates a comparison with the calculation in field theory since the leading singularities could be associated with zero external momentum. In the limit $k^2 = 0$, the solution of the radial Schrodinger equation with the power potential given above, can be written as

$$\psi_0(l, r) = \sqrt{r}\, k_{(2l+1)/(m-2)} \left[ \frac{2g^{1/2}}{m-2}\, r^{1-m/2} \right] \tag{5.8}$$

where $K$ is the Bessel function for imaginary argument and $l$ is the angular momentum.

Asymptotically we write the $s$-wave solution

$$\psi_0(0, r) \to \text{const}\,(r + a), \tag{5.9}$$

where on further inspection the zero energy amplitude $a$ is given by

$$a = -(vg^{1/2})^{2v}\,\{\Gamma(1-v)/\Gamma(1+v)\}, \quad v = \frac{1}{m-2}. \tag{5.10}$$

Since we know the wave functions and the zero energy amplitude for all classes of potentials considered here, we can compare them with the results obtained by peratization. This will be discussed below.

So far there has not been a definite criterion which states that this new mathematical technique (peratization) does in fact yield the correct phase shift as well as the scattering amplitude for any given singular potential satisfying conditions (i) and (ii). Only examples of special classes of singular potentials have been tested where the scattering amplitude in all cases is given explicitly.

For the class of potentials given by Eq. (5.1), the integral equation for the regular solution of the Schrodinger equation is given by

$$\psi(k, r) = \varphi_l(k, r) - gk^{-1}\overline{\varphi}_l(kr) \int_0^\infty \overline{\eta}_l(kr') V(r') \varphi(k, r') dr'$$
$$+ gk^{-1}\overline{\eta}_l(kr) \int_0^r \overline{\eta}_l(kr') V(r') \varphi(k, r') dr' \tag{5.11}$$

where

$$\overline{\varphi}_l(kr) = (\tfrac{1}{2}\pi kr)^{1/2} J_{l+1/2}(kr)$$

and

$$\overline{\eta}_l(kr) = -\tfrac{1}{2}(\pi kr)^{1/2} Y_{l+1/2}(kr).$$

We can then determine the scattering amplitude for all $l$ by writing,

$$\tan \delta_l = -(g/k) \int_0^\infty \overline{\varphi}_l(kr) V(r) \psi(k, r) dr. \tag{5.12}$$

If the potential is regular, (i.e. $rV(r) \to 0$ as $r \to \infty$ and $r^2 V(r) \to 0$). It is clear that for a small enough value of the coupling constant $g$, the Born series solution of (5.11) exists. For a more singular potential where $gV(r) \to gr^{-\beta}$, for $\beta > 0$, the $n$th iteration of Eq. (5.11) near the origin behaves as

$$\psi^{(n)} \xrightarrow[r \to 0]{} g^n k^{l+1} r^{l+1-n(\beta-2)} \tag{5.13}$$

The class of inverse power potential has been discussed by Tiktopoulos and Treiman (5) whose main results have been outlined in Section 4.2.4 above.

The iteration integrals diverge for order $n > \dfrac{(2l+1)}{(\beta-2)}$ ; no matter how small the value of $g$ is taken the Born series does not exist except for the value of $g = 0$. This indicates that there is a branch point at $g = 0$.

One way to construct the symptotic expansion of the solution is by imposing a cutoff on the potential,

$$V(r) = V(r) \theta(r - r_0). \tag{5.14}$$

With the cutoff, the existence of the Born series is guaranteed for a small value of the coupling constant $g$. Writing the series expansion of $\tan \delta_{r_0}$ in terms of the coupling constant $g$ and summing up the series, Tiktopoulos and Treiman were able to show that for $V(r) = g/r^4$ after letting the cutoff $r_0 \to 0$,

$$\tan \delta_{r_0} = -kg^{1/2} \tag{5.15}$$

A finite value for the phase shift may then be obtained from the expansion in powers of $g$ and a finite cutoff $r_0$ ($g$ must be small).

If the series of leading terms is given by an expansion such as

$$\tan \delta_{r_0} = k \sum_{n=0}^{\infty} a_n g^n r_0^{-nb+1} \tag{5.16}$$

which could also be written as

$$\tan \delta_{r_0} = kg^{1/b} \left[ (gr_0^{-b})^{-1/b} \sum_n a_n (gr_0^{-b})^b \right], \tag{5.17}$$

then we could infer that

$$\tan \delta \to \text{const } kg^{1/b} \tag{5.18}$$

provided the bracketed part in Eq. (5.17) exists for $r_0 \to 0$.

The introduction of a cutoff in potential theory is similar to that in non-renormalizable field theories, where for example the kernel of the two body Bethe–Salpeter equation may be very singular at very small distances so that its Fourier transform (momentum space) does not exist. In this case a cutoff procedure could be used to circumvent the difficulties.

We now look for an expansion of the solution of the radial Schrodinger equation for a singular power potential as $r \to 0$.

Equation (5.8) gives the solution in the limit of $k^2 = 0$. However, for a value of $k^2 \neq 0$, we write the solution as

$$\psi(r) = \psi_0 R(r) \tag{5.19}$$

where $R(r)$ satisfies the equation,

$$\frac{d}{dr} \left[ \psi_0^2 \frac{dR}{dr} \right] + k^2 \psi_0^2(r) R(r) = 0 \tag{5.20}$$

A Volterra type integral equation is obtained imposing the boundary condition $R(r) \to 1$ as $r \to 0$,

$$R(r) = 1 - k^2 \int_0^r \int_{r'}^r \frac{\psi_0^2(r')}{\psi_0^2(t)} R(r') \, dt \, dr'. \tag{5.21}$$

Using the behaviour of the modified Bessel function $K$ for $r \to 0$ we have

$$\psi_0(r) \xrightarrow[r \to 0]{} r^{(2\nu+1)/4\nu} e^{-2\nu\alpha} r^{-1/2\nu}$$

and

$$W(r, r') \equiv - \int_{r'}^r \frac{\psi_0^2(r')}{\psi_0^2(t)} \, dt \tag{5.22}$$

is bounded, i.e. there exists a positive number $B$ such that

for
$$\left.\begin{array}{l} |W(r, r')| < Br' \\[4pt] |\text{Arg } g| \quad \leqslant \pi \end{array}\right\} \tag{5.23}$$

From the previous discussion we have seen that $\tan \delta$ has a branch point at $g = 0$. In order to calculate the leading terms of an expansion for small $g$, we write

$$v(r) \equiv r^{\lambda+1} R(r) \tag{5.24}$$

so that

$$\frac{dv^2}{dr^2} + \left[ k^2 - \frac{\lambda^2 - \frac{1}{4}}{r^2} \right] v = \left[ \frac{2\lambda + 1}{r} - \frac{2\psi_0'}{\psi_0} \right] r^{\lambda+1/2} \frac{d}{dr} \left( \frac{v}{r^{\lambda+1/2}} \right)$$

$$\equiv B(g, r). \tag{5.25}$$

We also notice that the iterated solution of (5.25) can be differentiated term-wise to yield a uniformly convergent series. So for $\dfrac{dR}{dr}$ we may write an expression in terms of $v$ as

$$\frac{d^2v}{dr^2} + \left[ k^2 - \frac{\lambda^2 - \frac{1}{4}}{r^2} \right] v = k^2 \left[ 2\psi_0'/\psi_0 - \frac{2\lambda + 1}{r} \right] r^{\lambda+1/2} \psi_0^{-2}(r) \times$$

$$\times \int_0^r \psi_0^2(r') (r')^{-(\lambda+1/2)} v(r') \, dr' \equiv B(g, r). \tag{5.26}$$

The regular solution satisfies the integral equation

$$v(r) = \overline{\varphi}_{\lambda+1/2}(kr) - \frac{1}{k} \overline{\varphi}_{\lambda+1/2}(kr) \int_r^\infty \overline{\eta}_{\lambda+1/2}(kr') B(g, r') \, dr'$$

$$- k^{-1} \overline{\eta}_{\lambda+1/2}(kr) \int_0^r \overline{\varphi}_{\lambda+1/2}(kr') B(g, r') \, dr'. \tag{5.27}$$

When
(i) $\lambda v > 1$, the first iteration of Eq. (5.27) is proportional to $g$ and the leading term of $\psi$ in an expansion for small values of $g$ is the first Born iteration, which in this case is convergent.
(ii) For $\lambda v < 1$, it is shown (5) that the leading term for $\tan \delta_\lambda$ is

$$\tan \delta_\lambda \xrightarrow[g \to 0]{} \text{const } g^{\lambda^2 v}.$$

(iii) For $2\lambda v = 1$

$$\tan \delta_\lambda \xrightarrow[g \to 0]{} \text{const } g \ln g^{1/2}.$$

## 5.2   Logarithmic potentials

Following this analysis, various authors (18) investigated the singular logarithmic potentials of Eq. (5.2), (5.9).

The integral equation for the regular solution for $s$-waves and $k^2 \neq 0$ is

$$\psi(r) = r - gr \int_r^\infty \frac{(\ln r')^2}{r'^4} \, \psi(r') \, dr' - g \int_0^r r' \, \frac{(\ln r')^2}{r'^4} \, \psi(r') \, dr' \quad (5.28)$$

We now obtain the solution $\psi(r)$ and consequently $\tan \delta/k$ by using the peratization technique. For this purpose we replace the actual potential (5.2) by a cutoff potential

$$V_{r_0}(r) = \theta \, (r - r_0) \, V(r) \qquad (5.29)$$

so that the Born series in terms of the coupling constant $g$, exists. For $r > r_0$ we get

$$\psi_1(r) = gr \left\{ \left[ \frac{(\ln r)^2}{2r^2} + \frac{3 \ln r}{2r^2} + \frac{1}{4r^2} \right] - \frac{1}{r} \left[ \frac{\ln r_0}{r_0} + \frac{2 \ln v_0}{r_0} + \frac{2}{r_0} \right] \right\}$$

$$\psi_2(r) = g^2 r \left\{ \left[ \frac{(\ln r)^4}{24 r^4} + \frac{2}{9} \frac{(\ln r)^3}{r^4} + \cdots \right] \right.$$

$$- \left[ \left( \frac{(\ln r_0)^2}{r_0} + \frac{2 \ln r_0}{r_0} + \frac{2}{r_0} \right) \left( \frac{1}{6} \frac{(\ln r)^2}{r^3} + \cdots \right) \right.$$

$$\left. + \frac{1}{r} \left[ \frac{1}{3} \frac{(\ln r_0)^4}{r_0^3} + \cdots \right] \right\} \qquad (5.30)$$

etc.

If we sum the leading singularities in Eq. (5.30) as $r_0 \to 0$, we obtain

$$(\psi r) \sim r \left[ \cosh \left( g^{1/2} \frac{\ln r}{r} \right) - \frac{\ln r_0}{\ln r} \sinh \left( g^{1/2} \frac{\ln r}{r} \right) \tan \left( g^{1/2} \frac{\ln r_0}{r_0} \right) \right]$$

$$(5.31)$$

It is clear that $\psi(r)$ does not exist when $r_0 \to 0$. This result can be compared to the wave function obtained near $r = 0$ by the W.K.B. method which has the form

$$\psi(r) \underset{r \to 0}{\sim} \frac{r}{(\ln r)^{1/2}} \exp \left( -g \int_{r_0}^r \frac{\ln r'}{r'^2} \, dr' \right) \qquad (5.32)$$

and thus exists.

We turn now to calculate the scattering amplitude for this potential

$$k^{-1} \tan \delta|_0 = -g \left[ r_0^{-1} (\ln r_0)^2 + 2r_0^{-1} \ln r_0 + 2r_0^{-1} \right],$$

$$k^{-1} \tan \delta|_1 = g^2 \left[ \tfrac{1}{3} r_0^{-3} (\ln r_0)^4 + \tfrac{7}{9} r_0^{-3} \ln r_0 + \tfrac{17}{18} r_0^{-3} (\ln r_0)^2 \right.$$

$$\left. + \tfrac{17}{27} r_0^{-3} \ln r_0 + \tfrac{17}{81} r_0^{-3} \right]$$

$$k \tan \delta|_2 \quad = -g^3 \left[ \tfrac{2}{15} r_0^{-5} (\ln r_0)^6 + \cdots \right] \tag{5.33}$$

for $r_0 \to 0$, we obtain

$$k^{-1} \tan \delta = -g^{1/2} \ln r_0 \left[ g^{1/2} \ln r_0/r_0 - \tfrac{1}{3} g^{3/2} \ln r_0/r_0^3 + \tfrac{2}{15} (\ln r_0)^5/r_0^5 + \cdots \right]$$

$$= -g^{1/2} (\ln r_0) \tan \left[ g^{1/2} \frac{\ln r_0}{r_0} \right] \tag{5.34}$$

For $r_0 \to 0$,

$$\tan \delta/k \sim -g^{1/2} \ln r_0$$

which is not defined. This indicates that the peratization technique in the sense discussed above does not give the correct results.

However, Wu (14) dealing with a similar problem, pointed out that if all the singular terms are summed up, the final answer can be made meaningful as the cutoff $r_0 \to 0$. We must point out that in both scattering processes considered by Aly *et al.* (18) and Wu (14), the zero energy scattering amplitude is not known explicitly, since the radial wav eequation has no known solution which can be given in a closed form. It is then difficult to draw any definite conclusion with regard to the applicability of the peratization technique.

We now consider the logarithmic singular potential given by Eq. (5.5). For the case of $k^2 = 0$, we define the scattering amplitude $a(0)$ as

$$a(0) = \lim_{r \to \infty} [\psi(r) - r]$$

Again introducing a cutoff, we have the potential

$$V(r) = \theta (r - r_0) (-gr^{-4} \ln r). \tag{5.35}$$

Then the wave function $\psi$ satisfies the integral equation

$$\psi (r_1 r_0) = r + g \int_{r_0}^{\infty} dr' \, r'^{-4} \ln r' \, \min (r, r') \, \psi (r, r_0) \tag{5.36}$$

Iteration of (5.36) gives

$$\psi(r) = \sum_{n=0}^{\infty} \psi_n(r)$$

where

$$\psi_{n+1}(r, r_0) = g \int_{r_0}^{\infty} dr' r'^{-4} \ln r' \, \min(r, r') \, \psi_n(r', {}^0) \qquad (5.37)$$

and

$$\psi_0(r, r_0) = r.$$

Define $A_n(r_0)$ as

$$A_n(r_0) \equiv \psi_n(\infty, r_0^{-1}) = g \int_{r_0}^{\infty} dr' r'^{-3} \ln r' \, \psi_{n-1}(r', r_0^{-1}), \qquad (5.38)$$

we see that the $n$th iteration of the scattering amplitude i.e. of

$$A(r) = \sum_{n=1}^{n} \psi_n(r, r_0)$$
$$\text{as } t \to 0$$

has the following form

$$A_n(r_0^{-1}) = g^n r_0^{-(2n-1)} \sum_{m=0}^{n} a_{nm} (\ln r_0^{-1})^{n-m}. \qquad (5.39)$$

Let

$$B_m(r_0^{-1}) \equiv \sum_{n=m}^{\infty} a_{nm} g^n r_0^{-(2n-1)} (\ln r_0^{-1})^{n-m} \qquad (5.40)$$

then formally we have

$$A(r_0^{-1}) = \sum_{m=0}^{\infty} B_m(r_0^{-1}). \qquad (5.41)$$

We then ask what is the behaviour of $A(r_0^{-1})$ when $r_0 \to 0$?

From the analysis of Wu (14) it follows that the zero energy scattering amplitude

$$B_0(r_0^{-1}) = -g^{1/2} (\ln(r_0^{-1}))^{1/2} \tan(g r_0^{-2} \ln r_0^{-1})^{1/2}. \qquad (5.42)$$

It is interesting to notice that Aly *et al.* (18) obtained a similar result for the scattering amplitude by a direct Born expansion. However, Wu developed a scheme to get a general term for the amplitude, so it is not surprising that an agreement is obtained between the exact scattering amplitude and that obtained by summing *all* terms of the expansion in terms of $g$.

It is difficult to draw any definite conclusion with regard to the applicability of the peratization technique since the radial equation with logarithmic singular potential cannot be solved exactly.

Consider the potential given by Eq. (5.3)

$$V(r) = g \left( \frac{(\ln r)^2}{r^4} - g^{1/2} r^{-3} \right) \theta (r_0 - r).$$

The selection of this potential is motivated by the following consideration: It is the logarithmic term which dominates at the origin, since the $g^{1/2} r^{-3}$ term is less singular. The Schrodinger wave equation for $s$-wave and $k^2 = 0$ is solved exactly for this potential. Finally we must point out that the term $g^{1/2} r^{-3}$ does not fall off fast enough at infinity so as to define the phase shift at zero energy. To avoid this difficulty, we have introduced the $\theta$-function.

The introduction of the cutoff $\theta$-function does not affect the potential considered here, since it is the behaviour near the origin, i.e. $r = 0$ which concerns us. In the vicinity of the origin, the behaviour of the potential is determined by the more singular logarithmic term, i.e. $g \, \dfrac{\ln^2 r}{r^4}$.

For this potential the exact solution of the Schrodinger equation for $k^2 = 0$ and $s$-wave is

$$\psi(r) = r \exp \{g^{1/2} (\ln r + 1)/r\} \quad \text{for } r < r_0, \tag{5.43}$$

$$= ar + b \qquad\qquad\qquad \text{for } r > r_0.$$

This solution satisfies the normal boundary conditions for scattering,

$$\text{(i)} \quad \psi(r) \to 0 \quad \text{as } r \to 0$$

$$\text{(ii)} \quad \psi(r) \to r \quad \text{as } r \to \infty.$$

From the continuity conditions for $\psi(r)$ and $\dfrac{d\psi}{dr}$ at $r = r_0$, we have

and

$$\left. \begin{aligned} \alpha &= \exp \left\{ g^{1/2} \left[ \frac{\ln r_0 + 1}{r_0} \right] \right\} \left[ 1 - g^{1/2} \, \frac{\ln r_0}{r_0} \right] \\[2mm] \beta &= \exp \left\{ g^{1/2} \left[ \left( \frac{\ln r_0 + 1}{r_0} \right) \right] \right\} g^{1/2} \ln r_0 \end{aligned} \right\} \tag{5.44}$$

from Eq. (5.43) and (5.44) we write an expression for the zero-energy amplitude $a(0)$ as

$$a(0) = g^{1/2} \ln r_0 \, (|1 - g^{1/2} \ln r_0| \, r_0)^{-1}. \tag{5.45}$$

### 5.3  Calculation of $a(0)$ by peratization

Since we know the value of $a(0)$ in a closed form we now want to calculate it by the peratization technique. For this purpose we follow the method of ref. (17) applied to the case of inverse power potentials (Eq. (5.1)).

We write

$$\Psi(r) = \frac{1}{r}\,\psi(r) \tag{5.46}$$

and, regulating the potential by introducing a parameter $\alpha$, this leads to a regulated wave function $\Psi(r, \alpha)$

$$\Psi(r, \alpha) = \Psi_1(r, \alpha) + \Psi_2(r, \alpha). \tag{5.47}$$

$\Psi_1$ satisfies a singular integral equation for $\alpha \to 0$;

$$\Psi_1(r, \alpha) = -\frac{1}{r}\int_0^\infty y^2 dy\, V(y, \alpha)\,[\Psi_1(y, \alpha) + \Psi_2(y, \alpha)] \tag{5.48a}$$

while $\Psi_2(r, \alpha)$ satisfies the regular integral equation

$$\Psi_2(r, \alpha) = 1 - \frac{1}{r}\int_r^\infty y\, dy\, (r - y)\, V(y, \alpha)\, \Psi_1(y, \alpha)$$

$$- \frac{1}{r}\int_r^\infty y\, dy\, (r - y)\, V(y, \alpha)\, \Psi_2(y, \alpha) \tag{5.48b}$$

for $\alpha \neq 0$ there exists always a solution of Eq. (5.48a) of the form

$$\Psi_1(r, \alpha) = \frac{a(\alpha)}{r}. \tag{5.49}$$

Here $a(\alpha)$ is given by

$$a(\alpha) = -\int_0^\infty y^2\, dy\, V(y, \alpha)\,[\Psi_1(y, \alpha) + \Psi_2(y, \alpha)]. \tag{5.50}$$

On substituting (5.49) in (5.48b), a Volterra type integral equation for $\Psi_2$ is obtained for which we write the solution as

$$\Psi_2(r, \alpha) = \Psi_2^{(1)}(r, \alpha) + \Psi_2^{(2)}(r, \alpha). \tag{5.51}$$

We now assume that

$$\lim_{\alpha \to 0} a(\alpha) < \infty$$

implies that Eq. (5.48b), is non-singular when $\alpha \to 0$.

Let

$$\Psi_2^{(2)}(r) = F(r) - \frac{1}{r}, \quad \text{then} \quad \Psi_2^{(1)}(r) \text{ and } F(r)$$

satisfy the integral equations

$$\Psi_2^{(1)}(r) = 1 - \frac{1}{r}\int_r^\infty y\, dy\,(r - y)\, V(y)\, \Psi_2^{(1)}(y) \tag{5.52}$$

and

$$F(r) = \frac{1}{r} - \frac{1}{r}\int_r^\infty y\, dy\,(r - y)\, V(y)\, F(y). \tag{5.53}$$

From (5.52) it follows that

$$\Psi_2^{(1)} = A\exp\left\{g^{1/2}\left[\frac{\ln r + 1}{r}\right]\right\} + B\exp\left\{Dg^{1/2}\left[\frac{\ln r + 1}{r}\right]\right\} L(r) \quad \text{for } r < r_0,$$

$$\Psi^{(1)}(r) = 1 \quad \text{for } r > r_0. \tag{5.54}$$

Here $L(r)$ is defined as

$$L(r) = \int_\xi^r \frac{1}{r'^2}\exp\left\{-2g^{1/2}\left[\frac{\ln r' + 1}{r'}\right]\right\} dr'.$$

$\xi$ is an arbitrary point.

The continuity condition for $\Psi_2^{(1)}(r)$ and $\dfrac{d\Psi_2^{(1)}}{dr}$ at $r = r_0$ gives,

$$A = \exp\left\{-g^{1/2}\left[\frac{\ln r_0 + 1}{r_0}\right]\right\} - BL(r_0)$$

and

$$B = g^{1/2}\ln r_0 \exp\left\{g^{1/2}\left[\frac{(\ln r_0 + 1)}{r_0}\right]\right\}\left(g^{1/2}\frac{\ln r_0}{r_0} - 1\right)\right\}. \tag{5.55}$$

From (5.53) it follows that

$$F(r) = A'\exp\left\{g^{1/2}\left[\frac{\ln r + 1}{r}\right]\right\} + B'\exp\left\{g^{1/2}\left[\frac{\ln r + 1}{r}\right]\right\} L(r) \quad \text{for } r < r_0,$$

$$= \frac{1}{r} \quad \text{for } r > r_0. \tag{5.56}$$

By continuity condition for $F(r)$ and $F'(r)$ at $r = r_0$ we write

$$\left.\begin{array}{l} A' = \dfrac{1}{r_0}\,\exp\left\{-g^{1/2}\left(\dfrac{\ln r_0 + 1}{r_0}\right)\right\} - B'L\,(r_0) \\[2em] B' = \exp\left\{g^{1/2}\left[\dfrac{\ln r_0 + 1}{r_0}\right]\right\}\left(g^{1/2}\,\dfrac{\ln r_0}{r_0} - 1\right). \end{array}\right\} \tag{5.57}$$

and

Using the expression for $A(0)$ we get

$$a = \lim_{\sigma \to 0} \frac{-\displaystyle\int_\sigma^\infty y^2\,dy\,V\,(y,\alpha)\,\Psi_2^{(1)}\,(y,\alpha)}{1 + \displaystyle\int_\sigma^\infty y^2\,dy\,V\,(y,\alpha)\left(\Psi_2^{(2)}\,(y,\alpha) + \dfrac{1}{y}\right)}$$

Substituting for $\Psi_2^{(1)}\,(y,\alpha)$ and $\Psi_2^{(2)}\,(y,\alpha)$ and taking the leading singularities both in the numerator and denominator we obtain

$$a = -B/B'$$

which because of (5.55) and (5.57) reproduces the exact value (5.45) for the zero energy amplitude.

It is interesting to notice that the peratization technique seems to depend on the functional nature of the potential in the sense that for a pole type and branch point plus a pole type singularity, one can recover the zero energy scattering amplitude, while for a pure branch point singularity ($g \ln^2 r/r^4$) we cannot ensure the applicability of the technique.

### 5.4  Potentials with an essential singularity

We consider the repulsive singular potential Eq. (5.4)

$$V(r) = g\,\frac{e^{2/r} + v_1^2}{r^4} = g\,\frac{e^{2/r}}{r^4} + \frac{g'}{r^4}$$

Here $gv_1^2 = g'$.

We write again the radial Schrodinger equation for $k^2 = 0$ and $s$-waves. For this potential the wave equation

$$\psi''(r) - g\,\frac{e^{2/r} + v_1^2}{r^4}\,\psi = 0$$

with the transformations $r = z^{-1}$, $\psi(z) = z^{-1}\varphi = r\varphi$ and $t = -ig^{1/2}e^z$, we get the Bessel equation

$$\varphi''(t) + t^{-1}\varphi'\,(t) + (1 - v^2/t^2)\,\varphi(t) = 0. \tag{5.58}$$

For simplicity consider the case $v = \frac{1}{2}$ or $g' = \frac{1}{4}$ which leads to the two fundamental solutions of Eq. (5.58)

$$\varphi_{1/2}\left(-ig^{1/2}\exp\left(r^{-1}\right)\right) = \exp\left(-\tfrac{1}{2}r\right)\left(\exp\left[g^{1/2}\exp\left(r^{-1}\right)\right]\right.$$
$$\left. - \exp\left[-g^{1/1}\exp\left(r^{-1}\right)\right]\right)$$

and

$$\varphi_{-1/2}\left(-ig^{1/2}\exp\left(r^{-1}\right)\right) = \exp\left(-\tfrac{1}{2}r\right)\left(\exp\left[g^{1/2}\exp\left(r^{-1}\right)\right]\right.$$
$$\left. + \exp\left[-g^{1/2}\exp\left(r^{-1}\right)\right]\right).$$

The solution of Eq. (5.58) for $v = \frac{1}{2}$, which if regular at $r = 0$, is

$$\psi = r\left[\varphi_{1/2}\left(-ig^{1/2}\exp\left(r^{-1}\right)\right) - \varphi_{-1/2}\left(ig^{1/2}/\exp\left(r^{-1}\right)\right)\right] \qquad (5.59)$$

$$= 2r\exp\left(-\frac{1}{2}r\right)\exp\left(-g^{1/2}\exp\left(\frac{1}{r}\right)\right). \qquad (5.60)$$

Asymptotically this solution behaves as

$$\varphi(r) \underset{r\to\infty}{\longrightarrow} 2\exp\left(-g^{1/2}\right)\left(r - \left(g^{1/2} + \tfrac{1}{2}\right)\right). \qquad (5.61)$$

The zero-energy scattering amplitude is

$$a(0) = -\left(g^{1/2} + \tfrac{1}{2}\right). \qquad (5.62)$$

Following the same procedure as for the logarithmic potential given by Eq. (5.3) we can recover the zero-energy amplitude for this potential also.

It is clear that if we know the closed form of the scattering amplitude, of course there is no need for peratization. Spector (19) has also discussed this problem and gives a certain criterion for a general class of singular potentials, and quotes examples, where the peratization does in fact give the correct $s$-wave scattering amplitude.

Consider a potential of the form

$$V(r) = \frac{A(r)}{r^n}, \quad n > 3 \qquad (5.63)$$

where $A(r) > 0$ in some finite region $E > r > 0$, so that the potential is repulsive near the origin. The degree of singularity in the term $A(r)$ is chosen such that $V(r)$ falls off fast enough as $r \to \infty$, and a phase shift is defined for zero energy (i.e. conditions (i) and (ii) in section (5.2) are satisfied).

Then if $A(r)$ is more singular than $\dfrac{1}{r}\dfrac{dA(v)}{dr}$ or approaches zero less rapidly,

then the peratized zero energy scattering amplitude for a potential

$$V(r, r_0) = V(r)\,\theta\,(r - r_0); \quad r_0 > 0 \tag{5.64}$$

is given by

$$a(r_0) = -\frac{\pi v^{2v-1}}{[\Gamma(v)]^2 \sin v\pi}\, A^v(r_0)\, \frac{I_{1-v}(\Omega)}{I_{v-1}(\Omega)}$$

where

$$v = (n-2)^{-1}, \quad \Omega = \frac{2v}{r_0^{\,n/2-1}}\, A^{1/2}(r_0),$$

$I$ being a modified Hankel function.

For the potential (5.64) the scattering length $a'$ is given by

$$a'(r_0) = -g \int_{r_0}^{\infty} r'\, V\,(r')\, \psi\,(r', r_0)\, dr'. \tag{5.65}$$

Where $\psi\,(r', r_0)$ is the zero energy wave function obtained from an integral equation of the form of Eq. (5.28).

If we were to consider the lowest order term in $g$,

$$a'(r_0) = -g \int_{r_0}^{\infty} r'^{(2-n)}\, A(r')\, dr'$$

$$= \frac{-g}{(3-n)}\, \frac{A(r')}{r'^{(n-3)}}\, \Bigg|_{\alpha}^{\infty} + \frac{g}{(3-n)} \int_{t_0}^{\infty} r'^{(3-n)} \left(\frac{dA}{dr}\right)_{r=r'} dr'$$

$$= \frac{-g}{3-n}\, \frac{A(\alpha)}{\alpha^{n-3}} + \frac{g}{(3-n)} \int_{\alpha}^{\infty} r'^{(3-n)} \left(\frac{dA}{dr}\right)_{r=r'} dr', \tag{5.66}$$

where we have integrated by parts. The infinite limit in the integrated part does not contribute, because of the finiteness of the second moment of the potential. In general the $(m + 1)$th order in $g$ of $a'(r_0)$, call it $a'_{m+1}(r_0)$, is evaluated using $\psi_m\,(r_1 r_0)$, the $m$th order term of the wave function.

To evaluate $a'_{m+1}(r_0)$, only the most singular term in $\psi_m\,(r_1 r_0)$ is retained. Together with this term we consider the most singular terms in $r_0$.

Spector (19) also gives an expression for the sum of the most singular terms in $r_0$ of the Born series:

$$a(r_0) = \sum_{m=0}^{\infty} \gamma_m(n)\, [gA\,(r_0)]^m, \tag{5.67}$$

where $\gamma_m(n)$ does not depend on $A(r)$.

As a special case let $A(r) = 1$ and $n = 4$. Here the zero energy scattering amplitude is

$$a(r_0) = -g^{1/2} \tanh\left(\frac{g^{1/2}}{r_0}\right). \tag{5.68}$$

This is the exact form of $a(r_0)$ as given by Tiktopoulos and Treiman (5).

For the potential considered in Eq. (5.2), where $A(r) = g \ln^2 r$ and $n = 4$, we get

$$a(r_0) = -g^{1/2} \ln r_0 \tanh\left(g^{1/2} \frac{\ln r_0}{r_0}\right).$$

This is again the same result as given above, Eq. (5.34). Similarly for $A(r) = -g \ln r$, and any value of $n$ we get,

$$a'(r_0) = \frac{-\pi v^{2v-1}}{[\Gamma(v)]^2 \sin v\pi} g^{1/2} (-\ln r_0)^{1/2} \frac{I_{1-v}(\Omega)}{I_{v-1}(\Omega)}$$

$v, \Omega$, as defined above.

## 5.5   The potential $r^{-4}$

Among the various classes of singular potentials considered in this discussion, the singular potential $r^{-4}$ was the most widely discussed and investigated. This is because the radial Schrödinger equation is exactly solvable for all partial waves and energies. For full details see the set of references (20).

We shall briefly summarize the main steps to obtain the $S$-matrix and then discuss the asymptotic behaviour of the Regge trajectories for the $r^{-4}$ potential.

We consider the radial Schrödinger equation

$$\psi''(r) + \left[ k^2 - \frac{l(l+1)}{r^2} + \frac{V_0}{(\mu r)^4} \right] \psi(r) = 0$$

where

$$V(r) = -V_0/(\mu r)^4.$$

$V(r)$ is repulsive for $V_0 < 0$, attractive for $V_0 > 0$; $V_0$ has the dimension of energy, while $\mu$ has dimension of reciprocal length.

We set $\psi(r) = r^{1/2}\varphi(r)$, $\lambda = (l + \tfrac{1}{2})^2$ and $x = r/r_a = e^z$, we obtain

$$\varphi''(z) - [\lambda - 2h^2 \cosh 2z] \varphi(z) = 0, \tag{5.69}$$

where the range $0 \leqslant r \leqslant \infty$ corresponds to $-\infty \leqslant z \leqslant +\infty$, and where

$$r_a^2 k^2 = h^2, \quad r_a^2 = V^{1/2}/k\mu^2.$$

Equation (5.69) is known to be a modified Mathieu equation. This equation had received considerable attention and the transformations above arrived at by various authors (20). We shall consider both the attractive and repulsive potential.

### 5.6   The attractive potential

We briefly discuss the case where $V > 0$. Here the behaviour of solutions near $r = 0$ is obtained from the behaviour of the modified Mathieu function $\phi(z)$ for $z \to -\infty$, whereas for $r \to +\infty$, we have to use corresponding solution of (5.69) having the correct behaviour for $z \to +\infty$.

We directly apply the W.K.B. method to the radial Eq. above near $r = 0$ we obtain

$$\text{as } r \to 0: \quad \psi(r) \sim r \exp\left[\pm \frac{V^{1/2}}{r^2} i \int \frac{dr}{r^3}\right]$$

$$= r \exp\left[\mp \frac{iV^{1/2}}{\mu^2} \frac{1}{r}\right]. \tag{5.70}$$

We define for $r \to 0$:

$$\psi(r) \sim r \exp\left[ikr_a^2/r\right] \tag{5.71}$$

as an ingoing wave representing the solution which is regular at the origin. We now seek a solution of the type (5.71) with the asymptotic behaviour

$$r \to \infty: \quad \psi(r) \sim e^{\pm ikr}. \tag{5.72}$$

We now note the following solutions for Eq. (5.69) defined by Wannier (21)

$$h_e^{(1)}(z) \sim \frac{1}{(2h \cosh z)^{1/2}} \exp\left[-2ih \cosh z - \frac{i\pi}{4}\right]$$

$$\sim \frac{1}{(kr)^{1/2}} \exp\left[-ikr - \frac{i\pi}{4}\right] \tag{5.73}$$

$$h_e^{(2)}(z) \sim \frac{1}{(2h \cosh z)^{1/2}} \exp\left[+2ih \cosh z + \frac{i\pi}{4}\right]$$

$$\sim \frac{1}{(kr)^{1/2}} \exp\left[+ikr + \frac{i\pi}{4}\right] \tag{5.74}$$

for $z > 0$ and $z \to \infty$; and

$$h_e^{(3)}(z) \sim \frac{1}{(2h \cosh z)^{1/2}} \exp\left[-2ih \cosh z - \frac{i\pi}{4}\right]$$

$$\sim \left(\frac{r}{kr_a^2}\right)^{1/2} \exp\left[-\frac{ikr_a}{r} - \frac{i\pi}{4}\right] \tag{5.75}$$

$$h_e^{(4)}(z) \sim \frac{1}{(2h \cosh z)^{1/2}} \exp\left[2ih \cosh z + \frac{i\pi}{4}\right]$$

$$\sim \left(\frac{r}{kr_a^2}\right)^{1/2} \exp\left[+\frac{ikr_a}{r} + \frac{i\pi}{4}\right], \tag{5.76}$$

for $z < 0$ and $z \to -\infty$.

To obtain the $S$-matrix we require a relation connecting $h_e^{(1)}$ and $h_e^{(2)}$ with $h_e^{(4)}$. This can be obtained from Wannier's paper. The function $h_e^{(4)}$ ($z$ defined above for $z < 0$ may be re-expressed as a linear combination of two other linearly independent solutions $j_e^{\pm}(z)$, which are identical to $M_{e_{\pm}}(z)$ in the notation of ref. (21). We then write

$$h_e^{(4)}(z) = e^{(i\pi/2)(\gamma \pm \beta)} j_e^{(-)}(z) - e^{-(i\pi/2)(\gamma \pm \beta)} j_e^{(+)}(z)]/\sin \pi\beta. \tag{5.77}$$

Here $\beta$ is the Floquet parameter (usually denoted by $\nu$) and $\gamma$ is another parameter introduced by Wannier to play an analogous role in regions of instability of Mathieu functions as $\beta$ in regions of stability. These parameters are functions of $\lambda$ and $h$. We continue the functions $j_e^{\pm}(z)$ across $z = 0$ (i.e., $r = r_a$) with the help of the relations

$$j_e^{(\pm)}(z) = \tfrac{1}{2}i\, e^{\mp i\pi\gamma/2} \left[e^{\mp i\pi\beta/2} h_e^{(1)}(z) - e^{\pm i\pi\beta/2} h_e^{(2)}(z)\right]. \tag{5.78}$$

We put (5.78) into (5.77) we obtain

$$h_e^{(4)}(z) = -\frac{\sin \pi (\gamma + \beta)}{\sin \pi\beta} h_e^{(1)}(z) + \frac{\sin \pi\gamma}{\sin \pi\beta} h_e^{(2)}(z). \tag{5.79}$$

Hence the $S$-matrix is given by

$$s = -i \sin \pi\gamma / \sin \pi (\gamma + \beta) \tag{5.80}$$

$\gamma$ and $\beta$ are related by

$$e^{\Phi} = i \sin \pi\gamma / \sin \pi\beta. \tag{5.81}$$

where $\beta \neq$ integer. For further details see (19, 20, 21, 22).

### 5.7  The repulsive potential

In the case of $Vr^{-4}$ being repulsive, $V < 0$; i.e.

$$ir_a^2 = \frac{i\,|V|^{1/2}}{(k\mu^2)} \quad \text{so that} \quad z = \ln\left(\frac{r}{r_a}\right) - \frac{i\pi}{4}$$

at the point $r = r_a$, $z$ changes from $+\dfrac{i\pi}{4}$ to $-\dfrac{i\pi}{4}$. The matching relationship corresponding to the Wronskians of the solutions of the Mathieu equation (MS. p. 171) reads

$$
\begin{aligned}
W\,[M_\nu^{(i)}, M_\nu^{(k)}]\, M_\nu^{(j)}(-z) = {}& \left\{ M_\nu^{(i)}\left(-\frac{i\pi}{4}\right) M_\nu^{(k)}\left(\frac{i\pi}{4}\right) \right. \\
& \left. + M_\nu^{(i)}\left(-\frac{i\pi}{4}\right) M_\nu^{(k)}\left(\frac{i\pi}{4}\right) \right\} M_\nu^{(i)}(z) \\
& - \left\{ M_\nu^{(i)}\left(-\frac{i\pi}{4}\right) M_\nu^{(i)}\left(\frac{i\pi}{4}\right) \right. \\
& \left. + M_\nu^{(i)}\left(\frac{i\pi}{4}\right) M_\nu^{(i)}\left(-\frac{i\pi}{4}\right) \right\} M_\nu^{(k)}(z).
\end{aligned}
\tag{5.82}
$$

Evaluating the ratio of the coefficients for $j = 3$, $k = 4$, we obtain

where
$$
\left.
\begin{aligned}
\frac{A'}{B'} &= \frac{R^2 - 1}{R^2 - e^{-2i\pi\nu}} \\[2mm]
R &= M_{-\nu}^{(1)}(0)/M_\nu^{(1)}(0)
\end{aligned}
\right\}
\tag{5.83}
$$

In terms of the Wannier functions we write

$$M_\nu^{\left(\frac{3}{4}\right)}(z; h) \sim h_e^{\left(\frac{2}{1}\right)}(z), \quad R(z) \to +\infty.\tag{5.84}$$

This behaviour establishes the connections between the functions $M^{(j)}$ and $h_e$.

The $S$-matrix is finally given by

$$S = -i\left(\frac{A'}{B'}\right) e^{-i\pi\nu} \cdot e^{i\pi l}.\tag{5.85}$$

### 5.8  Regge poles and the potentials $r^{-n}$, $n \geqslant 4$

In this section we briefly discuss the asymptotic behaviour of the $\mathrm{Re}\,\alpha(s)$ as $s \to +\infty$. This feature of the problem is interesting since considerable attention was focused on the rise of $\mathrm{Re}\,\alpha(s)$ in relativistic processes. For simplicity we consider the potential $r^{-4}$ since this is exactly solvable and the $S$-matrix is already given in Section 5.

In a recent paper Mandelstan (23) proposed a dynamical scheme for rising Regge trajectories in the limit of very large energy, $s$. Four features form the basic Mandelstan schemes: (1) the asymptotic phase shift increases with energy, (2) the violation of the Levinson theorem, (3) $\mathrm{Im}\,\alpha(S) = 0$, and (4) the residue function $\beta(S) \sim s^{-1}$ as $s \to +\infty$.

He then constructs a linearly rising $\mathrm{Re}\,\alpha(s)$.

Interestingly we notice that the class of singular potentials $r^{-n}$ satisfies all these conditions while condition (3) is not required and yet it is clear that $\mathrm{Re}\,\alpha(S)$ is a monotonically increasing function of the energy $s$.

The radial Schrodinger equation with the repulsive potential $r^{-4}$ gives the $S$-matrix in the form (24).

$$S(\lambda, k) = \frac{e^{i\pi\lambda}}{\cos\pi\beta + i\,[e^{-2\Phi} + \sin^2\pi\beta]^{1/2}}, \qquad (5.86)$$

where

$$\cos\pi\beta = 1 - 2\,\Delta(0)\sin^2\frac{\pi\lambda}{2} \qquad (5.87)$$

and the function $\Phi$ is found by using the W.K.B. method

$$\Phi(\lambda, k) = \frac{\pi\,(\lambda^2 - 2ikg)}{2\,(\lambda^2 + 2ikg)^{1/2}} = \left(\frac{3}{2}, \frac{1}{2}; 2; \frac{\lambda^2 - 2ikg}{\lambda^2 + 2ikg}\right)$$

$$+ O\,[(\lambda^2 + 2ikg)^{-1/2}]. \qquad (5.88)$$

In Eq. (5.87), $\Delta(0)$ is a certain determinant function of $g$, momentum $k$ and $\lambda = l + \frac{1}{2}$. This approximation is good if $|\lambda^2 + 2ikg| \gg 1$.

The Regge poles and their residues can be found from the poles of the $S$-matrix Eq. (5.86), given by

$$e^{-2\Phi} = -1, \qquad (5.89)$$

and they lie along a line

$$\mathrm{Re}\,\Phi(\lambda, k) = 0. \qquad (5.90)$$

The residue functions $\beta(s)$ for the Regge poles in the case of $r^{-n}$ were calculated by Domby and Jones (25). We write

$$\beta_m(S) = \lim_{\lambda \to \lambda_m}\ (\lambda - \lambda_m)\,S(\lambda, k)$$

$$= \frac{\exp(2\pi\lambda_m)\,\cos\pi\beta\lambda_m}{\phi'(\lambda_m)}. \qquad (5.91)$$

Here we let $s = k^2$. In the limit of large $m$, i.e. for the leading trajectory, we have

$$\beta_n(s) = \left[ 2 \ln \left( \frac{4\lambda_m^2}{ig} \right) - \ln s \right]^{-1} \tag{5.92}$$

and asymptotically,

$$\beta_m(s) \sim -(\ln s)^{-1}. \tag{5.93}$$

If the trajectory function $\alpha(s)$ is assumed to obey real analyticity, then we can write the following dispersion relation,

$$\mathrm{Re}\,\alpha(s) = as^{n-1} + \frac{s^n}{\pi} P \int_{s_0}^{\infty} \frac{\mathrm{Im}\,\alpha(s')\,ds'}{(s')^n\,(s'-s)}. \tag{5.94}$$

In Eq. (5.94), $s_0$ is the threshold, $a$ is a constant and the dispersion relation has been subtracted $n$ times.

With the help of unitarity, we write

$$\mathrm{Im}\,\alpha(s) = s^{\alpha(0)+1/2}\,\beta(s). \tag{5.95}$$

Eq. (5.94) is a consequence of real analyticity and Eq. (5.95) is a consequence of unitarity. Together with the asymptotic form of $\beta(S)$ we obtain (26)

$$\mathrm{Re}\,\alpha(s) = \left[ \exp\left\{ \alpha(0) + \tfrac{3}{2} - n \right\} - 1 \right] s^{n-1} \ln s. \tag{5.96}$$

But for one subtraction we notice that

$$\mathrm{Re}\,\alpha(s) \sim \ln s \tag{5.97}$$

and two subtractions

$$\mathrm{Re}\,\alpha(s) \sim S \ln s \tag{5.98}$$

since

$$\mathrm{Re}\,\alpha(0) > -\tfrac{1}{2}.$$

We notice here that apart from corssing symmetry, this potential model has all the required features advanced by Mandelstam for his relativistic rising trajectories model.

## References


1. K. Meetz, *Nuovo Cimento*, **34**, 690 (1964).
2. N. Limic, *Nuovo Cimento*, **26**, 581 (1962); **28**, 1066 (1963); **36**, 100 (1965), and **A42**, 516 (1966).
3. H. Poincaré, *Acta Math.*, **4**, 213 (1884).
4. A. Pais and T. T. Wu, *Phys. Rev.*, **134**, B1303 (1964).

5. G. Tiktopoulos and S. B. Treiman, *Phys. Rev.*, **134**, B 844 (1964).

6. A. Pais and T. T. Wu, *J. Math. Phys.*, **5**, 799 (1964).

7. A. Bollino, A. Longoni and T. Regge, *Nuovo Cimento*, **23**, 954 (1962).

8. E. del Guidice and E. Galzenati, *Nuovo Cimento*, **38**, 443 (1965); **40**, 739 (1965).

9. T. F. O'Malley, L. Spruch and L. Rosenberg, *J. Math. Phys.*, **2**, 491 (1961); *Phys. Rev.*, **134**, A 1188 (1964).

10. B. R. Levy and J. B. Keller. *J. Math. Phys.*, **4**, 54 (1963).

11. L. Bertocchi, S. Fubini and G. Furlan, *Nuovo Cimento*, **35**, 633 (1965); **35**, 599 (1965).

12. N. Dombey, *Nuovo Cimento*, **37**, 1741 (1965).

13. F. Calogero, *Nuovo Cimento*, **37**, 756 (1965); *Phys. Rev.*, **139**, B 602 (1965); *Phys. Rev.*, **135**, B 693 (1964).

14. T. T. Wu, *Phys. Rev.*, **136**, B 1176 (1964).

15. K. M. Case, *Phys. Rev.*, **80**, 797 (1950).

16. G. Feinberg and A. Pais, *Phys. Rev.*, **131**, 2724 (1963); **133**, B 477 (1964).

17. N. N. Khuri and A. Pais, *Rev. Mod. Phys.*, **36**, 590 (1964).

18. H. H. Aly, Riazuddin and A. H. Zimerman, *Phys. Rev.*, **136**, B 1174 (1964); *J. Math. Phys.*, **6**, 1115 (1965). (See this paper for various references.)

19. R. M. Spector, *J. Math. Phys.*, **8**, 2357 (1967). (This is a very interesting paper and it contains further references.)

20. R. M. Spector, *J. Matth. Phys.*, **5**, 1185 (1964); H. H. Aly and H. J. W. Müller, *J. Math. Phys.*, **7**, 1 (1966), and further references in these two papers.

21. J. Meixner and F. W. Schäfke, *Mathieresche Funktionen und Sphäreidfunktionen*, Springer, Berlin (1954), and G. H. Wannier, *Quart. Appl. Math.*, **11**, 33 (1953).

22. E. Vogt and G. Wannier, *Phys. Rev.*, **95**, 1190 (1954).

23. S. Mandelstam, *Phys. Rev.*, **166**, 1539 (1968).

24. H. H. Aly and H. J. W. Müller, *J. Math. Phys.*, **7**, 1 (1966).

25. N. Dombey and R. H. Jones, *J. Math. Phys.*, **9**, 986 (1968).

26. H. H. Aly and P. Narayanaswamy, *Phys. Letters*, **28B**, 603 (1969).

# Singular Interaction in Quantum Field Theory


H. H. ALY

*Southern Illinois University*

W. GUTTINGER and H. J. W. MÜLLER

*Universität München, Germany*


## Contents



## 1  INTRODUCTION

About thirty years ago, Heisenberg (1) divided interactions between quantized fields into two classes which nowadays are termed renormalizable and unrenormalizable, respectively. Renormalizable interactions are characterized by dimensionless coupling constants, their interaction strength varies slowly with energy and the transition elements do not increase with increasing energy so that a finite number of subtractions suffices to render these theories finite. On the other hand, unrenormalizable interactions, such as 4-fermion theories, are characterized by coupling constants having dimension of some power of length, their interaction strength changes strongly with energy and the transition elements of $n$-th order increase with increasing energy as $|E|^n$ so that an infinite number of subtractions would seem to be necessary to make such theories finite. To avoid an indefinite increase of cross sections with increasing c.m. momentum, corresponding to a violation of unitarity, Heisenberg suggested that some sort of elementary length defining a kind of nonlocality should be introduced. For example, in a 4-fermion with coupling constant $g$ theory the unitarity limit was guessed to take place at those where $E \cdot g \approx 1$ i.e., at $E = 300$ GeV and thus an elementary length should be of the order of $l \approx 10^{-17}$ cm. Although in the meantime the picture of unrenormalizable theories has changed and those theories became more tractable, it is still an open question what should be done to make them physically meaningful and whether they are not already meaningful as they stand.

The division of field theories into renormalizable and unrenormalizable theories is closely linked with the singular structure of the interaction kernels characterizing the exchanged particles: The scattering of two particles is governed by a Bethe–Salpeter equation of the type

$$(\Box - m^2 - V(x))\,\psi = E\psi$$

with an interaction kernel $V(x)$ which as the mutual distance $x^2$ of the particles tends to zero behaves as $V(x) \sim 1/(x^2)^\alpha$. A theory is said to be renormalizable if $\alpha \leq 1$, it is called unrenormalizable if $\alpha < 1$; as in potential theory, the value $\alpha = 1$ discriminates between stable and unstable situations, respectively.

## 2  MATHEMATICAL TECHNIQUES FOR RELATIVISTIC SINGULAR INTERACTIONS

In this section we outline the generalized function techniques indispensable for dealing with singular relativistic interactions.

### 2.1  Lorentz invariant generalized functions

A generalized function $f$ on $R^4$, with $x = (x_0, \mathbf{x})$, $\mathbf{x} = (x_1, x_2, x_3)$, $x^2 = x_y^2 = x_0^2 - \mathbf{x}^2$, is a linear and continuous functional on a space $\phi$ of test functions $\phi$: $f\langle\varphi\rangle$, symbolically, $f\langle\varphi\rangle = \int f(x)\,\varphi(x)\,d^4x$. The test functions are assumed to be elements of the space $D$ or $Z$ (2, 3). $f$ is said to be invariant under the proper homogeneous Lorentz group

$$L_+^\uparrow \quad \text{if} \quad L_+^\uparrow f = f, \text{ i.e.,} \quad \text{if} \quad L_+^\uparrow f\langle\varphi\rangle = f\langle\varphi\,(\lambda x)\rangle = f\langle\varphi(x)\rangle; \quad x' = \lambda x.$$

Let $f(x)$ be an invariant function having an algebraic singularity on the light cone $x^2 = 0$. Then the generalized function $f$ associated with $f(x)$ is given by (2), (3)

$$\varphi(x)\,\langle\varphi(x)\rangle \;=\; \operatorname*{Res}_{z\to 0}\left[\frac{1}{z}\int_{-\infty}^{\infty} d^4x\,f(x)\left(\frac{a^2}{x^2 - i0}\right)^z \varphi(x)\right] \tag{2.1}$$

where $a$ is an arbitrary scaling parameter of dimension of length on which the l.h.s. depends, if and only if the classical integral $F(0) = \displaystyle\int_{-\infty}^{\infty} d^4x\,f(x)\,\varphi(x)$ exhibits logarithmic divergences. In Eq. (2.1) one first has to evaluate the $x$-integral for Re $(z)$ sufficiently large and negative, where it converges, then to continue the resulting function of $z$, $F(z) = \displaystyle\int_{-\infty}^{\infty} d^4x\,f(x)\,(a^2/(x^2 - i0))^z$ $\varphi(x)$, into a neighbourhood of the origin $z = 0$ and finally to take the residue. In case of functions of the type $(x^2)^{-\lambda}\,\theta(\pm x_0)\log^\mu|x^2|$ etc., Eq. (2.1) yields a generalization of the conventional $Pf$-notion. If $f(x)$ has an essential singularity at $x^2 = 0$, Eq. (2.1) is to be applied to every term in an expansion of $f(x)$. If the classical integral $F(0) = \displaystyle\int_{-\infty}^{\infty} d^4x\,f(x)\,\varphi(x)$ does not exhibit logarithmic divergences, then the l.h.s. of (2.1) is precisely the analytic continuation of $F(z)$ to $z = 0$. Thus, for example, if

$$f(x) = (x^2 \pm i0)^{-\lambda} = \lim_{\eta_{\nu\mu}\to 0} (x^2 \pm i\eta_{\mu\nu}x^\mu x^\nu)^{-\lambda} \tag{2.2}$$

where $\eta_{\mu\nu}x^\mu x^\nu$ is positive-definite, then

$$(x^2 \pm i0)^{-\lambda}\,\langle\varphi\rangle \;=\; \operatorname*{Res}_{z\to 0}\left[\frac{a^{2z}}{z}\int_{-\infty}^{\infty} d^4x\,(x^2 \pm i0)^{-\lambda-z}\,\varphi(x)\right] \tag{2.3}$$

is, for $\lambda \neq 2, 3, 4, \ldots$, the analytic continuation in $\lambda$ of $\displaystyle\int_{-\infty}^{\infty} d^4x\,(x^2 \pm i0)^{-\lambda}\varphi(x)$

$$\text{from Re } \lambda < -1 \text{ to Re } \lambda \geqslant -1.$$

If $\lambda = n = 2, 3, 4, \ldots$ the r.h.s. of Eq. (2.3) still exists and equals the pole-free part of $F(z)$ for $z = 0$ (but not, of course, the analytic continuation of

$F(z)$ to $z = 0$). In that case, the classical integral $\int_{-\infty}^{\infty} d^4x \, (x^2 \pm i0)^{-n} \, \varphi(x)$

indeed diverges logarithmically. $(x^2 \pm i0)^{-n}$, $n = 2, 3, \ldots$ depends explicitly on the arbitrary parameter $a$: With $(x^2 \pm i0)^{-n} = (x^2 \pm i0)^{-n}|_a$ one finds that

$$(x^2 \pm i0)^{-n}|_a - (x^2 \pm i0)^{-n}|_{a'} = c. \log \frac{a^2}{a'^2} \, \Box^{n-2} \, \delta(x); \quad n = 2, 3, \ldots.$$

$$(2.4)$$

This is easily seen by means of the Fourier transform of Eq. (2.1) which, if $f(x)$ decreases sufficiently rapidly as $|x| \to \infty$, follows simply by replacing the test function $\varphi(x)$ by $\varphi(x) = \exp(ikx)$:

$$\tilde{f}(k) = \mathscr{F} f(x) = \operatorname*{Res}_{z=0} \left[ \frac{1}{z} \int_{-\infty}^{\infty} d^4x \, f(x) \left( \frac{a^2}{x^2 - i0} \right)^z e^{ikx} \right]. \qquad (2.5)$$

In particular, for (2.2), $k$ real, it follows that

$$\mathscr{F} (x^2 - i0)^{-\lambda} = -i (4\pi)^2 (-4)^{-\lambda} \, \Gamma(2 - \lambda) (-k^2 - i0)^{\lambda-2}/\Gamma(\lambda),$$

$$\lambda \neq 2, 3, \ldots$$

$$\mathscr{F} (x^2 - i0)^{-n} = \frac{i (4\pi)^2 (-k^2 - i0)^{n-2}}{4^n (n-1)! (n-2)!} \left[ \log \left( \frac{a^2 (-k^2 - i0)}{4} \right) \right.$$

$$\left. - \psi(n) - \psi(n-1) \right], \quad n = 2, 3, \ldots, \qquad (2.6)$$

where $\psi(n) = \Gamma'(n)/\Gamma(n)$.

More generally, with $f(x^2 \pm i0) = \lim_{\eta \to 0} (x^2 \pm i\eta_{\mu\nu} x^\mu x^\nu)$, where $\eta_{\mu\nu} x^\mu x^\nu$ is a positive-definite quadratic form and means that the coefficients $\eta_{\mu\nu}$ of the form tend to zero, we have the generalized Bochner formula

$$\mathscr{F} f(x^2 - i0) = -4i\pi^2 \operatorname*{Res}_{z \to 0} \left[ \frac{1}{z \, (-k^2 - i0)^{1/2}} \int_0^{\infty} \tau^2 \left( -\frac{a^2}{\tau^2} \right) f(-\tau^2) \times \right.$$

$$\left. \times \mathscr{F} (\tau (-k^2 - i0)^2 \, d\tau \right]$$

which generalizes the "Reduction Formula" of Pais and Feinberg (4).

The notation $f(x^2 \pm i0) = \lim_{\varepsilon \to +0} f(x^2 \pm i\varepsilon)$ is nothing but a loose form of the above definition. It follows, therefore, that the classical function $(x^2$

$\pm i0)^{-\lambda}$ has a singularity only at the origin of the light cone and not on the whole cone as usually stated.

The result (2.4) means that two determinations of $(x^2 \pm i0)^{-n}$ corresponding to two different scaling parameters $a$ and $a'$ differ from one another by a term $\log \dfrac{a^2}{a'^2} \square^{m-2}\delta(x)$ localized at the origin of the light cone $x_0 = \mathbf{x} = 0$.

Consequently, two determinations of the infinite series

$$T = \sum_{n=1}^{\infty} C_n (x^2 - i0)^{-n} \tag{2.7}$$

differ from one another by the generalized function

$$T_0 = \log \frac{a^2}{a'^2} \sum_{0}^{\infty} C_n \square^n \delta(x). \tag{2.8}$$

$T_0$ is localized at the origin of the light one if and only if the coefficients $C_n$, satisfy the condition

$$\limsup_{n \to \infty} (|C_n| \, 2n!)^{1/n} = 0. \tag{2.9}$$

If the terms $(x^2 + i0)^{-n}$ in (2.7) are associated with different scaling parameters and for different orders $n$, one can easily prove that the condition (2.9) still remains satisfied provided that the $a_n$ do not increase too fast as $n \to \infty$.

The introduction of the scaling parameter $a$ into (2.1), and (2.3) is necessary for dimensional reasons: The argument in the logarithm in (2.6) would not be dimensionally correct if $a$ had not dimension of length. $a$ plays the role of a unit of length. Eq. (2.4) tells that, contrary to the classical function $(x^2 - i0)$ the generalized function $(x^2 - i0)^{-n} \langle \varphi \rangle$ is homogeneous only if a change of scale $x \to \alpha x$ is accompanied with a simultaneous change of scale $a \to \alpha a$:

$$\left. ((\alpha x)^2 - i0)^{-n} \right|_{\alpha a} = \frac{1}{\alpha^{2n}} \left. (x^2 - i0) \right|_{a} . \tag{2.10}$$

For details, see references (1), (2), (3). The most general distribution associated with $(x^2 \pm i0)^{-\lambda}$ follows by adding to the r.h.s. of (2.3) a term $\sum_{0}^{\lambda} b_n \square^n \delta$ which, however, destroys homogeneity. Also such a term would conflict the result of defining $(x^2 \pm i0)^{-\lambda}$ $(\lambda \neq 2, 3, \ldots)$ by analytic continuation.

The product $f_1(x) f_2(x)$ of two Lorentz invariant generalized functions $f_1(x)$ and $f_2(x)$ singular at $x_0 = \mathbf{x} = 0$ or at $x^2 = 0$ is defined by the functional

$$(f_1(x) \cdot f_2(x)) \langle \varphi \rangle = \operatorname*{Res}_{z=0} \left[ \frac{1}{z} \int_{-\infty}^{\infty} d^4x \, f_1(x) f_2(x) \left( \frac{a^2}{x^2 - i0} \right)^z \varphi(x) \right].$$

$$(2.11)$$

From this it follows, in particular, that

$$(\theta(\pm x) \cdot g(x^2)) \langle \varphi \rangle = \operatorname*{Res}_{z=0} \left[ \frac{1}{z} \int_{-\infty}^{\infty} d^4x \, \theta(\pm x_0) \, g(x^2) \left( \frac{a^2}{x^2 - i0} \right)^z \varphi(x) \right].$$

$$(2.12)$$

Thus one finds that

$$(x^2 - i0)^{-\lambda} (x^2 - i0)^{-\mu} = (x^2 - i0)^{-\lambda-\mu} \, \theta(\pm x_0) \, \delta^4(x^2)$$

is given by

$$\theta(\pm x_0) \, \delta^{(n)}(x^2) \langle \varphi \rangle = \operatorname*{Res}_{z=0} \left[ \frac{1}{z} \, \theta(\pm x_0) \left( \frac{a^2}{x_0^2} \right)^z \delta^{(n)}(x^2) \langle \varphi(x) \rangle \right]$$

and it is easily seen that $\delta^{(n)}(x^2) = \theta(x_0) \, \delta^{(n)}(x^2) + \theta(-x_0) \, \delta^{(n)}(x^2)$ depends explicitly on $a$, while $\varepsilon(x_0) \, \delta^{(n)}(x^2)$ does not (and indeed is a regular functional).

The generalized function $f(x^2 - ix_0 0) = f_+$ is defined by

$$f(x^2 - ix_0 0) \langle \varphi \rangle = \lim_{\eta \to 0} f((x - i\eta)^2) \langle \varphi \rangle$$

$$= \lim_{\eta \to 0} \int_{-\infty}^{\infty} f((x - i\eta)^2) \, \varphi \, (x - i\eta) \, d^4x \qquad (2.13)$$

with $\eta$ time-like. In particular, $(x^2 - ix_0 0)^{-\lambda}$ is a regular functional, and is the boundary value of an analytic function contrary to $(x^2 - i0)^{-\lambda}$. Products of these generalized functions are defined through (2.11). They may also be defined by the convolution formula $f_+ \cdot g_+ = \mathscr{F}^{-1} ((\mathscr{I}f_+) \times (\mathscr{I}g_+))$ which, however, yields the same result as (2.11), e.g.

$$(x^2 - ix_0 0)^{-\lambda} (x^2 - ix_0 0)^{-\mu} = (x^2 - ix_0 0)^{-\lambda-\mu}.$$

Series of the type

$$\sum_0^\infty (x^2 - ix_0 0)^{-\eta_n}, \quad \sum_0^\infty (x^2 - i0)^{-\eta_n} \qquad (2.14)$$

can be shown to converge in the space $Z'$. It should be noted that $f_+ \langle \varphi \rangle$ cannot be rotated to Euclidean space.

With the above techniques it is easy to construct convolution formulas e.g., by Fourier transformation one finds at once

$$(x^2 - i0)^{-\lambda} \times (x^2 - i0)^{-1} = \int_{-\infty}^{\infty} d^4y \, \frac{1}{[(x-y)^2 - i0]} \, \frac{1}{(y^2 - i0)}$$

$$= i\pi^2 \operatorname*{Res}_{z=0} \left[ \frac{1}{z} \, \frac{(x^2 - i0)^{-\lambda-1-z} \, a^{2z}}{(1 - \lambda - z)(2 - \lambda - z)} \right]. \qquad (2.15)$$

## 2.2  Generalized Lehmann representation

According to (2.1) the formal spectral representation

$$\Delta_F'(x) = \int_0^\infty dm^2 \, \varrho(m^2) \, \Delta_F(x; m^2) \qquad (2.16)$$

has the following meaning within the frame of generalized functions:

$$\Delta_F'\langle\varphi\rangle = \operatorname*{Res}_{z=0} \left[ \frac{1}{z} \int_{-\infty}^{\infty} \left\{ \int_0^\infty \varrho(m^2) \, \Delta_F(x; m^2) \right\} \left( \frac{-a^2}{x^2 - i0} \right)^z \varphi(x) \, d^4x \right].$$

$$(2.17)$$

For any $\varrho(m^2)$ of the type $\varrho(m^2) = (m^2)^\lambda \log^\mu m^2$, $\lambda \geqslant 0, \mu = 0, 1, 2, \ldots$ or series thereof it follows from (2.17) that

$$\Delta_F'\langle\varphi\rangle = \operatorname*{Res}_{z=0} \left[ \frac{1}{z} \int_0^\infty \varrho(m^2) \, (a^2 m^2)^z \, \{ \Delta_F(x, m^2) \, \langle\varphi\rangle \} \, dm^2 \right]. \qquad (2.18)$$

Taking the Fourier transform of (2.18) by means of (2.5) we arrive at the momentum space spectral representation

$$\Delta_F'(k) = 2i \operatorname*{Res}_{z=0} \left[ \frac{1}{z} \int_0^\infty \frac{\varrho(m^2) \, (a^2 m^2)^z}{(k^2 - m^2 - i0)} \, dm^2 \right]. \qquad (2.19)$$

This unsubtracted dispersion relation is called the generalized Lehmann representation. It contains only a single scaling parameter $a$ however $\varrho$ increases as $m \to \infty$. In case of infinite series $\varrho = \sum_0^\infty C_n(m^2)^{\eta_n}$, with $\eta_n \to \infty$ as $n \to \infty$, the residue operation is to be applied term-by-term. The representation of $\Delta_F'(x)$ preserves its usual form, viz.

$$\Delta'_F(x) \, \langle\varphi\rangle = \int_0^\infty \varrho(m^2) \, \{ \Delta_F(x, m^2) \, \langle\varphi\rangle \} \, dm^2 \qquad (2.20)$$

and for the absorptive part of $\Delta'_F(k)$, Eq. (2.19), we find

$$\operatorname*{disc}_{k^2 \gtrless 0} \Delta'_F(k) \equiv 2i \operatorname*{Im}_{h^2 \gtrless 0} \Delta'_F(k) = 4\pi \varrho(k^2) \qquad (2.21)$$

which is independent of $a$. If $\varrho(m^2) = m^\lambda$ and $\lambda \neq 0, 1, 2, \ldots$, then the l.h.s. of (2.19) is just the analytic continuation of $2i \displaystyle\int_0^\infty \frac{m^\lambda dm^2}{k^2 - m^2 - i0}$ from $\operatorname{Re} \lambda < 0$ to $\operatorname{Re} \lambda > 0$. The connection of (2.19) with the subtracted form is discussed in reference (2), (3), (5).

### 2.3  Integral equations

Integral equations of the type

$$f(x) = f_0(x) + g \int_{-\infty}^\infty G\,(x - x')\,V(x')f(x')\,d^4x' \qquad (2.22)$$

with singular interaction kernels $V(x)$ may now readily be solved by means of the preceding techniques. In case of Wightman-type functions $f(x) = f_+(x) = f(x^2 - ix_00)$ we have $G(x) = G_{\mathrm{ret}}^{(x)}$ and $V(x) = V_+(x) = (x^2 - ix_00)^{-\lambda}$ as $x^2 \to 0$ in case of a singular interaction where $f_0\,(x^2 - ix_00)$ is the free function. Since now the quantities involved in (2.22) are boundary values of analytic functions, Eq. (2.22) is of Fredholm type and can be solved by iteration, once the generalized function product $G_{\mathrm{ret}} \cdot V_+$ in (2.22) is defined in agreement with (2.11). In case of time-ordered functions $f(x) = f_F(x) = f(x^2 - i0)$ we have $G(x) = G_F = (x^2 - i0)^{-1}$ and $V(x) = V_F(x) = (x^2 - i0)^{-\lambda}$ as $x^2 \to 0$ in case of a singular interaction, $f_0$ being now the free time-ordered function. In this case, notwithstanding the definition of $G_F \cdot V_F$ by means of (2.11), Eq. (2.22) is no longer of Fredholm type and in fact has no classical (or even distribution) analogue. However, interpreting then (2.22) as a convolution equation in the sense of generalized functions, this equation can always be solved by iteration, utilizing convolution formula of the type (2.15). This solution, a power series expansion in $g$, coincides for $x_0 > 0$ and for $x_0 < 0$ with the corresponding Wightman type function $f_+, f_F(x) = f_+ (\pm x)$ for $x_0 \gtrless 0$, in accordance with (2.21). The time-ordered equation for $f_F$ may, however, also be rotated to Euclidean space in the $x_0$ or $\vec{x}$ plane, respectively, both rotations yielding the same Euclidean equation,

$$f_F^e(x') = f_{F_0}^e(x') + g \int_{-\infty}^\infty d^4x''\,G^e\,(x' - x'')\,V^e(x'')f_F^e(x''). \qquad (2.23)$$

This 4-dimensional Euclidean equation ($\pm x'^2 = x_0'^2 + \mathbf{x}'^2$) can be interpreted in two ways: (i) as a generalized function convolution equation and (ii) as a classical, though non-Fredholm, equation. If interpretation (i) is adopted, the equation can be solved in exactly the same way as the Minkowski space equation for $f_F$. Passing from Euclidean to Minkowski space by either $x_0$ or $\mathbf{x}$ rotation, gives the previously obtained iterative solution. If interpretation (ii) is adopted, this means that we have to look for a classically integrable solution $f_F^e(x')$. Such a solution may exist or may not exist. If an integrable solution exists, it vanishes strongly at $x' = 0$ and indeed has an essential singularity there. Passing from this Euclidean solution to Minkowski space by $x_0$ or $x$ rotation gives rise to two, in general different, Minkowski space solutions which, as does the Euclidean solution, have a singularity at $g = 0$. Already for this reason these solutions do not coincide with the Wightman type function for $x_0 \gtrless 0$ which, as we know already, is necessarily analytic at zero coupling. There are, moreover many cases where no integrable solution to (2.23) exists at all. In this case only the interpretation (ii) survives.

## 2.4  Peratization

The only way to obtain nonperturbative solutions to time-ordered equations is the one looking as (2.22) in terms of interpretation (ii) of the associated Euclidean equation (2.23). Peratization is a technique for obtaining such nonanalytic solutions in an iterative way. It requires a Euclidean interpretation of (2.22) and the existence of a classically integrable solution of Eq. (2.23). If such a solution exists, it does, except perhaps in case of marginally singular interactions ($\lambda = 1$), not preserve the absorptive parts and in general also conflicts with the physical interpretation of the time-ordered function. In case of propagators, in general no integrable solution exists at all.

## 3  RENORMALIZABLE AND UNRENORMALIZABLE FIELD THEORIES

### 3.1  General considerations

A field theory is called renormalizable if $\varrho(m^2)$, $\varrho_{1,2}(m^2)$ are tempered as $m^2 \to \infty$, i.e., bounded by some power of momentum; a field theory is called unrenormalizable if $\varrho(m^2)$, $\varrho_{1,2}(m^2)$ are not tempered but grow exponentially as $m^2 \to \infty$. In renormalizable theories, with $\varrho(m^2) \sim \exp(m^2)^\alpha$ as $m^2 \to \infty$, it follows from (2.18) that $\Delta_F'(x) \sim (x^2 - i0)^{-\lambda-2}$, i.e. $\Delta_F'(x)$ has at most algebraic singularities on the light cone. In unrenormalizable theories, with $\varrho(m^2) \sim \exp(m^2)^\alpha$ as $m^2 \to \infty$, $\Delta_F'(x) \sim \exp(x^2)^{-\alpha/(1-2\alpha)}$ has an essential singularity at $x^2 = 0$. For, the causality axiom poses the following bound on the high energy behavior: In a causal unrenormalizable field theory the functions $\varrho(k^2)$, $\varDelta'_+(k^2)$, $\varDelta'_F(k^2)$ $e^k$ satisfy the high energy growth property

$$|\varrho(k^2), \ldots, \varDelta'_F(k^2)| \leqslant C\, e^{|k^2|^\alpha}, \quad \alpha < \tfrac{1}{2}. \tag{3.1}$$

This means, if $\varrho(k^2) = \sum_0^\infty C_n(k^2)^n$, then

$$\varlimsup_{n\to\infty} ((2n)!\,|C_n|)^{1/n} = 0. \tag{3.2}$$

Consequently, the scattering amplitude $\pi$ is bounded by $|\pi\,(s, t)| \leqslant \left(\exp\left(\varepsilon\,\sqrt{|s|}\right)\right.$. Equivalent to this is the statement: A field theory is local if and only if the commutator $\varDelta'(x)$ [or the anti-commutator $S'(x)$] possesses the representation $(\varphi \in Z)$

$$\varDelta'\langle\varphi\rangle = \int_c dx_0 \int d\mathbf{x}\, \varDelta'\varphi\,(x)$$

with a closed contour in the $x_0$ plane which can be shrinked to the time-like domain.

### 3.2  Renormalizable theories

In a renormalizable theory, the interaction kernel in Eq. (2.22), $V_F \simeq (x^2 - i0)^{-\lambda}$, is characterized by $\mathrm{Re}\,\lambda \geqslant 1$. If $\lambda = 1$, the theory is called marginally singular. In those cases the solutions to propagator and Bethe-Salpeter equations of the type (2.22) (viz. local approximation) are analytic at zero coupling. The solutions can be obtained by straightforward application of the iteration techniques outlined in Section 2. Alternatively one may compute the Feynman diagrams by using the product definition (2.11) for propagators. This implies finite results to each order, with the conventional divergent renormalization constants appearing now replaced by the finite but arbitrary scaling parameters which may be different for different orders of the diagrams. In accordance with Eqs. (2.1) and (2.19), the renormalized expression for $\prod_{i,k} \varDelta_F\,(x_i - x_k)$ is given by

$$\int dx\,\{\varPi\,\varDelta_F\,(x_1 - x_k)\}$$

$$= \prod_P \operatorname*{Res}_{z_{ik}=0} \frac{1}{z_{i,k}} \int \cdots \int \varPi\,\varDelta_P\,(x_i - x_k)\cdot\left(\frac{a_{ik}^2}{(x_i - x_k)^2 - i0}\right)^{z_{ik}} dx. \tag{3.3}$$

That this is true also for overlapping diagrams has been shown in Refs. (6), (7), (8). The results obtained in this way from Feynman diagrams coincide with the generalized function solutions of the associated integral equations discussed above.

### 3.3 Unrenormalizable theories

The point is that the generalized function techniques which produce the renormalized results in renormalizable theories also render unrenormalizable theories finite. The procedure has already been discussed in reference (5) (cf. also [7]). It suffices to consider a 4-fermion theory as an example.

We start from the interaction Lagrangian

$$L_{\text{int}} = g : \tilde{\psi}_A \Gamma \psi_A : \, :\tilde{\psi}_B \Gamma \psi_B :$$

with two fermion fields $\psi_A, \psi_B$ with scalar or pseudoscalar coupling ($\Gamma = 1$ or $\Gamma = \gamma_S$). Confining ourselves to massless particles, the propagator $S'_F(x) = -2 \langle 0| T\psi_A (x/2) \tilde{\psi}_A (-x/2) |0\rangle$ satisfies in the strong approximation the equation

$$S'_F(x) = S_F(x) - \frac{g}{4} \int d^4u \, d^4v \, S_F (x - u) \, \Gamma V_F (\mu - v) \, S'_F (\mu - v) \, \Gamma S_F (v)$$

$$(3.4)$$

with

$$S_F(x) = -i\gamma_\mu \partial_\mu D_\Gamma(x), \quad V_F(x) = \frac{4}{(x^2 - i0)^3} \tag{3.5}$$

while for $S'_+(x)$ we obtain

$$S'_+(x) = S_+(x) - g \int S_R (x - u) \, \Gamma V_+ (u - v) \, \Gamma S'_f (m - v) \cdot \Gamma S_A (v) \, d^4u \, d^4v$$

$$(3.6)$$

with

$$S_{R(A)}(x) = -ix_\mu \partial_\mu \theta \, (\pm x_0) \, \delta(x^2)/2\pi, \quad V_+(x) = \frac{4}{(x^2 - ix_0 0)^2}. \tag{3.7}$$

By the methods of Section 2.2 one finds the generalized function iterative solutions

$$S'_F(x) = S_F(x) \, (-\sqrt{g}/2 \, (x^2 - i0))^{-1} \, I_1 \, (-\sqrt{g}/(x^2 - i0))$$

$$S'_+(x) = S_+(x) \, (-\sqrt{g}/2 \, (x^2 - ix_0 0))^{-1} \, I_1 \, (-\sqrt{g}/(x^2 - ix_0 0)) \tag{3.8}$$

respectively with $S'_F(x) = \mp \, S'_+(x)$ for $x_0 \gtrless 0$ as it should be. If a Euclidean rotation is performed in (3.4) one easily discovers that the resulting Euclidean equation does not possess an integrable classical solution. Hence, peratization—as a means of obtaining a solution nonanalytic in the coupling—does not make sense. For the commutator $S'$ one obtains

$$S'(x) = S'_+(x) + S'_+(-x) = \frac{i\gamma_\mu\partial_\mu}{2\pi} \sum_0^\infty (g/4)^n \frac{\varepsilon(x_0)\,\delta^{(2n)}(x^2)}{n!\,(n+1)!\,(2n+1)!} \qquad (3.9)$$

which satisfies the causality requirement (3.2). The spectral representation of $S'_F$ is in accordance with our generalized Lehmann representation (2.21) which, in the Fermion case takes the form

$$S'_F(k) = -2i \operatorname*{Res}_{z=0}\left[\frac{1}{z}\int_0^\infty \frac{dm^2\,[(\gamma k + m)\,\varrho_1(m^2) - \varrho_2(m^2)]\,(a^2m^2)^z]}{(k^2 - m^2 + i0)}\right] \qquad (3.10)$$

with

$$\varrho_1 = \delta(m^2) + \sigma(m^2), \quad \varrho_2(m^2) = m\sigma(m^2)$$

$$\sigma(m^2) = \sum_1^\infty (\sqrt{g}/8)^{2n}\,(m^2)^{2m-1}/n!\,(n+1)!\,(2n-1)!\,(2n+1)!. \qquad (3.11)$$

For the Bethe–Salpeter amplitude $\psi_F(x)$ one finds in the case of equal spins, $p = 0$ and zero total energy $E$ the simple equation in ladder approximation

$$\psi_F(x) = 1 - \frac{ig}{\pi^2}\int_{-\infty}^\infty d_4x'' \frac{1}{(x-x'')^2 - i0}\frac{1}{(x''^2 - i0)^3}\psi_F(x'') \qquad (3.12)$$

whose iterative generalized function solution is, via (2.15),

$$\psi_F(x) = \sum_0^\infty g^n\,(x^2 - i0)^{-2n}/(2n)! \equiv \cosh\left(\frac{\sqrt{g}}{x^2 - i0}\right) \qquad (3.13)$$

which agrees for $x_0 \gtrless 0$ with the associated non-time-ordered function $\psi_+(x)$ satisfying a similar equation. The Euclidean equation associated with (3.12) has the same iterative generalized function solution if one returns to Minkowski space. Moreover—and in contrast with (3.4)—this Euclidean equation also has an integrable solution, viz.

$$\psi_F^e(x') = e^{-\sqrt{g}/x'^2}, \quad x'^2 = x_0'^2 + \mathbf{x}_2' \qquad (3.14)$$

with a singularity at zero coupling. Returning to Minkowski space via $x_0$ rotation yields $e^{-\sqrt{g}/(x^2-i0)}$ and via $\mathbf{x}$ rotation $e^{+\sqrt{g}/(x^2-i0)}$. Hence one finds a 1-parameter solution

$$\psi_F = \cosh\frac{\sqrt{g}}{(x^2 - i0)} + A\sinh\frac{\sqrt{g}}{(x^2 - i0)}. \qquad (3.15)$$

The solution (3.14) in Minkowski space, obtained via an Euclidean definition of the equation, is in fact obtainable through peratization and coincides with the solution of Eq. (3.8) for $m = 4$. The discontinuity of (3.15) across the physical cut in momentum space however coincides with the perturbative Wightman type function if and only if $A \equiv 0$, i.e., if one sticks to the perturbative Bethe–Salperter solution (3.13).

It is a simple matter to write down the Feynman diagrams corresponding to the above *BS* equation and to evaluate them by utilizing the product definition (2.11), i.e. by adopting the generalized function renormalization procedure (3.3). The result is precisely equation (3.13).

### 3.4   Conclusions

We have seen that the techniques developed permit one to deal with renormalizable and unrenormalizable field theories in the same way and with the same mathematical success. This does not, however, immediately imply that unrenormalizable theories are physically meaningful. Indeed, in the Bethe–Salpeter case there have so far been no solutions obtained for unrenormalizable theories and total energy $E \neq 0$, nor is anything known as to whether unitarity is conserved in those interactions or not. Even if unitarity is guaranteed, the physical sense of unrenormalizable theories in the high energy domain seems somewhat doubtful. Indeed, the interaction kernels $(x^2 - i0)^{-n}$ are not only singular on the light cone but change sign at the cone and, therefore, act as repulsive *and* attractive singular "potentials". In the attractive region, which in field theory cannot be separated from the repulsive one, the motion of the particles becomes highly unstable. This does not necessarily conflict with unitarity but it might not correspond to reality at all. The Euclidean interpretation of the Bethe–Salpeter equation actually projects out of the interaction just a repulsive part in Euclidean space thus yielding a well behaved theory or, at least simulating such a theory, which, however, conflicts the perturbative absorptive parts. On the other hand, no Euclidean interpretation is possible in the case of propagators.

Besides this, there are many open problems in the present context, in particular the extension to $n$-point functions, the confrontation between definitions of the quantities through power series and the closed-form definitions via analytic continuation, the meaning of the scaling parameters and whether they appear in observable quantities, and, last but not least, the actual calculation of higher order corrections in unrenormalizable theories and their confrontation with experiment.

Things might, however, become much more physical if one considers 4-fermion theories as a formal $Z \to 0$ limit ($Z$ = renormalization constant $Z_3$) of a renormalizable Yukawa theory such that $(\overline{\psi}\psi)$ in $L_{\text{int}} = g\overline{\psi}\psi\,(\overline{\psi}\psi)$ is considered as a bound state particle. If this could be done, and there are some attempts in this direction (9), the so defined 4-femion theory would be renormalizable and would not suffer from instabilities and related unrenormalizable troubles.

One of the authors (H. H. A.) woulo like to thank Professor A. Salam for the hospitality extended to him in the Summer of 1969.

## References


1. W. Heisenberg, *Z. für Physik*, **101**, 533 (1936); *Ann. Physik*, **32**, 20 (1938).
2. W. Guttinger, *Fort. der Physik*, **14**, 483 (1966).
3. W. Guttinger, SIAM, *J. App. Math.*, **15**, 964 (1967).
4. A. Pais and G. Feinberg, *Phys. Rev.*, **131**, 2724 (1963); **133**, B477 (1964).
5. W. Guttinger and E. Pfaffelhuber, *Nuovo Cimento*, **52**, 389 (1967).
6. E. Pfaffelhuber, University of Munich Preprint (1967).
7. E. Pfaffelhulber and W. Guttinger, University of Munich Preprint (1967).
8. E. Spcer, private communiaction to H. Mitter.
9. D. Lurie and A. J. Macfarlane, *Phys. Rev.*, **136B**, 816 (1964); H. H. Aly and H. J. W. Müller, *J. Math. Phys.*, **8**, 367 (1967).

# Weak Interactions in the Nuclear Shadow

J. S. BELL

*CERN, Geneva, Switzerland*

Contents



## 1  INTRODUCTION

Adler[1] observed that in certain circumstances cross-sections for neutrino and pion induced reactions would be proportional in a way independent of the target. This seemed surprising for nuclear targets[2] because of the very different rates of absorption of neutrinos and pions in nuclear matter. The paradox was resolved by an investigation of the propagation of nearly real pions in such a medium. It was suggested that similar phenomena could occur at higher energies in association with the propagation of heavier mesons. Subsequently, the role of these heavier mesons in nuclear matter has attracted particular attention in connection with photonuclear reactions[3].

This paper is mainly an extended and informal account of ideas already reported briefly and formally[2]. That paper considered only neutrino reactions, which will again be the main interest here. However, because the photonuclear case is simpler we will mention it first by way of introduction.

The idea is widely entertained that the interactions of photons with hadrons are dominated by the intervention of vector mesons[4]. Thus one draws Feynman diagrams for $\gamma + T \to F$, where $T$ is some target and $F$ a final hadronic state:

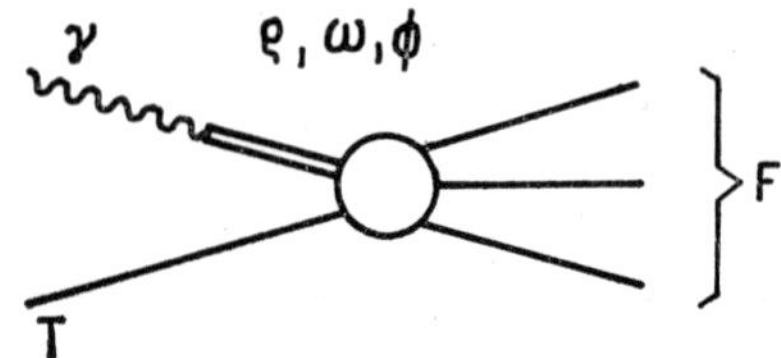


This suggests a rather simple relation between transition rates[4] with incident photons and vector mesons. If for simplicity we forget about $\omega$ and $\phi$ mesons, we could except for total transition rates[5]

$$\Gamma_\gamma = |g_{\gamma\varrho} m_\varrho^{-2}|^2 \, \Gamma_\varrho \tag{1.1}$$

where $g_{\gamma\varrho}$ is the $\gamma\varrho$ coupling constant, and we have put $q^2 = 0$ (for a real photon) in the propagator factor $(m_\varrho^2 + q^2)^{-1}$. Similar formulae arise for virtual photons in inelastic electron or muon scattering, and later we will take up in more detail related ideas in neutrino reactions.

The formula is at first sight paradoxial. The $\varrho$ meson is a strongly interacting particle. In a large target nucleus we would expect it to be strongly absorbed, so that only the face of the nucleus would contribute to the cross-section, i.e., $\Gamma_\varrho \propto A^{2/3}$. On the other hand, the photon should penetrate freely to all parts of the nucleus and one could have guessed relatively $\Gamma_\gamma \propto A$. It is clear therefore that the formula must be wrong for a large enough target. But it turns out that for large nuclei, with high-energy transfer it might be rather accurate—insofar indeed as light meson exchange dominates the process. Experimental confirmation would then be a nice demonstration of meson dominance ideas in a subtle situation.

The essential idea is the following: the difference between the virtual $\varrho$ meson, involved in the photonuclear reaction, and the real meson, is that the former is continually recreated by the photon at all depths in the medium. Thus there is undoubtedly at volume effect in the photonuclear reaction. However, it turns out that a high energy the virtual photon is (loosely speaking) absorbed in nuclear matter almost as soon as created, so that the strength of the virtual photon field is much reduced in nuclear matter as compared with its initial value. Thus the volume term in the reaction rate is diminished, and at sufficiently high energies is negligible.

The following discussion is based on a simple optical model analysis. For light nuclei a Glauber type multiple scattering theory[6] would be better. But the optical model contains the essential notions.

## 2   OPTICAL MODEL FOR VIRTUAL MESONS[2]

We will consider a single meson and for simplicity forget about spin. The propagation of the meson in free space is described by the Klein–Gordon equation

$$\left(m^2 - \frac{\partial}{\partial x_\mu}\frac{\partial}{\partial x_\mu}\right)\phi = 0.$$

To allow for the refractive and absorptive effect of nuclear matter we add an optical potential $V$ [1]

$$\left(V + m^2 - \frac{\partial}{\partial x_\mu}\frac{\partial}{\partial x_\mu}\right)\phi = 0. \tag{2.1}$$

Finally to allow for the fact that we are dealing with free propagation, but with continuous generation by a penetrating projectile, we add a source term

$$\left(V + m^2 - \frac{\partial}{\partial x_\mu}\frac{\partial}{\partial x_\mu}\right)\phi = g\,e^{iq_\mu x_\mu} \tag{2.2}$$

where $g$ is a coupling constant and $q$ is the four-momentum of the real or virtual photon, or intermediate $W$ meson, or whatever. We will always be concerned with steady conditions in time, so that time dependence $e^{-iq_0 t}$ can be assumed; replacing $\partial/\partial t = i\,(\partial/\partial x_4)$ by $-iq_0$, dropping the exponential factor and setting

$$K = +\sqrt{q_0^2 - m^2} \tag{2.3}$$

we have

$$(-\mathbf{V}^2 - K^2 + V)\phi = g\,e^{i\mathbf{q}\cdot\mathbf{x}}. \tag{2.4}$$

This gives the integral equation

$$\phi = \frac{g\,e^{i\mathbf{q}\cdot\mathbf{x}}}{|\mathbf{q}|^2 - K^2} - \int d\mathbf{x}'\,\frac{e^{ik|\mathbf{x}-\mathbf{x}'|}}{4\pi\,|\mathbf{x}-\mathbf{x}'|}\,V(\mathbf{x}')\,\phi(\mathbf{x}'). \tag{2.5}$$

If we write

$$\phi = \frac{g\psi}{|\mathbf{q}|^2 - K^2} = \frac{g\psi}{m^2 + q^2} \tag{2.6}$$

where $q^2 = |\mathbf{q}|^2 - q_0^2$, we have

$$(-\nabla^2 - K^2 + V)\psi = (m^2 + q^2)\,e^{i\mathbf{q}\cdot\mathbf{x}} \tag{2.7}$$

and

$$\psi = e^{i\mathbf{q}\cdot\mathbf{x}} - \int d\mathbf{x}'\,\frac{e^{ik|\mathbf{x}-\mathbf{x}'|}}{4\pi\,|\mathbf{x}-\mathbf{x}'|}\,V(\mathbf{x}')\,\psi(\mathbf{x}'). \tag{2.8}$$

Except that

$$|\mathbf{q}|^2 \neq q_0^2 - m^2$$

(in fact $|\mathbf{q}|^2 = q_0^2$ for real photons) (2.8) is precisely the integral scattering equation for incident real mesons. Hence the temptation to suppose that for $q_0 \gg m$ the transition amplitude is not much different from that for real mesons. However, the virtual mesons, unlike real mesons, are not completely attenuated in sufficiently deep matter (we always consider depths small compared with photon mean free path). They are sustained by the source term in (2.4) or (2.7). At sufficient depth $\psi$ will be a plane wave proportional to the source, whence from (2.7)

$$\psi = \frac{m^2 + q^2}{m^2 + q^2 + V}\,e^{i\mathbf{q}\cdot\mathbf{x}}. \tag{2.9}$$

If $V$ were small compared with $m^2 + q^2$, the virtual meson amplitude (in contrast with a real meson amplitude) would be little influenced by the presence of nuclear matter. Thus the validity of a simple formula like (2.1) depends on competition between off-shell propagation (represented by $m^2 + q^2$) and nuclear absorption and refraction (represented by $V$). We will return to this later.

First let us discuss in more detail the interpretation of the optical model wave function. Write

$$\psi = e^{i\mathbf{q}\cdot\mathbf{x}} + \chi \tag{2.10}$$

and note that (2.7) becomes

$$(-\nabla^2 - K^2 + V)\chi = -V e^{i\mathbf{q}\cdot\mathbf{x}}. \tag{2.11}$$

Outside the nucleus the scattered wave $\chi$ gives directly the coherent production of real mesons. The outward flux is

$$\Gamma_e = 2\,\mathrm{Im}\int d\mathbf{S}\cdot\chi^*\,\nabla\chi \tag{2.12}$$

integrated over an enclosing surface. Asymptotically this is just

$$2K \int d\Omega \, |f(\theta, \phi)|^2.$$

In addition there is the production of other final states, in this model represented only by absorption of the coherent wave. The absorption is determined by Im $V$ and has a rate

$$\Gamma_a = -2 \, \text{Im} \int d\mathbf{x} \, \psi^* V \psi. \tag{2.13}$$

As with real particles one can prove an "optical theorem" relating the total reaction rate to the imaginary part of a forward scattering amplitude. First, using Gauss' theorem, $\Gamma_e$ is transformed into a volume integral

$$\Gamma_e = 2 \, \text{Im} \int d\mathbf{x} \, \chi^* \nabla^2 \chi$$

[then using the wave equation (2.11)]

$$= 2 \, \text{Im} \int d\mathbf{x} \, \chi^* V \psi$$

$$= 2 \, \text{Im} \int d\mathbf{x} \, \psi^* V \psi - 2 \, \text{Im} \int d\mathbf{x} \, e^{-i\mathbf{q} \cdot \mathbf{x}} V \psi$$

or

$$\Gamma = \Gamma_e + \Gamma_a = -2 \, \text{Im} \int d\mathbf{x} \, e^{-i\mathbf{q} \cdot \mathbf{x}} V \psi$$

$$= 8\pi \, \text{Im} \, f'(0). \tag{2.14}$$

For the on-mass-shell case $(m^2 + q^2 = 0)$ $f'(0)$ would reduce to the usual forward scattering amplitude $f(0)$.

These transition amplitudes $\Gamma$ refer to an incident wave $e^{i\mathbf{q} \cdot \mathbf{x}}$ of unit amplitude. They can therefore be used in formulae like (2.1) in which the coupling constant and propagator factor (2.6) are explicitly supplied.


## 3 NUCLEAR ABSORPTION

It is necessary to form some idea about the size of the optical potential $V$ [7]. Note therefore that a free meson wave in nuclear matter would have the exponentially attenuated form

$$\psi \propto e^{i Z \sqrt{K^2 - V}}.$$

So $\left(2 \, \text{Im} \, \sqrt{K^2 - V}\right)^{-1}$ is mean free path. This we tentatively identify with the usual expression $(\varrho\sigma)^{-1}$ where $\varrho$ is nucleon density and $\sigma$ is total cross-section per nucleon. For $V \ll K^2$,

$$\text{Im} \, V = -K\varrho\sigma. \tag{3.1}$$

According to the optical theorem

$$K\sigma = 4\pi \, \mathrm{Im} \, f(0)$$

where $f(0)$ is forward elastic scattering amplitude. Then (3.1) is just the imaginary part of the usual relation between optical potential (essentially refractive index) and forward scattering amplitude

$$V = -4\pi\varrho \, f(0). \tag{3.2}$$

Empirically it seems that forward elastic scattering is mainly imaginary at high energy. Neglecting then the real part of $V$, we estimate $|V|$ from (3.1). Taking

$$\sigma \approx 25 \, \mathrm{mb} = 2.5 \times 10^{-26} \, \mathrm{cm}^2$$

$$\varrho \approx (1.8 \times 10^{-13})^{-3} \, \mathrm{cm}^3$$

one finds

$$\varrho\sigma K \approx (0.6) \, m_\pi K.$$

Then the critical parameter, measuring the competition between the off-shell regeneration ($m^2 + q^2 \neq 0$) and the absorption ($V$) is

$$\left| \frac{m^2 + q^2}{V} \right| \approx \frac{m^2 + q^2}{(0.6) \, m_\pi K}.$$

With real photons, $q^2 = 0$; taking for $m$ the $\varrho$ mass $\approx 5 \, m_\pi$, and 1 GeV $= 7 \, m_\pi$

$$\left| \frac{m^2}{V} \right| \approx \frac{6}{K \text{ in GeV}}. \tag{3.3}$$

Thus at SLAC energies ($K \approx 20$ GeV) the absorption effect is dominant, the electromagnetic interactions in the nuclear shadow are greatly suppressed ($(\frac{5}{20})^2 \approx \frac{1}{10}$) and the $A^{2/3}$ effect well developed. At the energies of other machines (a few GeV) the situation is intermediate. The effect should be there to some extent. Interpretation of the experiments will require at least an approximate solution of the equations in which both absorption and off-shell propagation are taken into account.


## 4 EIKONAL APPROXIMATION

The inhomogeneous Schrödinger equation can be solved exactly by expansion in spherical harmonics. For this see the paper of Løvseth and Frøyland[8]. However for high energy an "eikonal" type of approximation[9] should be very useful. Let us take $\mathbf{q}$ in the $z$ direction. Then we expect the solution to be of the form

$$e^{iKz} \ (\text{or } e^{i|\mathbf{q}|z})$$

$$\times \ (\text{relatively slowly varying factor}).$$

So we write

$$\left(\frac{\partial}{\partial z}\right) = iK + \left(\frac{\partial}{\partial z} - iK\right)$$

and ignore terms of second order in the relatively small quantities

$$\frac{\partial}{\partial x}, \ \frac{\partial}{\partial y}, \ \left(\frac{\partial}{\partial z} - iK\right).$$

The resulting equation, from (2.11), is

$$\left(\frac{\partial}{\partial z} - iK - \frac{V}{2iK}\right)\chi = \frac{V}{2iK} e^{iqz} \tag{4.1}$$

where $q \equiv |\mathbf{q}|$.

The homogeneous equation, obtained by putting zero on the right-hand side of (4.1), would be solved by

$$\chi = C e^{iKz + \int_{-\infty}^{z} dz'' \, (V(z'')/2iK)}$$

with $C$ constant. To solve the inhomogeneous equation $C$ must vary

$$\frac{\partial C}{\partial z} e^{iKz + \int_{-\infty}^{z} dz' \, (V(z')/2iK)} = \frac{V}{2iK} e^{iqz}$$

from which $C$ is obtained immediately by integration

$$C(z) = \int_{-\infty}^{z} dz' \, \frac{V(z')}{2iK} e^{i\Delta z' - \int_{-\infty}^{z'} dz'' \, (V(z'')/2iK)}$$

where $\Delta = q - K$. So finally

$$\chi(x, y, z) = e^{iqz} \int_{-\infty}^{z} dz' \left(\frac{V(x, y, z')}{2iK}\right) e^{i\Delta(z' - z) + \int_{z'}^{z} \frac{dz''}{2iK} V(x,y,z'')} \tag{4.2}$$

If $\Delta$ is ignored here the integration can be done at once, with the result

$$\chi = e^{iqz}\left(e^{\int_{-\infty}^{z} dz'' \, V(z'')/2iK} - 1\right)$$

                                                    Lectures on Particles and Fields

or

$$\psi = e^{iqz} + \chi$$

$$= e^{iqz} \, e^{\int_{-\infty}^{z} dz'' \, V(z'')/2iK}. \tag{4.3}$$

Apart from the replacement of $K$ by $q$, this is just the absorbed and refracted plane wave appropriate for an incident real meson. The condition that $\Delta$ be negligible in comparison with the rest of the exponent in (4.2) is

$$|\Delta| \ll |V/2K|. \tag{4.4}$$

This is again the condition of Section 2

$$|m^2 + q^2| \ll |V|. \tag{4.5}$$

We see again that the difference between real and virtual mesons depends on competition between absorption and off-shell propagation.

The eikonal approximation is not of course adequate to obtain the asymptotic form of $\chi$; it must be continued behind the nucleus by means of diffraction theory[10]—Huygen's principle. The result is

$$\chi \sim r^{-1} \, e^{iKr} f$$

$$f(\theta, \phi) = -\frac{2iK}{4\pi} \int dx \, dy \, e^{-i\mathbf{K}\cdot\mathbf{K}} \chi(x, y, z) \tag{4.6}$$

where the $(x, y)$ integration is to be done at a value of $z$ sufficiently far behind the nucleus, and $(\mathbf{K})$ is a vector of magnitude $K$ and the required asymptotic direction $(\theta, \phi)$.

Substituting $\chi$ from (4.1) and considering only small angles so that to a sufficient approximation $K_z = K$

$$f = \frac{-1}{4\pi} \int dx \, V(\mathbf{x}) \, e^{\int_z^{\infty} dz' \, (V(x,y,z')/ziK)} \, e^{i\Delta z - iK_x x - iK_y y}. \tag{4.7}$$

If $V$ has cylindrical symmetry, and one goes to cylindrical coordinates,

$$dx \, dy \, dz = d\phi \, b \, db \, dz$$

the azimuthal integration can be done

$$f(\theta) = \frac{-1}{2} \int b \, db \, dz \, J_0 \, (Kb \sin \theta) \, V(b, z) \, e^{\int_z^{\infty} \frac{dz'}{ziK} V(b,z')} \, e^{i\Delta z - iKb \sin \theta}. \tag{4.8}$$

A similar formula has been quoted by Drell and Trefil[3] for the photo-production of vector mesons. They did not assume the vector meson dominance of the primary interaction of the photon with the nuclear matter. Thus in their formula the two potentials $V$ appearing in (4.7) are not the same. One represents the creation of a meson in a photon-nucleon collision, and the other (in the exponent) the subsequent absorption and refraction of this meson. In the meson dominance model the two processes are essentially the same.

## 5   APPLICATION TO PHOTOPRODUCTION OF VECTOR MESONS

In general the vector meson problem is complicated by spin. However for particles created by real photons, and at high energy where propagation is rectilinear, we are concerned with indepedently propagating transverse components of the vector field. Then each such component separately can be discussed along the lines indicated. Such a theory has been applied to the photoproduction of neutral mesons by a DESY group[11], following suggestions of Ross and Stodolsky[3] and Drell and Trefil[3].

Assuming a purely imaginary $V$, related by (3.1) to the nucleon cross-section, they analyse their results for this latter cross-section and for the photon-$\varrho$ meson coupling constant $g_{\gamma\varrho}$ which fixes the over-all magnitude of the process. From experiments at several energies in the range of a few GeV, and on various nuclei, they obtain

$$\sigma_{\varrho N} = 31.3 \pm 2.3 \text{ mb}$$

for the cross-section, and for the coupling constant $g_{\gamma\varrho}$ with

$$g_{\gamma\varrho} = m_\sigma^2/2\gamma_\varrho$$

they quote

$$\gamma_\varrho^2/4\pi = 0.49 \pm 0.12 \quad \text{(carbon)}$$

$$0.42 \pm 0.10 \quad \text{(copper)}$$

$$0.40 \pm 0.10 \quad \text{(lead)}.$$

The agreement of these values with one another and with the quite independently measure value

$$\gamma_\varrho^2/4\pi = 0.41 \pm 0.10$$

from the process

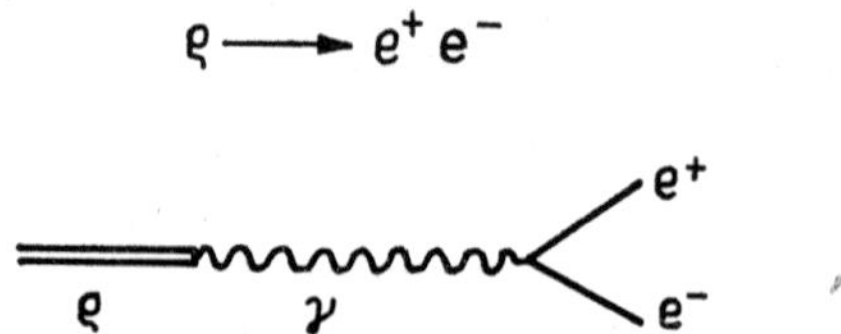


suggests that the theory has some validity.

From a theoretical viewpoint the following observation of Stodolsky[12] should be borne in mind. Just as the propagation of the photon in nuclear matter is complicated by the coupled propagation of the $\varrho$ meson, so the propagation of this latter can be complicated by the coupled propagation of other mesons. If we neglect the isospin asymmetry of nuclear matter (i.e. neutron excess and Coulomb field) then the first candidate to couple with the $\varrho$ would be the $A_1$ meson (if it exists). We will describe shortly the formalism for coupled propagation. On conventional ideas of high-energy collisions the coupling amplitude

$$\varrho + \text{nucleon} \rightarrow A_1 + \text{nucleon}$$

would be expected to be small compared with the elastic amplitudes

$$\varrho + \text{nucleon} \rightarrow \varrho + \text{nucleon}$$

$$A_1 + \text{nucleon} \rightarrow A_1 + \text{nucleon}.$$

Then the coupling phenomenon could be negligible. A more careful analysis of experiment could allow a test of these ideas.

If one wishes to apply these ideas to total photonuclear cross-sections, then $\omega$ and $\phi$ mesons have to be considered. The coupled propagation of $\omega$ and $\phi$ in nuclear matter has been studied by Ross and Stodolsky[13].


## 6   COUPLED PROPAGATION

When it is necessary to consider the coupled propagation of more than one meson, Eq. (2.2) has to be generalized to a set of coupled equations. Let us consider the simple case of two particles with waves $\phi_1$ and $\phi_2$. We go straight away to the case of a plane wave in a homogeneous medium $((\partial/\partial\chi) \equiv q)$; the more general form of the equations will be obvious. Then we write instead of (2.2)

$$(m_1^2 + q^2 + V_{11})\phi_1 + V_{12}\phi_3 = g \tag{6.1}$$

$$(m_2^2 + q^2 + V_{22})\phi_2 + V_{21}\phi_1 = 0.$$

We have supposed here that only particle 1 is created directly by the source; the more general case is obtained by linear superposition. The diagonal potentials, $V_{11}$ and $V_{22}$ can be supposed to be related in the usual way to forward elastic scattering amplitudes, and $V_{12}$ and $V_{21}$ in a similar way to forward transition amplitudes.

From the second of Eqs. (6.1)

$$\phi_2 = -\frac{V_{21}}{m_2^2 + q^2 + V_{22}}\phi_1. \tag{6.2}$$

Substituting into the first equation, we eliminate $\phi_2$ at the expense of complicating the optical potential for $\phi_1$

$$V_{11} \rightarrow V_{11} - \frac{V_{12}\,V_{21}}{m_2 + q^2 + V_{22}}. \tag{6.3}$$

This complication will be small if the second meson is sufficiently massive, or if the coupling amplitudes are sufficiently small, loosely speaking.

Note that these effects arise also with real mesons incident. One could for example investigate with pions incident on nuclei for evidence of propagation of the longitudinal component of the $A_1$ meson. At high energy one would expect $V_{22}$ and $V_{11}$ to be of similar magnitude, large compared with $m_1^2$ and $m_2^2$, while $V_{12}$ and $V_{21}$ would be relatively much smaller. So the effect would not be large.


## 7   NEUTRINO REACTIONS

We come finally to neutrino reactions. Note at once that we have now to deal with charged rather than neutral mesons. The Coulomb field of heavy nuclei could then be important in quantitative comparison of theory and experiment. We explicitly neglect this in what follows, setting the question aside for future investigation.

For neutrino reactions vector meson exchange is of course part of the story. When these mesons are sufficiently energetic (i.e., for high-energy transfer) the phenomena described already will be encountered. However, it was observed by Adler[1] that in the case of *forward* neutrino reactions the *pion* can play the dominant role. Because it is much less massive interesting effects then occur at much lower energies.

The obvious way for the pion to enter is through the one-pion exchange diagram

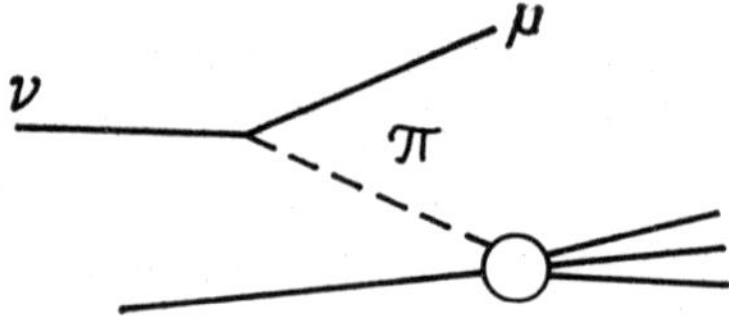


Denoting the contribution of the lower (pion-hadron) vertex by $M$, this total diagram gives

$$f_\pi \left( \frac{G}{\sqrt{2}} \right) l_\lambda i q_\lambda M \, (m_\pi^2 + q^2)^{-1} \tag{7.1}$$

where $l_\lambda$ is the pair of lepton wave functions

$$l_\lambda = \bar{\mu} i \gamma_\lambda (1 + \gamma_5) V$$

and $f_\pi G / \sqrt{2}$ (where $G$ is the universal weak coupling constant) is known from pion decay

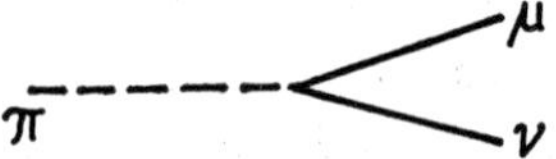


Denoting initial and final lepton four-momenta by $k$ and $k'$, $q$ denotes the difference $(k - k')$. Using the Dirac equations

$$i\gamma_\lambda k_\lambda v = \bar{\mu} \, (i\gamma_\lambda k'_\lambda + m_\mu) = 0$$

one finds

$$q_\lambda l_\lambda = m_\mu \bar{\mu} \, (1 + \gamma_5) v. \tag{7.2}$$

The charged lepton mass, even for the muon and much more so for the electron, is small compared with other relevant masses and energies. This, together with the rapid fall-off with $q^2$ suggested by the pion propagator in (7.1), indicates that this one-pion pole diagram is relatively unimportant.

The main role of the pion is not then the direct one just discussed. It is important rather because of the generalized Goldberger–Treiman relation[14].

Let the complete process

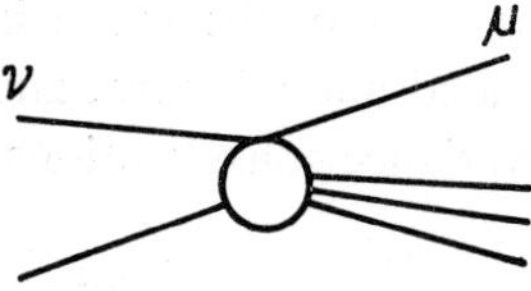


have amplitude

$$\frac{G}{\sqrt{2}}\, l_\lambda M_\lambda\,.$$

A particular contribution to $M_\lambda$ is the one-pion pole term discussed above. Let $M'_\lambda$ denote the remainder

$$M_\lambda = M'_\lambda + f_\pi i q_\lambda\, (m_\pi^2 + q^2)^{-1} M\,. \tag{7.3}$$

The generalized Goldberger–Treiman relation asserts a connection between pionic and non-pionic parts

$$i q_\lambda M'_\lambda = f_\pi M\,. \tag{7.4}$$

This reduces to the original Goldberger–Treiman relation in the special case of the "elastic" neutrino reaction

$$\nu + n \to \mu + p\,.$$

In this case

$$M_\lambda = \bar{p} i \gamma_\lambda\, (f_V + f_A \gamma_5)\, n + \bar{p} \sigma_{\lambda\nu} i q_\nu f_1 n - \bar{p} \gamma_5 q_\lambda f_P n \tag{7.5}$$

$$M = \sqrt{2}\, g \bar{p}\, i \gamma_5 n \tag{7.6}$$

where the $f$'s are form factors (functions of $q^2$) and $g$ is the strong pion-nucleon coupling constant. We identify the induced pseudoscalar term $(f_p)$ with the pole term of (7.5). Then from the remainder of $M_\lambda$ we find, using the Dirac equation

$$i q_\lambda M'_\lambda = -2M f_A\, \bar{p} i \gamma_5 n\,.$$

Then from (7.4) and (7.6)

$$f_\pi = -\sqrt{2}\, \frac{M f_A}{g}$$

which is the original Goldberger–Treiman relation.

The special significances of (7.4) for forward inelastic reactions is that for them $l_\lambda \propto q_\lambda$ when the mass of the charged lepton is neglected. The neutrino has always left-hand helicity, but the muon may, in general, be right- or left-handed and we have to add the squares of the two corresponding matrix elements. However, when the mass of the muon is neglected the transition amplitude for the right-handed muon is zero, and for the other—either by direct calculation or by invariance considerations

$$l_\lambda = 4 \sqrt{k_0 k_0'} \left( \frac{q_\lambda}{q_0} + \theta l_\lambda' \right) \left( \cos \frac{1}{2} \theta \right) e^{-1/2 i\varphi} \tag{7.7}$$

where $q_0$ and $q_0'$ are laboratory energies of neutrino and muon, and $(\theta, \phi)$ is the angle between $\mathbf{k}'$ and $\mathbf{k}$; for small $\theta$, $l_\lambda'$ is a constant vector such that $l_0'$ and $\mathbf{l}' \cdot \mathbf{q}$ are zero.

It is perhaps of interest to see in more detail how the pionic amplitude $M$ becomes sufficient in the optical model. Consider some reaction initiated by a pion deep inside nuclear matter, and let us represent it by the diagram

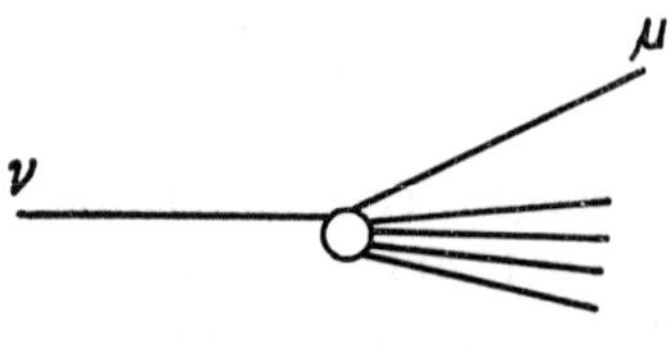

The circles indicate that the pion may interact coherently with the nuclear matter (i.e., with the optical potential $V$) any number of times before causing the reaction. We do not concern ourselves with what happens subsequently to the particles and holes created. If $g$ is the amplitude for this reaction on an isolated nucleon, and if we allow for only the absorptive and refractive effects of nuclear matter

$$M = \frac{m_\pi^2 + q^2}{m_\pi^2 + q^2 + V} g. \tag{7.8}$$

This reduction factor due to $V$ may be obtained by summing the set of Feynman diagrams, or from (2.9).

Consider now the corresponding neutrino reaction. We leave aside the process in which the lepton spontaneously emits a pion which subsequently interacts with a nucleon—the pole term in (7.3). There are two other processes to consider. First the reaction may be induced directly by the lepton

giving an amplitude, say, $\qquad\qquad l_\lambda g_\lambda'.$

Or the lepton may first create a pion, not spontaneously but in collision with a nucleon

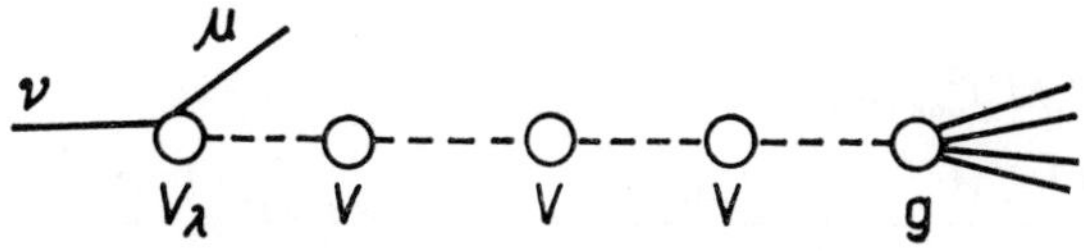

$$\mu \quad \nu \qquad V_\lambda \quad V \quad V \quad V \quad g$$

Denoting the amplitude for this forced pion creation by $l_\lambda V'_\lambda$, the over-all amplitude is[15]

$$- V'_\lambda l_\lambda \frac{1}{m_\pi^2 + q^2 + V} g \tag{7.9}$$

where the pion propagator in nuclear matter appears. Altogether

$$l_\lambda M'_\lambda = l_\lambda g'_\lambda - l_\lambda V'_\lambda \frac{1}{m_\pi^2 + q^2 + V} g. \tag{7.10}$$

Now the various amplitudes $V'_\lambda$, $g'_\lambda$, $g$ are not independent, but are connected by Goldberger–Treiman relations; whenever the lepton pair wave function is proportional to $q_\lambda$, it is like an incident pion; thus

$$iq_\lambda V'_\lambda = f_\pi V; \quad iq_\lambda g'_\lambda = f_\pi g. \tag{7.11}$$

Putting $l_\lambda = iq_\lambda$ in (7.10), and using (7.11) and (7.8), on confirms (7.4). We have gone through this story to emphasize the following point; the "non-pionic" term $M'_\lambda$ by definition excludes the contribution of the pion spontaneously emitted by the lepton. However, the contribution from a pion created in lepton-nucleon collision enters in an essential way; through generalized Goldberger–Treiman relations it is responsible for the quenching of the non-pionic $M'_\lambda$ term for forward neutrino reactions in nuclear matter.

Let us concentrate for a moment on the pole term in (7.3). If here also the muon mass is regarded as negligible, this term makes no contribution to the reaction amplitude-according to (7.2). However because of the rapidly varying factor $(m_\pi^2 + q^2)^{-1}$ it is worth-while[1] to handle this term more accurately. To lowest non-vanishing order in $m_\mu$ one finds

$$iq_\lambda l_\lambda = im_\mu \bar{u} (1 + \gamma_5) V$$

$$\approx 2im_\mu^2 \sqrt{k_0/k'_0} \cos \frac{\theta}{2} e^{-1/2 i\phi} \text{ (non flip)}$$

$$\approx 4im_\mu \sqrt{k_0 k'_0} \sin \frac{\theta}{2} e^{-1/2 i\phi} \text{ (flip)} \tag{7.12}$$

Using these for the pole term, and (7.7) for the non-pole term, we have for small angles

$$|l_\lambda M_\lambda|^2_{\text{flip}} + |l_\lambda M_\lambda|^2_{\text{non flip}}$$

$$= 16 k_0 k_0' \left\{ \left| \frac{f_\pi M}{q_0} - \frac{f_\pi M m_\mu^2}{2 k_0' (m_\pi^2 + q^2)} + \theta\, e^{i\phi} l_\lambda' M_\lambda' \right|^2 + \frac{\theta^2}{4} \left| \frac{m_\mu f_\pi M}{m_\pi^2 + q^2} \right|^2 \right\}.$$

$$\tag{7.13}$$

In this expression $M$ is the matrix element with an incident pion on the given target to whatever final hadron state is considered. To obtain the total reaction rate summation over final states is made. Summation over hadron states is effected simply by replacing $|M|^2$ by the total reaction rate $\Gamma(q^2, q_0)$ for incident pions of indicated energy and mass. The summation over final lepton states gives a factor (neglecting $m_\mu$)

$$\frac{d^3\mathbf{k}'}{(2\pi)^3\, 2 k_0'} = \frac{k_0'^2\, dk_0'\, d\cos\theta}{8\pi^2 k_0'} = \frac{k_0'\, dq_0\, d\cos\theta}{8\pi^2}.$$

Thus for incident neutrinos, the differential reaction rate with respect to energy transfer and angle of the secondary lepton is[16]

$$\frac{d^2\Gamma_\nu}{dq_0\, d\cos\theta} = \Gamma_\pi k_0 \left( \frac{G f_\pi k_0'}{\pi q_0} \right)^2 \left\{ \left( 1 - \frac{q_0 m_\mu^2}{2 k_0' (m_\pi^2 + q^2)} \right)^2 \right.$$

$$\left. + \frac{\theta^2}{4} \left( \frac{m_\mu q_0}{m_\pi^2 + q^2} \right)^2 \right\}.$$

$$\tag{7.14}$$

This formula is derived for $\theta$ small. Note that terms linear in $\theta^{\pm i\phi}$ must disappear by symmetry. The only $\theta^2$ terms retained are those associated with a factor $(m_\pi^2 + q^2)^{-2}$. It is in formula (7.14) that we substitute our optical model $\Gamma_\pi(q^2, q_0)$, i.e. $\Gamma\,(= \Gamma_e + \Gamma_a)$ of Section 2.


## 8   FINAL REMARKS

Using the pion mass rather than the $\varrho$ mass in (3.3) we obtain

$$\left| \frac{m^2}{V} \right| \approx \frac{200}{K \text{ in MeV}}.$$

$$\tag{7.15}$$

Thus the "$A^{2/3}$ effect" in neutrino reactions should begin to set in at quite modest energy transfers of a few hundred MeV. That this is indeed what the theory says has been confirmed in detailed calculations by Løvseth and Frøyland[8].

The question arises of how large is the region around the forward direction for which the effect should be expected. It could be feared that because of the essential role of the pion the relevant region could be determined by the pion propagator $(m_\pi^2 + q^2)^{-1}$. Actually it is clear from the theory that it is the effective pion propagator $(m_\pi^2 + q^2 + V)^{-1}$ which enters [e.g. in Eq. (7.8)], and from (7.15) $V$ is large at high energy. Then it is the increase with $\theta$ of the $l'_\lambda M'_\lambda$ term in (7.13), rather than the decrease with $\theta$ of $M$, which conceals the effect and dictates the required angular resolution. To estimate this one will introduce the vector mesons explicitly, and such a model will also allow discussion of incipient "$A^{2/3}$ effects" with these particles which will become fully developed only at higher energies. At sufficiently high energies, insofar as light mesons are in a simple sense responsible for the process, even large angle neutrino reactions will show the effect. Note that even if meson propagation is not the *whole* story, it must surely *occur*, and give effects of the general kind discussed.

Let it be admitted finally that the hole story *could* be much more complicated. We have simply made minimal modifications to the free propagation of mesons. More violent modifications could be necessary in the dense and strongly interacting medium with which we are concerned[17]. In the end we can only propose it as an interesting empirical question. Is the notion of meson propagation useful, at the phenomenological level, even in nuclear matter?

## References


1. S.L.Adler, *Phys. Rev.*, **135**, B963 (1964).
2. J.S.Bell, *Phys. Rev. Letters*, **13**, 57 (1964).
3. M.Ross and L.Stodolsky, *Phys. Rev.*, **149**, 1172 (1966); S.D.Drell and J.S.Trefil, *Phys. Rev. Letters*, **16**, 552 (1966); and *Phys. Rev. Letters*, **16**, 832 (1966).
4. Note that we are not here concerned with whether photons couple through vector mesons in some "fundamental" Lagrangian. The existence of Lagrangian models [e.g., that of Kroll, Lee and Zumino, *Phys. Rev.*, **157**, 1376 (1967)] is of course of great interest. However, with large coupling constants the structure of the Lagrangian may not be very apparent at the phenomenological level. The possibility that interests us here is that at that level, or in the *solution* of the fundamental equations, the vector mesons might play a role analogous to that of the photon.
5. For generalization to virtual particles the notion of transition rate is somewhat more natural than that of cross-section. For free mesons, covariantly normalized to $2q_0$ per unit volume, where $q_0$ is energy, transition rate $\Gamma$ and cross-section $\sigma$ for a stationary

target are related by

$$\Gamma = 2\,|\mathbf{q}|\,\sigma$$

where $|\mathbf{q}|^2 = q_0^2 - m^2$.

6. R.J.Glauber, *Lectures in Theoretical Physics*, Vol. 1, Interscience, New York, 1959.

7. It will be noted that in this section we disregard any $q^2$ dependence of $V$, i.e., we employ the same potential for real and virtual mesons. The assumption is that the significant $q^2$ dependence arises from the extension of the target, and is allowed for by solving the off-shell Schrödinger equation.

8. J.Løvseth and J.Frøland, to be published.

9. See, for example, L.D.Landau and E.M.Lifshitz, *Quantum Mechanics*, p. 489, Pergamon, London, 1965.

10. From the asymptotic form of (2.8) one finds

$$f = -\frac{1}{4\pi}\int d\mathbf{x}\, e^{-i\mathbf{K}\cdot\mathbf{x}}\, V(\mathbf{x})\,\psi(\mathbf{x}).$$

From (2.11)

$$e^{-i\mathbf{K}\cdot\mathbf{x}}V\psi = e^{-i\mathbf{K}\cdot\mathbf{x}}(\nabla^2 + K^2)\chi = \nabla\cdot(e^{-i\mathbf{K}\cdot\mathbf{x}}\nabla\chi) + i\mathbf{K}\cdot\nabla(e^{-i\mathbf{K}\cdot\mathbf{x}}\chi).$$

Substituting this in the above expression for $f$ the surface integral can be converted into a volume integral over an enclosing surface. Taking for the significant part of this surface a plane behind the target at position $z$

$$f = -\frac{1}{4\pi}\int_z dx\,dy\, e^{-i\mathbf{K}\cdot\mathbf{x}}\left\{\frac{\partial\chi}{\partial z} + iK_z\chi\right\}.$$

This reduces to (4.6) with the approximations $K_z = K$, $(\partial/\partial z) = iK$.

11. J.G.Asbury *et al.*, *Phys. Rev. Letters*, **19**, 865 (1967) and DESY preprint.

12. L.Stodolsky, *Phys. Rev. Letters*, **18**, 135 (1967).

13. M.Ross and L.Stodolsky, *Phys. Rev. Letters*, **17**, 563 (1966).

14. For the generalized Goldberger–Treiman relation see any recent account of weak interaction theory under the heading PCAC or PDDAC; for example J.S.Bell in *High Energy Physics*, Les Houches 1965, Gordon and Breach, 1965.

15. This can be obtained from the usual Feynman rules and the sign conventions we have implicitly used; or it can be obtained from (2.11) replacing in the source term $V$ (appropriate for an incident pion wave) by $V_\mu l_\mu$ for the incident lepton.

16. The result is due to Adler (unpublished) and to Løvseth and Frøyland[8].

17. For example a modification of the Goldberger–Treiman relation due to the medium has been proposed[8]. Some modification must indeed occur at some level of accuracy. However, I do not follow the arguments[8] for the particular large modification advocated.

# On the Composite Structure of Leptons and the Existence of the $\mu$-Meson


A. O. BARUT

*University of Colorado*


---

If we wish to understand the existence of the $\mu$-meson, which has exactly the same properties, except mass, as the electron and which is coupled to the electron only via the weak interactions, within the general context of leptons, we must see wether both the electron and the muon states, together with their associated neutrinos, can be obtained from a general dynamical principle. One can give, of course, *a posteriori* group theoretical *descriptions* of the known lepton states. We do not attempt to list all the group theoretical classifications of leptons because our concern is to have a dynamical framework in which one can also discuss the interactions of the leptons.

There are several ways of approaching this problem. One can start from the bare electron and muon fields interacting with the electromagnetic field (but not with each other) and ask whether non-perturbative solutions of this system contain two distinct masses, i.e., whether mass renormalization can result in the electron and muon. There are no clear-cut answers to this question and the considerations depend on cut-offs.[1] Another way is to start with the four-component spinors but to postulate again in the spirit of non-pertubative solutions, higher-order equations.[2] Yet another way is to start from larger representations of the Lorentz group, or even to use representations of larger groups to describe rest frame states of leptons and then write new wave equations.[3]

Any approach to the electron muon puzzle has to satisfy a number of requirements. The $\mu$-meson seems to arise in a very subtle way:

a) Lorentz transformations and electromagnetic interactions should not couple $e$ and $\mu$. We must forbid electromagnetic transitions $\mu \to e \to \gamma$.

"

b) The theory should not contain any new unknown lepton states, nor should it be a trivial direct sum of two non-interacting fermions.

c) The rules of the usual quantum electrodynamics should hold for electron and muon separately.

These are tight requirements. It is possible that these requirements are not quite exact; that is there may be some very small $\mu \to e \to \gamma$ transitions, or some new leptonic states. But at the present state, it seems safe to impose these requirements.

In this note we *propose* a solution to the problem by giving up one of the standard (tacit) assumptions in the Dirac theory of the electron: the proportionality between the conserved quantum mechanical matter current (or probability current in the first quantized sense) and the conserved electromagnetic current. In other words we assume that the distribution of the leptonic matter and the distribution of the electric charge is *not* the same. That these two distributions are in general not the same is clear if one thinks of neutral constituents. In Schrodinger and Dirac theories we are used to making no distinctions between the two. The leptonic matter distribution will contain, we expect, in addition to electric charges, also the weak charges, for example. The conserved matter current describes the internal dynamics and yields the mass spectrum. The conserved electromagnetic current tells us what electromagnetic transitions are possible.

We shall formulate the theory in a more general context to point out that the generalization mentioned above (i.e. the inequality of matter and charge distribution) occurs also in the case of other physical systems.

We start with a fairly general form of the matter current. The equation

$$(J_\mu P^\mu + \beta L_{46} + \tfrac{1}{2}\gamma)\, \tilde{\psi}(P) = 0 \tag{1}$$

with the minimal conserved current

$$J_\mu = \alpha_1 \Gamma_\mu + \alpha_2 P_\mu + \alpha_3 P_\mu L_{46} + \alpha_4 L_{\mu\nu}q^\nu + \cdots \tag{2}$$

$$(+ \text{ axial currents})$$

$$J_\mu q^\mu = 0 \tag{3}$$

describes for special values of $\alpha_i, \beta, \gamma$ and for special representations of the $O\,(4, 2)$ generators $\Gamma_\mu, L_{46}$, the properties and interactions of at least three distinct physical systems: (a) a Dirac particle, (b) the H-atom, and (c) the hadrons. In this equation the rest frame states $\tilde{\psi}(0)$, satisfying $[J_0 m + \beta L_{46}$

$+ \gamma] \tilde{\psi}(0) = 0$, belong to an irreducible representation of the group $O\,(4, 2)$ whose generators are denoted by $L_{ab}$, $a, b = 1, 2, \ldots, 6$, with $\Gamma_\mu = (L_{56}, L_{i6})$, $i = 1, 2, 3$; $L_{ij}$ are the spin operators ($L^2 = j\,(j + 1)$; $L_{12} = m$) and $L_{i5}$ are the generators of pure Lorentz transformations. A basis in the representation space may be denoted by $|njm, \pm\rangle$, where the principal quantum number $n$ is the eigenvalue of $L_{56}$ and $\pm$ refer to the parity of the state. The $j$-values are either all integers (boson case), or all half integers (fermion case). Table 1 shows the specialization of (1) in the three known models.

Table 1

| System | Representation | $\Gamma_\mu$ | $L_{46}$ | $\alpha_1$ | $\alpha_2$ | $\alpha_3$ | $\alpha_4$ | $\beta$ | $\gamma$ |
|---|---|---|---|---|---|---|---|---|---|
| Dirac particle | 4-dimensional[4] | $-\tfrac{1}{2}\gamma_\mu$ | $-\tfrac{1}{2}\gamma_5$ pseudo-scalar | $\neq 0$ | 0 | 0 | 0 | $\neq 0$ 0 | $0 \neq 0$ |
| H-atom | most degenerate[5] (boson) | $\Gamma_\mu$ | scalar | $\neq 0$ | 0 | $\neq 0$ | 0 | $\neq 0$ | 0 |
| Hadron model | most degenerate[6] (boson or fermion) | $\Gamma_\mu$ | scalar | $\neq 0$ | $\neq 0$ | $\neq 0$ | $\neq 0$ | $\neq 0$ | $\neq 0$ |

The reason the $L_{46}$ terms do not occur in the usual form of the Dirac equation is that in this case $L_{46}\,\gamma_5$ is odd under parity (which is represented by $\gamma_0$), hence the equation would not conserve parity. In the previous paper[4] it was noted that if the full form of eq. (1) is postulated, i.e. with the $\alpha_2, \alpha_3, \ldots$ terms present, it leads to two unequal possible mass values. In the hadron models the matter current may even contain more terms than indicated in (1) and the electromagnetic current is in general not the same as (1), for example, in the case of the neutron and the neutral excited states of the baryons.

In the special case of the parity conserving equation ($\alpha_3 = \alpha_4 = 0$) (four-dimensional case) the two mass values obtained either from (1) or (2) are

$$m_1 \text{ and } m_2 = \frac{\alpha_1}{\alpha_2} - m_1. \tag{4}$$

The current (2) is interpreted as the matter current.

Next we introduce the electromagnetic current and assume it to have the usual form, but different from (2), i.e.

$$j_\mu^{\text{em}} = e\gamma_\mu \mp J_\mu^{\text{matter}}. \tag{5}$$

Evidently $j_\mu^{em}$ is conserved only between states of equal mass. Thus there are no electromagnetic transitions of the form $m_2 \to m_1 + \gamma$. Had we taken $j_\mu^{em}$ proportional to $J_\mu^{matter}$ of eq. (1) then there would be electromagnetic transitions between the two different mass states 1 and 2 with a transition vertex having the form $(1|\,\sigma_{\mu\nu}q^\nu\,|2)$. In this case the second mass state must be interpreted as an excited (or heavy) electron. Also, both the particles would have in this case anomalous intrinsic magnetic moments. This case is discussed elsewhere where the complicated current (2) has been viewed as describing a "dressed" particle, describing part of the radiative corrections in a first quantized form.[7] In the case of the electromagnetic current (5), both of the particles have zero anomalous magnetic moments with no electromagnetic transitions between them. From (4) it is seen that a very small ratio of $\alpha_2/\alpha_1 \approx 4.8 \times 10^{-3}$ would result in the observed $\mu$-mass:

$$m_1 = m_e, \quad m_2 = m_\mu. \tag{6}$$

In this case there are, however, transitions between the $e$- and $\mu$-states due to the non-electromagnetic part of the current which are proportional to $P_\mu$, as we have noted, and due to the axial currents $\gamma_5\gamma_\mu$, etc. It is thus tempting to set the complete matter current (1) as the sum of electromagnetic and weak currents. i.e., the total distribution of matter consists of the distribution of electromagnetic and weak charges.

One feature of the second states ($m_2$) is that, as in the Klein–Gordon equation, they have a negative norm $(2|\,j_0\,|2) = -1$. However, a consistent and complete second quantized theory exists.[7]

The generalization of the mass spectrum (4) in the case of the more general current (1) containing axial vectors is straightforward. The zero mass leptons could arise either from such a generalization,[8] or from a transition of the four-component equation into a higher-order two-component one. For it has been known for some time that the consistent quantization of the second-order Gell–Mann–Feynman equation implies that every fermion must have its own neutrino.[9]

These considerations again show the ambiguities in the coupling schemes of seemingly equivalent wave equations. The ambiguities do *not* arise if we specify the currents (matter, electromagnetic, weak, …) that cause the transitions. This is the procedure which we have followed.

## References


1. M.Baker and S.L.Glashow, *Phys. Rev.*, **128**, 2462 (1962); P.Pascual, *Phys. Letters*, **6**, 184 (1963); R.Haag and Th.A.J.Maris, *Phys. Rev.*, **132**, 2325 (1963).
2. For higher-order field equations see A.Pais and G.Uhlenbeck, *Phys. Rev.*, **79**, 145 (1950). A theory of this type using arguments of the classical radiation theory has been given by G.Rosen, *Nuovo Cimento*, **32**, 1037 (1964) (this author would also like to interpret the $\mu$-meson as an "excited" state of the electron, i.e., an electromagnetic origin of the $\mu$-mass). See also M.A.Markov, *Nucl. Phys.*, **55**, 130 (1964).
3. An example has recently been given by B.Kurşunoğlu, *Phys. Rev.*, **167**, 1452 (1968) using the six-dimensional representation of the group $SU(3, 1)$. His equations also predict two new spin 3/2 leptons.
4. A.O.Barut, *Phys. Rev. Letters*, **20**, 893 (1968).
5. Y.Nambu, *Phys. Rev.*, **160**, 1171 (1967); C.Fronsdal, *Phys. Rev.*, **156**, 1665 (1967); A.O.Barut and H.Kleinert, *Phys. Rev.*, **156**, 1541; **157**, 1180; **160**, 1149 (1967).
6. For baryons, A.O.Barut, D.Corrigan and H.Kleinert, *Phys. Rev. Letters*, **20**, 167 (1968); for bosons, A.O.Barut, *Nucl. Phys.*, **B4**, 455 (1968); K.C.Tripathy, *Phys. Rev.*, **170**, 1626 (1968); C.Fronsdal, *Phys. Rev.*, **171**, 1811 (1968).
7. A.O.Barut, P.Cordero and G.C.Ghirardi, ICTP, Trieste, preprints IC/68/69 and IC/68/96, to be published in Phys. Rev.
8. A.O.Barut, P.Cordero and G.C.Ghirardi, Trieste preprint April 1969 (to be published in Nuovo Cimento).
9. The quantization of the second-order Gell–Mann–Feynman equation without separating it into two coupled first-order equations is given in A.O.Barut and G.H.Mullen, *Ann. Physik*, **20**, 184 and 203 (1962).

# Phenomenology of $K_{l_3}$ Decay

C. RYAN

*University College, Dublin, Ireland*

## Contents

## 1  INTRODUCTION

The three body semileptonic decays of $K$-mesons (referred to as $K_{l_3}$ decays) are processes of considerable theoretical and experimental interest. On the theoretical side these decays constitute examples of weak interaction processes which are at once non trivial and at the same time sufficiently simple that one feels a fairly satisfactory theoretical treatment of them ought to be possible. On the experimental side, at least up to recently, the situation was that the basic measurements of rates, spectra and polarization had not been thoroughly carried out due to experimental difficulties and low statistics. At present a fairly consistent experimental picture of these decays appears to be emerging though the matter is by no means completely settled yet.

The present article aims at giving a general phenomenological treatment of $K_{l_3}$ decays. The emphasis throughout will be on generality for the reason that it seems useful at this particular time to try to provide as wide a framework as possible within which one may (a) deduce the maximum possible information from the increasing experimental information now becoming available and (b) make a critical evaluation of the various theories which have been advanced in this area.

We begin in Section 2 with a discussion of the basic experimental facts, the kinematics, the general form of the matrix element together with the restrictions imposed upon it on the one hand by various symmetries and on the other by the principle of the local action of leptons. In Section 3 we deduce the general decay-polarization distribution and the various special distributions which can be obtained from it. Next in Section 4 we make a comparison between theory and experiment. Since it emerges that all present experiments are consistent with the $V,A$ theory of weak interactions we devote the remaining section to considering the form factors $f_+(q^2)$ and $f_-(q^2)$ in this theory.

## 2  PRELIMINARY QUESTIONS

### 2.1  General experimental situation

Experimentally $K_{l_3}$ decays comprise the six processes

$$K^- \rightarrow \pi^0 + l^- + \bar{\nu}_e, \tag{2.1}$$

$$K^+ \rightarrow \pi^0 + l^+ + \nu_e, \tag{2.2}$$

$$K_L^0 \to \pi^+ + l^- + \bar{\nu}_e, \tag{2.3}$$

$$K_L^0 \to \pi^- + l^+ + \nu_e, \tag{2.4}$$

$$K_S^0 \to \pi^+ + l^- + \bar{\nu}_e, \tag{2.5}$$

$$K_S^0 \to \pi^- + l^+ + \nu_e, \tag{2.6}$$

where $K_S^0$ and $K_L^0$ denote the short-lived and long-lived neutral kaons and $l$ stands for either $e$ or $\mu$. With regard to the decays $K_S^0 \to \pi^\pm + l^\mp \pm \nu_e$ [Eqs. (2.5) and (2.6)] the situation is that we expect their rates to be similar to those for the decays $K_L^0 \to \pi^\pm + l^\mp \mp \nu_e$ ($\sim 3 \times 10^6$ sec$^{-1}$) but since these are very much smaller than those of the dominant $2\pi$ modes of the $K_S^0$ ($\sim 5 \times 10^9$ sec$^{-1}$) the $K_S^0$ semileptonic decays will be very difficult to detect. In fact we have at present no direct experimental information about these processes though some knowledge of them can however be deduced from regeneration experiments. For the observed $K_{l_3}$ decays we give in Table 1 the experimental branching ratios and decay rates[1]. Despite the big difference in branching ratios as between the charged and neutral cases we see that the decay rates in all cases are very similar which of course strongly suggests that they have a common origin. From TCP invariance, which we shall assume throughout this article, it follows that if we neglect final state interactions the branching ratio and decay rates for $K_{l_3}^-$ are equal to those for $K_{l_3}^+$ and the same is true for $K_{L_{l_3}^-}^0$ and $K_{L_{l_3}^+}^0$ to the extent that we also neglect the effect of CP violation which is of order $10^{-3}$. These properties seem to be verified in the actual decays.

**Table 1**   Branching ratios and Decay Rates of $K_{l_3}$ decays. The $K_L^0$ branching ratios and rates are for the sum of the $l^-$ and $l^+$ modes.

| Decay | Branching ratio | Decay rate (sec$^{-1}$) |
| --- | --- | --- |
| $K_{e_3}^\pm$ | $(4.86 \pm 0.07)\,\%$ | $(3.93 \pm 0.06)\,10^6$ |
| $K_{\mu_3}^\pm$ | $(3.18 \pm 0.14)\,\%$ | $(2.58 \pm 0.11)\,10^6$ |
| $K_{L_{l_3}}^0$ | $(37.7 \pm 0.9)\,\%$ | $(7.01 \pm 0.30)\,10^6$ |
| $K_{L_{\mu_3}}^0$ | $(28.1 \pm 0.8)\,\%$ | $(5.22 \pm 0.25)\,10^6$ |

## 2.2 General theoretical considerations

For purposes of theoretical analysis it is not the states $K_S^0$ and $K_L^0$ that one considers but rather the states $K^0$ and $\bar{K}^0$. These states are eigenstates of the Hamiltonian of strong interactions; they are connected by particle-antiparticle conjugation and in terms of them $K_S^0$ and $K_L^0$ are given as

$$K_{\genfrac{}{}{0pt}{}{S}{L}}^0 = \frac{pK^0 \mp q\bar{K}^0}{\sqrt{|p|^2 + |q|^2}}, \tag{2.7}$$

where $p$ and $q$ are complex constants. Unlike $K_S^0$ and $K_L^0$, $K^0$ and $\bar{K}^0$ possess definite values of strangeness and third component of isospin and this makes them the more appropriate states to deal with them considering the strangeness and isospin properties of $K_{l_3}$ decays. Thus in what follows we shall in fact consider the decays.

$$1. \quad K^- \to \pi^0 + l^- + \bar{\nu}_e, \tag{2.8}$$

$$2. \quad \bar{K}^0 \to \pi^+ + l^- + \bar{\nu}_e, \tag{2.9}$$

$$3. \quad K^0 \to \pi^+ + l^- + \bar{\nu}_e, \tag{2.10}$$

and their particle antiarticle conjugates

$$4. \quad K^+ \to \pi^0 + l^+ + \nu_e, \tag{2.11}$$

$$5. \quad K^0 \to \pi^- + l^+ + \nu_e, \tag{2.12}$$

$$6. \quad \bar{K}^0 \to \pi^- + l^+ + \nu_e. \tag{2.13}$$

Consequences for $K_S^0$ and $K_L^0$ decays can be deduced by using the relation (2.7).

### 2.3   Two component neutrino

Considering now the particles involved in these decays we remark that the only uncertainty which exists is connected with the neutrino particles. These particles are not observed directly and what is usually done is to assume that they have the same kinematic properties as $\beta$-decay neutrinos i.e. that they are two component massless objects, the particle state $\nu_l$ having helicity $-1$ (left-handed particle) and the antiparticle state $\bar{\nu}_l$ having helicity $+1$ (right-handed antiparticle). From high energy neutrino experiments we know that $K_{e_3}^\pm$ neutrinos interact with nuclear matter to produce inverse $\beta$-decay; hence if we assume that these decays involve a unique final state i.e. that only one kind of neutrino state is produced in each decay, then the above assumption is true for $K_{e_3}^\pm$ decay. For $K_{\mu_3}^\pm$, $K_{l_3}^0$ or $\bar{K}_{l_3}^0$ we have no similar arguments; however we shall make the usual assumption in the case of these decays also.

## 2.4   Kinematics

Turning now to the kinematics of $K_{l_3}$ decays see that the six processes (2.8) to (2.13) may be represented schematically as

$$K(k) \rightarrow \pi(\pi) + l(l) + \nu_l(\nu), \tag{2.14}$$

where the symbols in parentheses denote the respective four momenta. Kinematically, each such decay is completely specified by giving the values of the two invariant variables $s$ and $t$ defined by

$$s = -(k - l)^2 = -(\pi + \nu)^2, \tag{2.15}$$

and

$$t = -(k - \pi)^2 = -(l + \nu)^2. \tag{2.16}$$

One easily verifies from these definitions that

$$s = m_K^2 + m_l^2 - 2m_K E_l, \tag{2.17}$$

and

$$t = m_K^2 + m_\pi^2 - 2m_K E_\pi, \tag{2.18}$$

where $E_\pi$ and $E_l$ are the energies of the pion and the lepton resepectively in the rest frame of the kaon. The ranges of $s$ and $t$ are given by

$$m_\pi^2 \leqslant s \leqslant (m_K - m_l)^2, \tag{2.19}$$

and

$$m_l^2 \leqslant t \leqslant (m_K - m_\pi)^2. \tag{2.20}$$

If we define a third invariant variable $u$ by the relation

$$u = -(k - \nu)^2 = -(l + \pi)^2, \tag{2.21}$$

we find that

$$u = m_K^2 - 2m_K E_\nu, \tag{2.22}$$

$E_\nu$ being the energy of the neutrino in the kaon rest frame, with the limits on $u$ being given by

$$(m_\pi + m_l)^2 \leqslant u \leqslant m_K^2. \tag{2.23}$$

The conservation of four momentum leads to the result that $s, t$ and $u$ satisfy the linear relation

$$s + t + u = m_K^2 + m_\pi^2 + m_l^2 \tag{2.24}$$

and due to the same conservation law we find that the allowed values of $s$ and $t$ for the decay must lie inside the cubic curve

$$st(s+t) - st(3m_K^2 + m_\pi^2 + m_l^2) + sm_l^2(m_K^2 - m_\pi^2)$$

$$+ tm_\pi^2(m_K^2 - m_l^2) - m_l^2 m_\pi^2(m_K^2 - m_\pi^2 - m_l^2) = 0. \qquad (2.25)$$

[This latter may be shown by considering the equation of conservation of energy in the kaon rest frame when the neutrino momentum has been eliminated, namely,

$$m_K - E_\pi - E_l - \sqrt{\pi^2 + l^2 + 2\pi l \cos\theta_{\pi l}} = 0$$

and then using the inequality $-1 \leqslant \cos\theta_{\pi l} \leqslant 1$]. Thus we see that the various possible configurations in $K_{l_3}$ decay may be represented by the points in the $s$–$t$ plane lying inside the curve (2.25) Fig. 1.

### 2.5   Form of the matrix element

Coming now to the form of the matrix element for a process of the form (2.14) we remark that in view of the connection (through crossing) of this process to the process

$$K + \bar{\nu} \rightarrow \pi + l \qquad (2.26)$$

and the similarity of this second process to a pion-nucleon scattering process, for example, it is easy to convince oneself that the most general form of the reduced $T$-matrix element for the processes 1 to 3 [Eqs. (2.8) to (2.10)] is

$$\tilde{T}_k = \frac{G'}{\sqrt{2}} \frac{1}{(2\pi)^6} \sqrt{\frac{m_l m_\nu}{4k_0 \pi_0 l_0 \nu_0}} \, \bar{u}(l) [A_k + iB_k(\gamma \cdot \mathbf{P})](1 + \gamma_5) v(\nu),$$

$$\qquad (2.27)$$

$$P = k + \pi, \quad k = 1, 2, 3$$

while for the processes 4 to 6 [Eqs. (2.11) to (2.13)] it is

$$\tilde{T}_k = \frac{G'}{\sqrt{2}} \frac{1}{(2\pi)^6} \sqrt{\frac{m_l m_\nu}{4k_0 \pi_0 l_0 \nu_0}} \, \bar{u}(\nu)(1 - \gamma_5)[A_k + iB_k(\gamma \cdot \mathbf{P})](v(l),$$

$$\qquad (2.28)$$

$$P = k + \pi, \quad k = 4, 5, 6.$$

Here the reduced $T$-matrix element is related to the $S$-matrix element by the relation

$$S_{fi} = \delta_{fi}\delta^4\,(p_f - p_i) + i(2\pi)^4\,\delta^4\,(p_f - p_i)\,\tilde{T}_{fi}. \qquad (2.29)$$

In Eqs. (2.27) and (2.28) $G'/\sqrt{2}$ is a factor which we take out at this stage in order to facilitate out later discussion of the $V$, $A$ theory of $K_{l_3}$ decay (see Sections 4 to 7), the $\gamma$ matrices are chosen hermitian with $\gamma_5 = \gamma_1\gamma_2\gamma_3\gamma_4$, the factors $1 \pm \gamma_5$ appear because of the assumed two component nature of the neutrino and the $A_k$ and $B_k$ are scalar functions of the invariant variables $s$ and $t$. These last functions represent the most interesting part of our problem since they in fact contain all the dynamical information about the decay. The ultimate goal in the study of $K_{l_3}$ decay is, on the experimental side, to measure these functions accurately and on the theoretical side to develop a theory which predicts them correctly.

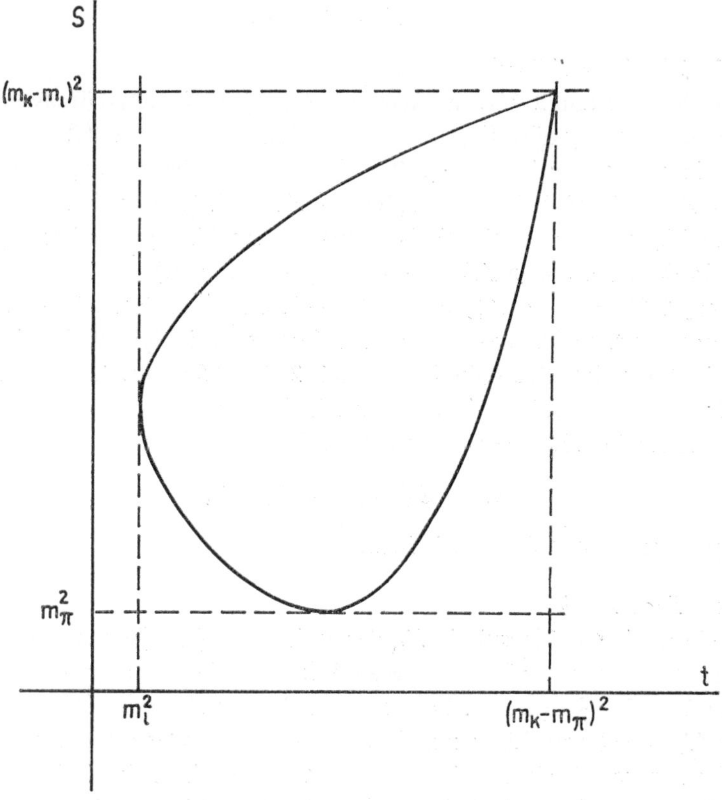


**Figure 1**  Boundary of Dalitz plot for $K_{l_3}$ decay

## 2.6 Symmetry considerations

Symmetry principles, if they are valid, enable us to deduce certain information about the functions $A_k$ and $B_k$ of Eqs. (2.27) and (2.28).

### 2.6.1 *TCP invariance*

If TCP invariance is valid and we neglect final state interactions (which is a reasonably good approximation since they are only electromagnetic or weak) then we have the relations

$$A_{4,5,6} = (A_{1,2,3})^*; \quad B_{4,5,6} = -(B_{1,2,3})^*. \tag{2.30}$$

### 2.6.2 *Time reversal*

If time reversal invariance is valid and again we neglect final state interactions the functions $A_k$ and $B_k$ for each decay are relatively real i.e.

$$\mathrm{Arg}\, A_k = \mathrm{Arg}\, B_k. \tag{2.31}$$

### 2.6.3 *The $\Delta y = \Delta Q$ rule*

In the decays 1 and 2 $\Delta I_3$, the change in $I_3$, is $\frac{1}{2}$ while in the decay 3 it is $\frac{3}{2}$. An equivalent way of stating this is to say that in decays 1 and 2 the changes in the hadronic electric charge, $\Delta Q$, and in the hypercharge, $\Delta Y$, are given by $\Delta Q = \Delta Y = 1$, while in decay 3 $\Delta Q = -\Delta Y = 1$. (The equivalence can be checked by reference to the Gell–Mann–Nakano–Nishijima relation.) Likewise in decays 4 and 5 $\Delta I_3 = -\frac{1}{2}$ or equivalently $\Delta Q = \Delta Y = -1$ while in decay 6 $\Delta I_3 = -\frac{3}{2}$ or $\Delta Q = -\Delta Y = -1$. On this basis the decays 3 and 6 appear to be the odd ones out and therefore as a principle of simplification we might postulate that only the decays 1, 2, 4 and 5 occur. A way of stating this is the so-called $\Delta Y = \Delta Q$ rule which is the postulate that only transitions for which $\Delta Y = \Delta Q = \pm 1$ occur. If this is the case then

$$A_3 = B_3 = A_6 = B_6 = 0, \tag{2.32}$$

and processes 3 and 6 are forbidden.

### 2.6.4 *The $\Delta I = \frac{1}{2}$ rule*

The change in total isospin in $K_{l_3}$ decays is either $\frac{1}{2}$ or $\frac{3}{2}$ or in other words the transition operator for these decays is in general a sum of an $I = \frac{1}{2}$ operator and an $I = \frac{3}{2}$ operator. Now if one rules out $\Delta Y = -\Delta Q$ transitions, only $\Delta I_3 = \pm\frac{1}{2}$ transitions remain and then it becomes attractive to make the further hypothesis that the change in total isospin is simply $\frac{1}{2}$, i.e. $I = \frac{3}{2}$ transitions are absent. This is the $\Delta I = \frac{1}{2}$ rule and in addition to the relations (2.32) it predicts that

$$A_1 = \frac{1}{\sqrt{2}} A_2, \quad B_1 = \frac{1}{\sqrt{2}} B_2,$$

$$A_4 = \frac{1}{\sqrt{2}} A_5, \quad B_4 = \frac{1}{\sqrt{2}} B_5. \tag{2.33}$$

To see the above rules in their general setting we remark that the transition operator $T$ for the processes 1, 2 and 3 in general consists of three parts

$$T = T^{1/2}_{1/2} + T^{3/2}_{1/2} + T'^{3/2}_{3/2} \tag{2.34}$$

the upper and lower indices on each part denoting its total isospin ($I$) and third component of isospin ($I_3$) tensor properties respectively. Only the first two parts in this decomposition contribute to processes 1 and 2 and only the third contributes to process 3. In terms of the $I = \frac{1}{2}$ and $I = \frac{3}{2}$ amplitudes $A_1$ and $A_2$ are given by

$$A_1 = \frac{1}{\sqrt{2}} A_{1/2} - \frac{1}{\sqrt{2}} A_{3/2},$$

$$A_2 = A_{1/2} + \frac{1}{2} A_{3/2}, \tag{2.35}$$

with similar relations for $B_1$ and $B_2$. Now if the $K_{l_3}$ decays satisfy the $\Delta Y = \Delta Q$ rule the term $T'^{3/2}_{3/2}$ is absent from $T$, whereupon the first two relations of Eq. (2.32) follow. If the $\Delta I = \frac{1}{2}$ rule is satisfied only the first term is present in $T$ and then in addition to the $\Delta Y = \Delta Q$ rule, which is automatically satisfied, we obtain $A_{3/2} = B_{3/2} = 0$ from which using Eq. (2.35) and a similar one for $B_1$ and $B_2$ the first two of relations (2.33) follow. Similar considerations follow for the processes 4, 5 and 6 by considering an operator similar to in Eq. (2.34) but with $I_3 = -\frac{1}{2}$ and $I_3 = -\frac{3}{2}$ tensor properties.

We shall discuss the experimental situation regarding these various symmetries when we have calculated the various distributions arising from these decays.

## 2.7  The local action of leptons

It is generally believed that in semileptonic weak interactions the leptons act locally and without derivatives. What this means is that whereas in general the transition operator $T$ for a process such as $K \rightarrow \pi + l^- + \bar{\nu}_l$ is of the form

$$T = \int O_{\alpha\beta}(x_1, x_2)\, \bar{l}_\alpha(x_1)\, v_{l\beta}(x_2)\, d^4x_1\, d^4x_2$$

where $l_\alpha(x)$ and $v_l(x)$ are the out fields of the leptons and $O_{\alpha\beta}(x_1, x_2)$ is a hadronic operator capable of transforming $K$ into $\pi$, in actual fact it takes the very special form resulting from $O_{\alpha\beta}(x_1, x_2)$ having the structure

$$O_{\alpha\beta}(x_1, x_2) = \bar{O}_{\alpha\beta}(x_1)\, \delta^4(x_1 - x_2).$$

Roughly speaking this has the consequence that both leptons in the decay are created at the same space-time point as is illustrated in Fig. 2. More precisely, the local action of leptons means that $T$ can be written as

$$T = \int \bar{O}_{\alpha\beta}(x)\, \bar{l}_\alpha(x)\, v_{l\beta}(x)\, d^4x$$

$$\equiv \int I(x)\, d^4x$$

where from Lorentz covariance $I(x)$ must have the form

$$I(x) = \frac{G'}{\sqrt{2}}\, [J^S(x)\, \bar{l}(x)\, (1 + \gamma_5)\, v_l(x) + J^v_\lambda(x)\, \bar{l}(x)\, i\gamma_\lambda\, (1 + \gamma_5)\, v_l(x)$$

$$+ J^T_{\lambda v}\bar{l}(x)\, \delta_{\lambda v}\, (1 + \gamma_5)\, v_l(x) + J^A_\lambda(x)\, \bar{l}(x)\, i\gamma_\lambda\, (1 + \gamma_5)\, v_l(x)$$

$$+ J^P(x)\, \bar{l}(x)\, (1 + \gamma_5)\, v_l(x)], \quad \delta_{\lambda v} = \frac{1}{2i}\, (\gamma_\lambda\gamma_v - \gamma_v\gamma_\lambda),$$

$$J^S(x), \quad J^v_\lambda(x), \quad J^T_{\lambda v}(x), \quad J^A_\lambda(x) \quad \text{and} \quad J^P(x)$$

being scalar, vector, antisymmetric tensor, axial vector and pseudoscalar hadronic operators respectively. The reduced $T$-matrix element for $K \to \pi + \bar{l} + \bar{v}_l$ decay is then given by

$$\bar{T} = \frac{G'}{\sqrt{2}}\, \frac{1}{(2\pi)^3}\, \sqrt{\frac{m_l m_v}{l_0 v_0}}\, \{\langle\pi|\, J^S(0)\, |K\rangle\, \bar{u}(l)\, (1 + \gamma_5)\, v(v)$$

$$+ \langle\pi|\, J^v_\lambda(0)\, |K\rangle\, \bar{u}(l)\, i\gamma_\lambda\, (1 + \gamma_5)\, v(v)$$

$$+ \langle\pi|\, J^T_\lambda(0)\, |K\rangle\, \bar{u}(l)\, \sigma_{\lambda v}\, (1 + \gamma_5)\, v(v)\}.$$

$$(2.36)$$

(possible pseudovector and pseudoscalar contributions are absent because $K$ and $\pi$ are spinless and possess the same intrinsic parity). Using Lorentz covariance we may now express the various hadronic matrix elements appearing in Eq. (2.36) in terms of known kinematic quantities and unknown scalar functions which we call form factors. Thus we obtain

$$\langle\pi|\, J^S(0)\,|K\rangle = \frac{1}{(2\pi)^3}\,\frac{1}{\sqrt{4k_0\pi_0}}\, m_K f_S^k(t), \qquad (2.37a)$$

$$\langle\pi|\, J_\lambda^v(0)\,|K\rangle = \frac{1}{(2\pi)^3}\,\frac{1}{\sqrt{4k_0\pi_0}}\,[f_+^k(t)(k+\pi)_\lambda + f_-^k(t)(k-\pi)_\lambda],$$
$$\qquad (2.37b)$$

$$\langle\pi|\, J_{\lambda v}^T(0)\,|K\rangle = \frac{1}{(2\pi)^3}\,\frac{1}{\sqrt{4k_0\pi_0}}\,\frac{1}{m_K}\, f_T^k(t)\,(k_\lambda\pi_v - k_v\pi_\lambda),$$
$$\qquad (2.37c)$$

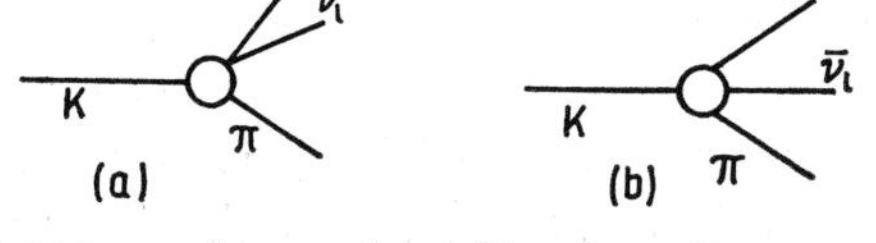


**Figure 2**   Local (a) as against non local (b) actions of leptons in semileptonic
decays

and hence from Eqs. (2.36) and (2.37) we deduce for the functions $A_k$ and $B_k$ in Eq. (2.27) the forms

$$A_k = m_K f_S^k(t) - m_l f_-^k(t) + \frac{1}{m_K}\,(2s + t - m_K^2 - m_\pi^2 - m_l^2)f_T^k(t) \qquad (2.38a)$$

$$B_k = f_+^k(t) - \frac{m_l}{m_K}\, f_T^k(t). \qquad (2.38b)$$

While the foregoing discussion applied only to the decays 1, 2 and 3 it is obvious that a completely analogous considerations hold for the decays 4, 5 and 6. We may therefore take Eqs. (2.38) as expressing the consequence of the local action of leptons for all six decays under consideration. It is seen from these equations that the hypothesis of the local action of leptons makes quite specific predictions about the forms of the functions $A_k$ and $B_k$. In particular, the $S$ dependence of these quantities is completely specified by this hypothesis.

With the foregoing preparation we may now come to the various distributions which can be compared with experiment.

## 3  DISTRIBUTIONS

The decay-polarization distribution for the general process $K \to \pi + l^- + \bar{\nu}_l$ with the matrix element given by Eq. (2.28) is given by

$$\frac{d\Gamma_k^-}{d\pi \, d\mathbf{l} \, d\mathbf{v}} = \frac{(G')^2}{2} \frac{1}{(2\pi)^5} \frac{\delta^4 (k - \pi - \mathbf{l} - \mathbf{v})}{4k_0 \pi_0 l_0 v_0} \times$$

$$\times [|A_k|^2 \{-\mathbf{v}(\mathbf{l} + m_l \mathbf{n})\}$$

$$+ (A_k B_k^* + A_k^* B_k) \{m_l (\mathbf{vP}) - (\mathbf{vn})(\mathbf{Pl}) + (\mathbf{vl})(\mathbf{Pn})\}$$

$$+ (A_k B_k^* - A_k^* B_k) \varepsilon_{\mu\nu\alpha\beta} P_\mu q_\nu l_\alpha n_\beta$$

$$+ |B_k|^2 \{2 (\mathbf{Pv}) [P(\mathbf{l} - m_l \mathbf{n})] - P^2 [\mathbf{v}(\mathbf{l} - m_l \mathbf{n})]\}]. \qquad (3.1)$$

For the decays $K \to \pi + l^+ + \nu_l$ the corresponding distribution is that obtained from (3.1) by making the replacements $m_l \to -m_l$, $n \to -n$. In this expression $n$ is a unit four vector satisfying $(n \cdot l) = 0$ which corresponds to the spin direction of the lepton. Since the neutrino is never detected directly we integrate over its momentum and in so doing we eliminate the three-dimensional delta function of momentum. In order to eliminate the energy delta function we go to the rest frame of the kaon, write

$$d\pi \, d\mathbf{l} = \pi^2 \, d\pi \, \mathbf{l}^2 \, d\mathbf{l} \, d(\cos \theta_{\pi l}) (8\pi^2)$$

and use the relation

$$\delta \left( m_k - E_\pi - E_l - \sqrt{\pi^2 + \mathbf{l}^2 + 2\pi l \cos \theta_{\pi l}} \right) d(\cos \theta_{\pi l}) = \frac{E_\nu}{\pi l}$$

whereupon we find

$$\frac{d\Gamma_k^-}{dE_\pi \, dE_l} = \frac{(G')^2}{2} \frac{1}{(2\pi)^3} \frac{1}{2m_k} [ \qquad ], \qquad (3.2)$$

the bracket term being the same as that appearing on Eq. (3.1). Then using Eqs. (2.15) to (2.18) we can write Eq. (3.2) in the covariant form

$$\frac{d\Gamma_k^-}{dt \, ds} = \frac{(G')^2}{2} \frac{1}{(2\pi)^3} \frac{1}{8m_k^3} \left[ \frac{1}{2} (t - m_l^2) |A_k|^2 - m_l \left\{ (s - m_\pi^2) \right. \right.$$

$$\left. + \frac{1}{2}(t - m_l^2) \right\} (A_k B_k^* + A_k^* B_k) - \frac{1}{2} \{(t - m_l^2)^2$$

$$+ (2s - t)(t - m_l^2) + 4(s - m_\pi^2)(s - m_K^2)\} |B_k|^2$$

$$- \left[ m_l \, |A_k|^2 + \left\{ (s - m_\pi^2) + \frac{1}{2} (t - m_k^2) \right\} (A_k B_k^* + A_k^* B_k) \right.$$

$$- m_l (t - 2m_K^2 - 2m_\pi^2) |B_k|^2 (\mathbf{q} \cdot \mathbf{n})$$

$$- \frac{1}{2} (t - m_l^2) (A_k B_k^* + A_k^* B_k) - m_l \left\{ 2 (s - m_\pi^2) \right.$$

$$\left. + (t - m_l^2) \right\} |B_k|^2 ] (\mathbf{P} \cdot \mathbf{n}) + (A_k B_k^* - A_k^* B_k) \, \varepsilon_{\pi\nu\alpha\beta} P_\mu q_\nu l_\alpha n_\beta \Bigg].$$

$$(3.3)$$

Eq. (3.3) contains all the information it is possible to deduce from $K_{l_3}$ decay. By measuring the rate of the process as a function of $s$, $t$ and $n$, the polarization direction, one can determine the functions $A_k$ and $B_k$ and so learn the structure of the decay.

When the muon polarization is not observed the decay distribution is that obtained from (3.3) by setting $n$ equal to zero and multiplying by 2. The resulting expression is

$$\frac{d\Gamma_k^-}{dt \, ds} = \frac{(G')^2}{2} \frac{1}{(2\pi)^3} \frac{1}{4m_K^3} \left[ \frac{1}{2} (t - m_l^2) |A_k|^2 - m_l \left\{ (s - m_\pi^2) \right. \right.$$

$$+ \frac{1}{2} (t - m_l^2) \right\} (A_k B_k^* + A_k^* B_k) - \frac{1}{2} \left\{ ((t - m_l^2)^2 \right.$$

$$+ (2s - t) (t - m_l^2) + 4 (s - m_\pi^2) (s - m_K^2) \right\} |B_k|^2 \Bigg] \qquad (3.4)$$

The expression on the right-hand side here represents the Dalitz plot distribution function.

In cases where the polarization of the charged lepton is measured, if the unit vector $\xi$ denotes the direction of the lepton's spin in its own rest frame, the four vector $n$ appearing in (3.3) is given by

$$\mathbf{n} = \xi + \frac{(\xi \cdot \mathbf{l}) \, \mathbf{l}}{m_l (l_0 + m_l)}, \qquad (3.5a)$$

$$n_0 = \frac{1}{m_l} (\xi \cdot \mathbf{l}). \qquad (3.5b)$$

($n$ is obtained from $\xi$ by applying to it a Lorentz transformation with velocity $-\mathbf{l}/l_0$. From this we obtain from (3.3) the distribution

$$\frac{d\Gamma_k^-}{dt\,ds} = \frac{(G')^2}{2}\,\frac{1}{(2\pi)^3}\,\frac{1}{8m_k^3}\,[\mathfrak{A} + \mathfrak{B}\cdot\xi] \tag{3.6a}$$

where $\mathfrak{A}$ denotes the expression appearing in the square brackets in Eq. (3.4) and $\mathfrak{B}$ is the vector

$$\mathfrak{B} = -[m_l\,|A_k|^2 + \{(s - m_\pi^2) + (t - m_l^2)\}\,(A_k B_k^* + A_k^* B_k)$$

$$- m_l\,(2t + 2s - 2m_k^2 - 4m_\pi^2 - m_l^2)\,|B_k|^2]\left[\mathbf{k} + \frac{\mathbf{l}}{m_l}\left\{\frac{\mathbf{k}\cdot\mathbf{l}}{l_0 + m_l} - k_0\right\}\right]$$

$$+ [m_l\,|A_k|^2 + (s - m_\pi^2)\,(A_k B_k^* + A_k^* B_k) + m_l\,(2s + 2m_k^2 - m_l^2)\,|B_k|^2]\times$$

$$\times\left[\boldsymbol{\pi} + \frac{\mathbf{l}}{m_l}\left\{\frac{\boldsymbol{\pi}\cdot\mathbf{l}}{l_0 + m_l} - \pi_0\right\}\right] + 2i\,(A_k B_k^* - A_k^* B_k)\times$$

$$\times\left[k_0\,(\boldsymbol{\pi}\times\mathbf{l}) + \pi_0\,(\mathbf{l}\times\mathbf{k}) + l_0\,(\mathbf{k}\times\boldsymbol{\pi}) + \frac{(\mathbf{l}\cdot\boldsymbol{\pi}\times\mathbf{k})\,\mathbf{l}}{l_0 + m_l}\right]. \tag{3.6b}$$

This expression coincides with that derived by Cabibbo and Maksymowicz for the special case of a local $V$–$A$ coupling[2].

A method of analysing $K_{l_3}$ decay distributions which has become fairly popular is that in which instead of the variables $t$ and $s$ one uses the variables $t$ and $\cos\theta$ where $\theta$ is the angle between the momentum of the pion and that of the neutrino in the eentre of mass system of the two leptons. The particular appropriateness of this method of analysis is best seen if one applies the helicity formalism to the process[3]. Since the neutrino $\bar{\nu}_l$ is assumed to be a two component particle with positive helicity $K \to \pi + l^- + \bar{\nu}_l$ involves just two possible helicity states which we denote by $++$ and $+-$. If then the matrix elements for decay into these states are denoted by $\tilde{T}_k(++)$ and $\tilde{T}_k(+-)$ respectively and the quantities $\tau_k(++)$ and $\tau_k(+-)$ are defined by

$$\tilde{T}_k(++) = \frac{1}{(2\pi)^6}\,\frac{1}{\sqrt{4k_0\pi_0 l_0 \nu_0}}\,\tau_k(++), \tag{3.7a}$$

and

$$\tilde{T}_k(+-) = \frac{1}{(2\pi)^6}\,\frac{1}{\sqrt{4k_0\pi_0 l_0 \nu_0}}\,\tau_k(+-), \tag{3.7b}$$

we find from Eq. (2.27)

$$\tau_k (++) = \frac{G'}{\sqrt{2}}\, m_l p \left(\frac{t - m_l^2}{t}\right)^{1/2} \{- C_k + B_k \cos \theta\}, \qquad (3.8a)$$

and

$$\tau_k (+-) = \frac{G'}{\sqrt{2}}\, p\, (t - m_l^2)^{1/2}\, B_k \sin \theta, \qquad (3.8b)$$

where

$$C_k = [(m_k^2 - m_\pi^2 + t)^2 - 4m_k^2 t]^{-1/2} \left[\frac{t}{m_l}\, A_k + (m_k^2 - m_\pi^2)\, B_k\right],$$

and $p$ is the magnitude of the kaon or pion momentum in the dilepton centre of mass system.

Angular momentum analysis shows that $\tau_k(++)$ has the expansion

$$\tau_k (++) = \sum_J (J + \tfrac{1}{2})\, H_{k++}^J(t)\, P_J (\cos \theta), \qquad (3.9a)$$

while for $\tau_k(+-)$ the corresponding expansion is

$$\tau_k (+-) = \sum_J \frac{J + \tfrac{1}{2}}{J(J+1)}\, H_{k+-}^J(t)\, P_J' (\cos \theta). \qquad (3.9b)$$

The differential rates of decay into the two states are given by

$$\frac{d\Gamma_k^- (+, \pm)}{dt\, d\Omega} = \frac{(G')^2}{4}\, \frac{1}{(2\pi)^4}\, \frac{p^3}{m_k^3}\, (t - m_l^2)\, t^{1/2} \left|\frac{\tau_k (+, \pm)}{P}\right|^2 \qquad (3.10)$$

In the case when the charged lepton polarization is not observed the differential decay rate is

$$\frac{d\Gamma_k^-}{dt\, d\Omega} = \frac{(G')^2}{4(2\pi)^4}\, \frac{P^3}{m_k^3}\, (t - m_l^2)\, t^{1/2} \left\{\left|\frac{\tau_k (+-)}{P}\right|^2 + \left|\frac{\tau_k (++)}{P}\right|^2\right\}$$

$$= \frac{(G')^2}{4(2\pi)^4}\, \frac{p^3}{m_k^3}\, \frac{(t - m_l^2)^2}{t^{1/2}} \left\{|B_k|^2 \sin^2 \theta + \frac{m_l^2}{t}\,|- C_k + B_k \cos \theta|^2\right\}.$$

$$(3.11)$$

The polarization vector $\mathscr{B}$ Eq. (3.7) can be expressed in terms of $t$ and $\cos \theta$ simply by using the relation

$$s = m_\pi^2 - \frac{t - m_l^2}{t} \{(t - m_k^2 + m_\pi^2) - [(t - m_k^2 + m_\pi^2)^2 - 4m_\pi^2 t]^{1/2} \cos \theta\}$$

$$(3.12)$$

The formulae (3.4), (3.6a, b) and (3.11) form the basis for the analysis of $K_{l_3}$ decay.

## 4   COMPARISON WITH EXPERIMENT. THE $V$–$A$ THEORY

### 4.1   Local action of leptons

Logically the first feature to study in $K_{l_3}$ decay would seem to be the local action of leptons and this is done by examining the dependence of the decay distribution on $s$ or $\cos\theta$. According to Eqs. (2.39a, b) the local action of leptons implies that the decay distribution (3.4) is a polynomial of degree 2 in $s$; i.e. it is of the form

$$\frac{d\Gamma_k}{dt\,ds} = a_k(t) + b_k(t)\,s + c_k(t)\,s^2, \tag{4.1}$$

where $a_k(t)$, $b_k(t)$ and $c_k(t)$ are functions of $t$ alone. An equivalent statement is that the decay distribution (3.11) is a second degree polynomial in $\cos\theta$ i.e.

$$\frac{d\Gamma_k}{dt\,d\Omega} = \tilde{a}_k(t) + \tilde{b}_k(t)\,\cos\theta + \tilde{c}_k(t)\,\cos^2\theta, \tag{4.2}$$

$\tilde{a}_k(t)$, $\tilde{b}_k(t)$ and $\tilde{c}_k(t)$ being functions of $t$ alone. Experiments performed so far[4] are certainly consistent with the forms (3.13) and (3.14) but the level of accuracy of these experiments is not very high. Future experiments should greatly improve our knowledge on this point.

It may be remarked that electromagnetic corrections effectively destroy the local action of leptons since they give rise to Feynman diagrams such as that shown in Fig. 3. This mechanism introduces an $s$ dependence in the

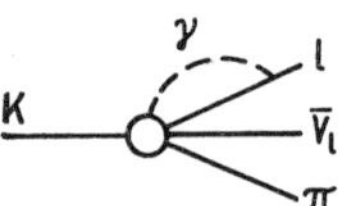


**Figure 3**   Feynman diagram for electromagnetic correction in $K_{l3}$

amplitudes $A_k$ and $B_k$ different from that given by Eqs. (2.39). This phenomenon occurs, of course, at order $\alpha$ in the matrix element but calculations show that it should be a detectable effect[5].

## 4.2  T C P invariance

According to the conditions (2.30) TCP invariance implies the equality of the decay (Dalitz plot) distributions for $K^+$ and $K^-$ decays if we neglect final state interactions. Because of the difficulty of observation of $K^-$ decays (due to $K^-$ absorption in nuclear matter) no test of this equality has yet been performed. For the total lifetimes of the $K^+$ and the $K^-$, which should be equal if TCP invariance holds, it is known that[6]

$$\frac{\tau(K^+)}{\tau(K^-)} - 1 = -0.0010 \pm 0.0017$$

This is the best indication we have that TCP invariance holds in $K$ decays.

## 4.3  T invariance

If time reversal invariance is valid then $A_k$ and $B_k$ are relatively real if we neglect final state interactions. A test of this invariance is the existence or non-existence of a component of lepton polarization perpendicular to the decay plane i.e. the presence or absence of the term proportional to $A_k B_k^* - A_k^* B_k$ in Eq. (3.3) or Eq. (3.6b). Three experiments undertaken to measure this transverse component of polarization in $K_{\mu_3}$ decay essentially yielded null results[7]. The accurancy of these experiments though is about the $10\%$ level whereas known $CP$ and probably $T$ violating effects occur at the level of $2 \times 10^{-3}$.

## 4.4  The $\Delta Y = \Delta Q$ and $\Delta I = \frac{1}{2}$ rules

The amplitudes for the semileptonic decays of the $K_S^0$ and $K_L^0$ states are obtained by combining the definition of these states in Eq. (2.7) with the expressions for the amplitudes for the decays 1 to 6 given in Eqs. (2.27) and (2.28). We obtain

$$\tilde{T}_S(l^-) = Nq\tilde{T}_2 + Np\tilde{T}_3, \tag{4.3a}$$

$$T_S(l^+) = Np\tilde{T}_5 - Nq\tilde{T}_6, \tag{4.3b}$$

$$\tilde{T}_L(l^-) = Nq\tilde{T}_2 + Np\tilde{T}_3, \tag{4.3c}$$

$$\tilde{T}_L(l^+) = Np\tilde{T}_5 + Nq\tilde{T}_6, \tag{4.3d}$$

where

$$N = \frac{1}{\sqrt{|p|^2 + |q|^2}}.$$

The $\Delta Y = \Delta Q$ rule corresponds to having

$$\tilde{T}_3 = \tilde{T}_6 = 0$$

and in this event we obtain from Eqs. (4.3a–d))

$$\tilde{T}_S(l^-) = \tilde{T}_L(l^-) = -Nq\tilde{T}_{\overline{K}^0}(l^-), \qquad (4.4a)$$

and

$$\tilde{T}_S(l^+) = T_L(l^+) = Np\tilde{T}_{K^0}(l^+). \qquad (4.4b)$$

We therefore have the predictions

$$\Gamma_S(l^-) = \Gamma_L(l^-) = N^2\,|q|^2\,\Gamma_{\overline{K}^0}(l^-), \qquad (4.5a)$$

$$\Gamma_S(l^+) = \Gamma_L(l^+) = N^2\,|p|^2\,\Gamma_{K^0}(l^+). \qquad (4.5b)$$

If in addition we neglect final state interactions, $TCP$ invariance implies that $\Gamma_{\overline{K}^0}(l^-) = \Gamma_{K^0}(l^+)$ and we get the additional relation

$$\Gamma_S(l^-) + \Gamma_S(l^+) = \Gamma_L(l^-) + \Gamma_L(l^+) = \Gamma_{\overline{K}^0}(l^-) = \Gamma_{K^0}(l^+). \qquad (4.6)$$

Unfortunately none of the relations (4.5) or (4.6) can be tested experimentally because the rates $\Gamma_S(l^\pm)$ are quite unknown at present. However if we invoke in addition the $\Delta I = \tfrac{1}{2}$ rule we have $\tilde{T}_1 = \dfrac{1}{\sqrt{2}}\,\tilde{T}_2$ which implies

$$\Gamma_{K^-}(l^-) = \tfrac{1}{2}\Gamma_{\overline{K}^0}(l^-). \qquad (4.7)$$

Then this relation together with relations (4.6) gives the testable relation

$$\Gamma_L(l^-) + \Gamma_L(l^+) = 2\Gamma_{K^-}(l^-)$$
$$= 2\Gamma_{K^+}(l^+). \qquad (4.8)$$

By reference to Table 1 one sees that this relation is very well satisfied experimentally for the case $l = \mu$ and satisfied within the errors for the case $l = e$. However more precise experiments would be desirable on this point. We should like to emphasize here that the relations (4.6) and (4.8) hold only to the extent that final state interactions may be neglected. These effects however are electromagnetic at most and probably not very large.

### 4.5  Charge asymmetry in the semileptonic decays of neutral kaons

It is well known that the charge asymmetry in the semileptonic decays of neutral kaons is an important source of information regarding $CP$ violation and the rule. By now four mutually consistent experiments have confirmed

the existence of such an asymmetry[8] and accordingly certain conclusions
have been drawn about $CP$ violation and the $\Delta Y = \Delta Q$ rule[9]. In order to
see how this comes about let us examine the matter in some detail.

We assume (1) that all interactions are $TCP$ invariant and (2) that semileptonic decays are first order effects of a semileptonic weak interaction
Hamiltonian $H^{SL} = \sum\limits_{l=e,\mu} H^l$ with $H^l = \int d^3 \times H^l(x)$. From assumption (1) it
follows that the short and long lived neutral kaon states are given by

$$|K_S\rangle_L = p\,|K^0\rangle \mp q|\overline{K}^0\rangle \quad (|p|^2 + |q|^2 = 1 \tag{4.9}$$

where $|K^0\rangle$ and $|\overline{K}^0\rangle$ are eigenstates of the Hamiltonian of strong and electro-
magnetic interactions with $|\overline{K}^0\rangle$ given by $C\,|K^0\rangle$ if this Hamiltonian is in-
variant under $C$, by $CP\,|K^\circ\rangle$ if it is invariant under $CP$ but not under $C$, and
by $TCP\,|K^0\rangle$ if it is invariant under $TCP$ but not under $C$ or $CP$. (Note that
this is a somewhat different definition of $K_S^0$ and $K_L^0$ from that employed previously.) Assumption (2) means that the reduced $T$-matrix element for the
decays $K_S \to n^-l^+\nu_l$ and $K_S \to n^+l^-\overline{\nu}_l$ are given respectively by

$$\langle n^-l^+\nu_{l\,\text{out}}\,|H_{(0)}^l|\,K_{S\,\text{In}}^0\rangle_L = p\,\langle n^-l^+\nu_{l\,\text{out}}|H_{(0)}^l|\,K_{\text{In}}^0\rangle$$
$$+ q\,\langle n^-l^+\overline{\nu}_{l\text{out}}|\,H_{(0)}^l|\,\overline{K}_{\text{In}}^0\rangle \tag{4.10}$$

and

$$\langle n^+l^-\nu_{l\,\text{out}}\,|H_{(0)}^l|\,K_{S\,\text{In}}^0\rangle_L = p\,\langle n^+l^-\nu_{l\,\text{out}}|\,H_{(0)}^l|\,K_{\text{In}}^0\rangle$$
$$+ q\,\langle n^+l^-\overline{\nu}_{l\,\text{out}}|\,H_{(0)}^l|\,\overline{K}_{\text{In}}^0\rangle \tag{4.11}$$

Here $n^-$ and $n^+$ stand for any collection of particles which can be produced
with an $l^+\nu_l$ or an $l^-\overline{\nu}_l$ pair respectively in a semileptonic decay and the states
in (4.10) and (4.11) are "in" and "out" states of the Hamiltonian of strong
and electromagnetic interactions.

Consider now the semileptonic decays of the neutral kaon state $K_R^0(t)$
which at time $t = 0$ is given by

$$K_R^0(0) = \frac{|K_L^0\rangle + R\,|K_S^0\rangle}{\sqrt{1 + (R)^2 + 2\,\text{Re}(R\langle K_L^0|\,K_S^0\rangle)}}. \tag{4.12}$$

Notice that $K_R^0(0)$ represents quite a variety of initial states; for $R = 0$ it
represents an initial $K_L^0$ state, for $R = \infty$ an initial $K_S^0$ state, for $R = 1$ an
initial $K^0$ state, for $R = -1$ an initial $\overline{K}^0$ state and for $R$ a regeneration amplitude it represents a regenerated neutral kaon state. Consequently the for-

mulas deduced below have quite a wide application. Now the state $K_R^0(t)$ is given in its own rest frame by

$$K_R^0(t) = \frac{e^{-im_L t - (\Gamma_L/2)t} |K_L^0\rangle + e^{-im_S t - (\Gamma_S/2)t} R |K_S^0\rangle}{\sqrt{1 + |R|^2 + 2\,\mathrm{Re}\,(R\,\langle K_L^0| K_S^0\rangle)}} \tag{4.13}$$

and hence using Eqs. (4.9) to (4.13) we obtain for the sum and difference of the rates for the decays $K_R^0(t) \to n^- l^+ \nu_l$ and $K_R^0(t) \to n^+ l^- \bar{\nu}_l$ the expressions

$$\Gamma_R (n^- l^+ \nu_l;\, t) \pm \Gamma_R(n^+ l^- \bar{\nu}_l;\, t) = \frac{1}{1 + |R|^2 + 2\,\mathrm{Re}\,(R\,\langle K_L^0| K_S^0\rangle)} \times$$

$$\times [e^{-\Gamma_L t} \{\Gamma_L (n^- l^+ \nu_l) \pm \Gamma_L (n^+ l^- \bar{\nu}_l)\} + e^{-\Gamma_S t} |R|^2 \{\Gamma_s (n^- l^+ \nu_l)$$

$$\pm \Gamma_s(n^+ l^- \bar{\nu}_l)\} + e^{-(\Gamma_L + \Gamma_S/2)t} (2\pi)^7 \int \delta^4 (k - p_n - l - \nu)\, d^3l\, d^3\nu\, d^3p_n \times$$

$$\times \{e^{-i\Delta m t} R^* [|p|^2 \langle K_{in}^0| H^I(0)| n^- l^+ \nu_{l\,out}\rangle \langle n^- l^+ \nu_{l\,out} |H^I(0)| K_{in}^0\rangle$$

$$- |q|^2 \langle \bar{K}_{in}^0 |H^I(0)| n^- l^+ \nu_{l\,out}\rangle \langle n^- l^+ \nu_{l\,out} |H^I(0)| \bar{K}_{in}^0\rangle$$

$$\pm |p|^2 \langle K_{in}^0 |H^I(0)| n^+ l^- \bar{\nu}_{l\,out}\rangle \langle n^+ l^- \bar{\nu}_{l\,out} |H^I(0)| K_{in}^0\rangle$$

$$\mp |q|^2 \langle \bar{K}_{in}^0 |H^I(0)| n^+ l^- \bar{\nu}_{l\,out}\rangle \langle n^+ l^- \bar{\nu}_{l\,out} |H^I(0)| \bar{K}_{in}^0\rangle$$

$$- pq^* \langle \bar{K}_{in}^0 |H^I(0)| n^- l^+ \nu_{l\,out}\rangle \langle n^- l^+ \nu_{l\,out} |H^I(0)| K_{in}^0\rangle$$

$$+ p^*q \langle K_{in}^0 |H^I(0)| n^- l^+ \nu_{l\,out}\rangle \langle n^- l^+ \nu_{l\,out} |H^I(0)| \bar{K}_{in}^0\rangle$$

$$\mp pq^* \langle \bar{K}_{in}^0 |H^I(0)| n^+ l^- \bar{\nu}_{l\,out}\rangle \langle n^+ l^- \bar{\nu}_{l\,out} |H^I(0)| K_{in}^0\rangle$$

$$\pm p^*q \langle K_{in}^0 |H^I(0)| n^+ l^- \bar{\nu}_{l\,out}\rangle \langle n^+ l^- \bar{\nu}_{l\,out} |H^I(0)| \bar{K}_{in}^0\rangle]$$

$$+ \text{complex conjugate}\}], \tag{4.14}$$

where

$$\Delta m = m_L - m_S.$$

By studying these quantities as functions of time one can learn about the mass difference $\Delta m$, the magnitude of the decay rates $\Gamma_{s,L}\,(nl\nu_l)$, $CP$ violation and the validity of the $\Delta Y = \Delta Q$ rule in the various decay modes. The matrix elements $\langle \pi^+ l^- \bar{\nu}_l |H^I(0)| K^0\rangle$ and $\langle \pi^- l^+ \nu_l| H^I(0) |\bar{K}^0\rangle$, for example, represent $\Delta Y = -\Delta Q$ transitions and so if they are non-zero they signify violation of the $\Delta Y = \Delta Q$ rule.

A quantity of particular interest is the charge asymmetry of the decays of the $K_L^0$ into the charge conjugate modes $n^- l^+ \nu_l$ and $n^+ l^- \nu_l$ (think especially of the case $n = \pi$). This quantity is given by

$$\delta_L^n = \frac{\Gamma_L(n^-l^+\nu_l) - \Gamma_L(n^+l^-\bar{\nu}_l)}{\Gamma_L(n^-l^+\nu_l) + \Gamma_L(n^+l^-\bar{\nu}_l)}. \tag{4.15}$$

The numerator and denumerator in this expression are given by

$$\Gamma_L(n^-l^+\nu_l) \pm \Gamma_L(n^+l^-\bar{\nu}_l) = (2\pi)\sum_{\text{pol}} \int \delta^4(k - p_n - l - \nu)\, d^3l\, d^3\nu\, d^3p_n \times$$

$$\times\ [|p|^2|\langle n^-l^+\nu_{l\,\text{out}}|\, H^l(0)\, |K_{\text{in}}^0\rangle|^2 + |q|^2\,|\langle n^+l^-\bar{\nu}_{l\,\text{in}}|\, H^l(0)\, |K_{\text{in}}^0\rangle|^2$$

$$\pm\ |p|^2\,|\langle n^+l^-\nu_{l\,\text{out}}|\, H^l(0)|\, K_{\text{in}}^0\rangle|^2 \pm |q|^2\,|\langle n^-l^+\nu_{l\,\text{in}}|\, H^l(0)\, |K_{\text{in}}^0\rangle|^2$$

$$+\ pq^*\ \{\langle \bar{K}_{\text{in}}^0|\, H^l(0)\, |n^-l^+\nu_{l\,\text{out}}\rangle\, \langle n^-l^+\nu_{l\,\text{out}}|\, H^l(0)\, |K_{\text{in}}^0\rangle$$

$$\pm\ \langle \bar{K}_{\text{in}}^0|\, H^l(0)\, |n^-l^+\nu_{l\,\text{in}}\rangle\, \langle n^-l^+\nu_{l\,\text{in}}|\, H^l(0)\, |K_{\text{in}}^0\rangle\}$$

$$+\ p^*q\ \{\langle K_{\text{in}}^0|\, H^l(0)\, |n^-l^+\nu_{l\,\text{out}}\rangle\, \langle n^-l^+\nu_{l\,\text{out}}|\, H^l(0)\, |\bar{K}_{\text{in}}^0\rangle$$

$$\pm\ \langle K_{\text{in}}^0|\, H^l(0)\, |n^-l^+\nu_{l\,\text{in}}\rangle\, \langle n^-l^+\nu_{l\,\text{in}}|\, H^l(0)\, |\bar{K}_{\text{in}}^0\rangle\} \tag{4.16}$$

Notice that every second term in this expression involves an "in" $n l \nu$ state rather than an "out" one. This is because we have made use of the relations

$$\langle n^-l^+\nu_{l\,\text{out}}|\, H^l(0)\, |K^0(\bar{K}^0)_{\text{in}}\rangle = \langle n^+l^-\bar{\nu}_{l\,\text{in}}|\, H^l(0)\, |\bar{K}^0(K^0)_{\text{in}}\rangle^*$$

and

$$\langle n^+l^-\bar{\nu}_{l\,\text{out}}|\, H^l(0)\, |K^0(\bar{K}^0)_{\text{in}}\rangle = \langle n^-l'^+\nu_{l\,\text{in}}|\, H^l(0)\, |\bar{K}^0(K^0)_{\text{in}}\rangle^*$$

($l'$ denotes the spin flipped state of $l$), which follow from $TCP$ invariance, to replace the terms which were originally in those places. Now so far, discussions of the quantity $\delta_L^n$ have concentrated solely on the case $n = \pi$ and in all these discussions the difference between the "in" and "out" states in (4.16) has been completely disregarded. The reason for so doing is presumably that in this particular case this difference can arise only due to the electromagnetic interaction and so is likely to be small of order $\alpha$. Whatever the reason, the fact is that with this assumption and the definition

$$\langle \pi^+l^-\bar{\nu}_{l\,\text{out}}|\, H^l(0)\, |K_{\text{in}}^0\rangle = X_\pi \langle \pi^-l^+\nu_{l\,\text{out}}|\, H^l(0)\, |K_{\text{in}}^0\rangle,$$

($X_\pi$ which is assumed constant, is obviously a measure of the violation of the $\Delta Y = \Delta Q$ rule in $K_{l_3}^0$ decay) it immediately follows that the charge asymmetry in $K_{L_{l_3}}^0$ is given by

$$\delta_L^\pi \equiv \frac{\Gamma_L(\pi^-l^+\nu_l) - \Gamma_L(\pi^+l^-\bar{\nu}_l)}{\Gamma_L(\pi^-l^+\nu_l) + \Gamma_L(\pi^+l^-\bar{\nu}_l)} = |p|^2 - |q|^2 \frac{|- |X_\pi|^2}{|1 + X_\pi|^2 + 0(\varepsilon)}, \tag{4.17}$$

where $\varepsilon = \dfrac{p - q}{p + q}$ is known to be smaller than $3 \times 10^{-3}$ in absolute value, so that the term $O(\varepsilon)$ may be safely neglected. It was on the basis of the expression (4.17) that the conclusions about $CP$ violation and the $\Delta Y = \Delta Q$ rule mentioned previously were drawn.

The difficulty is, however, that in deriving this relation one has neglected an effect which is likely to be of the same order of magnitude as the quantity one is computing; experimentally $\delta_L^\pi$, is of the order $2$–$4 \times 10^{-3}$ while the difference between "in" and "out" states is expected to be of the order of $\alpha = 1/137$ an estimate which is supported by the calculations of radiative corrections in $K_{l_3}^0$ decay[5]. Of course neglecting this difference is a matter of consequence only if $CP$ or $T$ invariance is violated ny the Hamiltonian of strong, electromagnetic or weak interactions. If these interactions are all $CP$ or $T$ invariant then we have the relations

$$\langle n^-(\mathbf{p}_n)\, l^+(l)\, \nu_l(\mathbf{v})_{\text{in}}|\, H_{(0)}^l\, |K^0(K^0)\,(\mathbf{k})_{\text{in}}\rangle$$
$$= \langle n^-(-\mathbf{p}_n)\, l'^+(-\mathbf{l})\, \nu_l(-\mathbf{v})|\, H_{(0)}^l\, |K^0(\overline{K}^0)\,(-\mathbf{k})_{\text{in}}\rangle^*,$$

and

$$\langle n^+(\mathbf{p}_n)\, l^-(l)\, \bar{\nu}_l(\mathbf{v})_{\text{in}}|\, H_{(0)}^l\, |K^0)\overline{K}^0)\,(\mathbf{k})_{\text{in}}\rangle$$
$$= \langle n^+(-\mathbf{p}_n)\, l'^-(-\mathbf{l})\, \bar{\nu}_l(-\mathbf{v})|\, H_{(0)}^l\, |K^0(\overline{K}^0)\,(-\mathbf{k})\rangle^*,$$

from $T$-invariance and because of these relations the difference between "in" and "out" states in (4.16) may be rigorously neglected. This would be the case in the super-weak theory, for example, but not in any of the theories where the $CP$ and $T$ violation is located either in the strong, electromagnetic or weak interactions. Since, at present, experiment has not decided as between the superweak and non-superweak explanations of $CP$ violation it is necessary to take seriously the difference between "in" and "out" states in (4.16). Elsewhere[10] we have shown that the corrections to Eq. (4.17) resulting from this difference are quite small; they are in fact of order $\alpha \lambda \Gamma(K_{l_4}^0)/\Gamma(K_{l_3}^0)$ ($\lambda$ is the strength of the $CP$ violating interaction) and this quantity is at most equal to $10^{-3}\alpha$.

A rather interesting way of circumventing the difficulty about the difference between "in" and "out" states it to measure not $\delta_L^r$ as given in Eq. (4.15) but rather

$$\delta_L^{\text{total}} = \frac{\sum\limits_n [\Gamma_L(n^- l^+ \nu_l) - \Gamma_L(n^+ l^- \bar{\nu}_l)]}{\sum\limits_n [\Gamma_L(n^- l^+ \nu_l) + \Gamma_L(n^+ l^- \bar{\nu}_l)]}, \tag{4.18}$$

i.e. the charge asymmetry in the decay of $K_L^0$ into *all* its semileptonic modes and not simply into a pair of modes charge conjugate to each other. If one now looks at Eq. (4.16) one sees that the effect of the summation over $n$ indicated in (4.18) is to replace the contribution of the single state $nl\nu$ by a summation over states which is equivalently a summation over a complete set of states. This fact enables one rigorously to neglect the difference between the "in" and "out" states in the equivalent of Eq. (4.16) because sums over complete sets of in or out states are both equal to the identity operator and hence equal to one another.

On performing the calculation one finds that

$$\delta_L^{\text{total}} = |p|^2 - |q|^2 \frac{\sum\limits_n [\Gamma_{K^0}(n^-l^+\nu_l) - \Gamma_{K^0}(n^+l^-\bar{\nu}_l)]}{\sum\limits_n [\Gamma_{K^0}(n^-l^+\nu_l) + \Gamma_{K^0}(n^+l^-\bar{\nu}_l)]} . \tag{4.19}$$

From this relation one sees that a non zero value of $\delta_L^{\text{total}}$ automatically implies $CP$ violation and furthermore the factor of $|p|^2 - |q|^2$ in this quantity is a measure of the validity of the $\Delta Y = \Delta Q$ rule since, if one disregards the minute $K^0 \to K^+ + l^- + \bar{\nu}_l$ decay and the probably non-existent $K^0 \to K^- + l^+ + \nu_l$ decay, this factor is equal to unity if the rule is exact and any deviation from this value indicates violation of the rule. In both these respects $\delta_L^{\text{total}}$ is seen to be an analogue of the discarded expression (4.17).

If we now use the same procedure for the decays of the state $K_R^0(t)$, Eq. (4.13), and measure the quantity

$$\delta_R^{\text{total}}(t) = \frac{\sum\limits_n [\Gamma_R(n^-l^+\nu_l; t) - \Gamma_R(n^+l^-\bar{\nu}_l; t)]}{\sum\limits_n [\Gamma_R(n^-l^+\nu_l; t) + \Gamma_R(n^+l^-\bar{\nu}_l; t)]}, \tag{4.20}$$

we can determine both of the quantities on the r.h.s. of (4.19). The numerator and denominator of (4.20) are respectively found from (4.14) to be

$$\sum_n [\Gamma_R(n^-l^+\nu_l; t) - \Gamma_R(n^+l^-\bar{\nu}_l; t)] = \frac{1}{1 + |R|^2 + 2\,\text{Re}\,(R\,\langle K_L^0|K_S^0\rangle)} \times$$

$$\times \left[ e^{-\Gamma_L t} \sum_n \{\Gamma_L(n^-l^+\nu_l) - \Gamma_L(n^+l^-\bar{\nu}_l)\} \right.$$

$$+ e^{-\Gamma_S t} |R|^2 \sum_n \{\Gamma_S(n^-l^+\nu_l) - \Gamma_S(n^+l^-\bar{\nu}_l)\}$$

$$\left. + e^{-1/2(\Gamma_L + \Gamma_S)t}\, 2|R| \cos(\Delta m t + \phi_R) \sum_n \{\Gamma_{K^0}(n^-l^+\nu_l) - \Gamma_{K^0}(n^+l^-\bar{\nu}_l)\} \right]$$

$$\tag{4.21}$$

and

$$
\sum_n \left[ \Gamma_R \left( n^- l^+ \nu_l ; t \right) + \Gamma_R \left( n^+ l^- \nu_l ; t \right) \right] = \frac{1}{1 + |R|^2 + 2 \, \mathrm{Re} \left( R \, \langle K_L^0 | K_S^0 \rangle \right)} \times
$$

$$
\times \left[ e^{-\Gamma_L t} \sum_n \left\{ \Gamma_L \left( n^- l^+ \nu_l \right) + \Gamma_L \left( n^+ l^- \bar{\nu}_l \right) \right\} \right.
$$

$$
+ \, e^{-\Gamma_S t} |R|^2 \sum_n \left\{ \Gamma_S \left( n^- l^+ \nu_l \right) + \Gamma_S \left( n^+ l^- \bar{\nu}_l \right) \right\}
$$

$$
+ \, e^{-1/2 (\Gamma_L + \Gamma_S) t} \, 2 \, |R| \times
$$

$$
\left. \times \left\{ \cos \left( \Delta m t + \phi_R \right) \mathrm{Re} \, \langle K_L^0 | \Gamma^l | K_S^0 \rangle - s_m \left( \Delta m t + \phi_R \right) \mathrm{Im} \, \langle K_L^0 | \Gamma^l | K_S^0 \rangle \right\} \right]
$$

$$
(4.22)
$$

where $\phi_R$ is the phase of $R$ and $\Gamma^l$ is the $l$-lepton part of the decay operator: i.e.

$$
\Gamma^l = \frac{(2\pi)^3}{2} \int d^4 x \, \left( H^l(x) \, H^l(0) + H^l(0) \, H^l(x) \right). \qquad (4.23)
$$

The determination of the quantity $|p^2| - |q^2|$ in the experiments reported in the last two of references 8 was in all probability done by measuring $\delta_R^{\mathrm{total}}(t)$.

In concluding this section we should like to draw attention to the expression (4.22) for the total semileptonic decay rate of the state $K_R^0(t)$. We believe that this quantity would repay considerable experimental study. From it, as one can see, it is possible to determine the four matrix elements of the decay operator $\Gamma^l$, Eq. (4.23), in the two dimensional subspace spanned by the $K_S^0$ and $K_L^0$ states and this knowledge would be a substantial help towards a fuller understanding of the $K^0$–$\bar{K}^0$ system.

### 4.6   The $V$, $A$ theory

The $V$, $A$ theory of weak interactions requires that if one neglects radiative corrections and higher order weak interactions the matrix elements for $K_{l_3}$ decay are those obtained from the local action of leptons together with the absence of the scalar and tensor terms. This means that $A_k$ and $B_k$ take on the very simple forms (see Eqs. (2.38))

$$
A_k = \left. \begin{array}{l} -m_l f_-^k(t) \; k = 1, 2, 3 \\[2mm] m_l f_-^k(t) \quad k = 4, 5, 6 \end{array} \right\}, \quad B_k = f_+^k(t). \qquad (4.24)
$$

In principle it is straightforward to test this prediction; one simply looks at the Dalitz plot using either expression (3.4) or expression (3.11) and sees whether the theoretical shape agrees with experiment. In practice the situation is that experiment is consistent with pure vector interaction i.e. prediction (4.24), but it is not sufficiently accurate to rule out sizeable amounts of scalar and tensor terms. A recent $K_{l_3}^+$ experiments[11], for example, set the limits

$$\frac{f_S}{f_+^{(0)}} < 0.23; \quad \frac{f_\tau}{f_+^{(0)}} < 0.58,$$

with 90% confidence, where it was assumed that $f_S$ and $f_T$ are constant and these are the best limits available still.

If we accept that the interaction is pure vector then $K_{l_3}$ decay reduces to a study of the pairs of form factors $f_+^k(t)$ and $f_-^k(t)$. (Of course if the $\Delta I = \frac{1}{2}$ rule is correct, which it seems to be, there are really only two form factors in all.) It is assumed that in the region of $K_{l_3}$ decay $f_-^k(t)$ is not many times greater than $f_+^k(t)$. With this assumption it follows that in $K_{l_3}$ the quantity $A_k = \pm m_l f_-^k(t)$ is negligible and so by studying $K_{l_3}$ decay we can determine the form factor $f_+^k(t)$.

From Eq. (3.11) we find that for $K_{l_3}$ decay

$$\frac{d\Gamma_k^\pm}{dt} = \frac{(G')^2}{(2\pi)^3} \frac{2m_k^2 p_\pi^3}{3} |f_+^k(t)|^2,$$

or in other words

$$\frac{d\Gamma_k^+}{dE_\pi} = \frac{(G')^2}{(2\pi)^3} \frac{2m_k^2 p_\pi^3}{3} |f_+^k(t)|^2, \tag{4.25}$$

where $E_\pi$ is the pion energy in the kaon's rest frame and of course $t = m_K^2 + m_\pi^2 - 2m_k E_\pi$. Thus by studying the energy spectrum of the pion in $K_{l_3}$ decay one can determine $f_+^k(t)$. Studies of this kind have established that $f_+^k(t)$ does not vary very strongly with $t$ in the physical region of $K_{l_3}$ decay and a linear form of the kind

$$f_+(t) = f_+(0)\left(1 + \frac{\lambda_+^k}{m_\pi^2} t\right) \tag{4.26}$$

with $\lambda_+^k = 0.03 \pm 0.01$ is indicated[12]. The lifetime derived from (4.25) and (4.26) is

$$\Gamma_k^\pm = \frac{(G')^2}{768\pi^3} m_k^5 |f_+^k(0)|^2 \, 0.573 \, (1 + 3.1\lambda_+^k). \tag{4.27}$$

A point of interest in this formula is the fact that the $t$ dependence of the form factor has considerable influence on the lifetime and may increase it by as much as 10%.

Now in the Cabibbo theory of weak interactions the coupling constant $G'$ is given by

$$G' = G \sin \theta, \qquad (4.28)$$

where $G$ is the bare $\mu$-decay constant having the value $1.026 \times 10^{-5} \times m_p^{-2}$ and $\theta$ is the Cabibbo angle. If we now use the relation (4.28) and the $K_{l_3}^+$ lifetime[1] we find that

$$\sqrt{2}\, f_+'(0) \sin \theta\, (1 + 3.1\lambda_+')^{1/2} = 0.227 \pm 0.006 \qquad (4.29)$$

From the $K_{L\,l_3}^0$ decay rate if we assume the $\Delta Y = \Delta Q$ rule we find

$$f_+^2(0) \sin \theta\, (1 + 3.1\lambda_+^2)^{1/2} = 0.215 \pm 0.010 \qquad (4.30)$$

These results are consistent with the $\Delta I = \frac{1}{2}$ rule. If we now invoke the Ademollo–Gatto theorem and set $\sqrt{2}\, f_+^1(0) = f_+^2(0) = 1$ then we can find the value of $\theta$ from Eq. (4.29) and (4.30). The result is consistent with the value derived from nucleon and nuclear $\beta$-decay but there are still considerable errors present in both determinations.

We next turn to a consideration of $K_{\mu_3}$ decay. Provided the form factor $f_+^k(t)$ has been determined in $K_{l_3}$ decay, $K_{\mu_3}$ decay reduces to a study of the form factor $f_-^k(t)$, but so far no very good determinations of this form factor are available. It is usual to follow the pattern set with regard to $f_+^k(t)$ and introduce the parametrization

$$f_-^k(t) = f_-^k(0) \left( 1 + \frac{\lambda_-^k}{m_\pi^2}\, t \right). \qquad (4.31)$$

Also used is the auxiliary quantity $\xi^k(t)$ given by

$$\xi^k(t) = \frac{f_-^k(t)}{f_+^k(t)} = \xi^k(0) \left( \frac{1 + \dfrac{\lambda_-^k}{m_\pi^2}\, t}{1 + \dfrac{\lambda_+^k}{m_\pi^2}\, t} \right). \qquad (4.32)$$

The straightforward way to measure $f_-^k(t)$ is through a Dalitz plot analysis but experiments of this type performed so far are very inconclusive. Partial information about $\xi^k(t)$ can be obtained from the ratios of the rates of $K_{\mu_3}$ and $K_{l_3}$. Using Eqs. (4.26) and (4.32) it is found that the ratio $\Gamma(K_{\mu_3}^\pm)/\Gamma(K_{l_3}^\pm)$

is given by

$$\frac{\Gamma(K_{\mu3}^{\pm})}{\Gamma(K_{e3}^{\pm})} = 0.649 + 1.32\lambda_+^1 + 0.127 \, \text{Re} \, \xi^1(0) + 0.019 \, |\xi^1(0)|^2$$

$$+ \, 0.008 \, \lambda_+^1 \, \text{Re} \, \xi^1(0) + 0.456\lambda_-^1 \, \text{Re} \, \xi^1(0)$$

$$+ \, 0.161\lambda_-^1 \, |\xi^1(0)|^2 - 0.375\lambda_+^1 \, |\xi^1(0)|^2. \tag{4.33}$$

Since we have no information about the quantity $\lambda_-^1$ this formula is not very useful. If we assume that $\lambda_-^1$ is negligible and in addition neglect $\lambda_+^1$ and take $\xi'(0)$ to be real we find

$$\frac{\Gamma(K_{\mu3}^{\pm})}{\Gamma(K_{e3}^{\pm})} - 0.649 + 0.127\xi^1(0) + 0.019 \, (\xi^1(0))^2 \tag{4.34}$$

Quite similar formulas hold for the ratio $\dfrac{\Gamma(K_{L\,\mu3}^{0})}{\Gamma(K_{L\,e3}^{0})}$.

On using Eq. (4.34) in conjunction with the experimental value $0.656^1$ we find for $\xi^1(0)$

$$\xi^1(0) = -0.06.$$

The difficulty with this result is that it is considerably different from that obtained from measurements of nuon polarization in $K_{\mu3}^{+}$ decay. These results have an average value of[12]

$$(\xi_{\text{pol}}^{1}(0))_{\text{expt}} = -0.98 \pm 0.20$$

One way to reconcile this result with that obtained from the branching ratio experiments is to use the full formula (4.33) for the branching ratio rather than the approximate one (4.34). However the value of $\lambda_-^1$ required for that purpose is $-0.27$ which seems unreasonably large when we recall that $\lambda_+^1$ is $0.03 \pm 0.01$.

The situation with regard to $K_{L\,l_3}^{0}$ decay is similar to that for $K_{l_3}^{\pm}$ decay; branching ratio experiments using the analogue of Eq. (4.34) give $\xi^L(0) = 1.3$ while measurements of muon polarization give $\xi^L(0) = -1.45 \pm 0.26$. Again the differences can be reconciled by giving to $f_-^L(t)$ a sufficiently strong dependence (i.e. sufficiently large $\lambda_-^L$) but the solution is no more satisfactory than in the previous case. What is required in this question is more accurate experimentation and particularly a detailed study of the Dalitz plot which simultaneously determines $\xi^k(0)$ and $\lambda_-^k$ thus eliminating the uncertainty of the previous analysis[13].

This survey of the phenomenology of $K_{l_3}$ decay shows that while the process is quite well understood in general, there are a number of points of detail on which more information would be extremely valuable. It is to be hoped that this information will be forthcoming in the not too distant future.

### References


1. N. Barash-Schmidt *et al.*, *Rev. Mod. Phys.*, **41**, 109 (1969).
2. N. Cabibbo and A. Maksymowicz, *Phys. Letters*, **9**, 352 (1964); **11**, 360 (1964); and **14**, 72 (1965).
3. W. R. Frazer and J. R. Fulco, *Phys. Rev.*, **117**, 1603 (1960).
4. E. Bellotti, E. Fiorin and A. Pullia, *Nuovo Cimento*, **52A**, 1287 (1967).
5. E. S. Ginsberg, *Phys. Rev.*, **142**, 1035 (1966); *ibid.* **171**, 1975 (1968).
6. F. Lobkowicz *et al.*, *Phys. Rev. Letters*, **17**, 548 (1966).
7. D. Bartlett *et al.*, *Phys. Rev. Letters*, **16**, 282 (1966); and *ibid.* (E) **16**, 601 (1966); P. S. Young *et al.*, *Phys. Rev.*, **156**, 1464 (1967); J. Bettels *et al.*, *Nuovo Cimento*, **56A**, 1106 (1968).
8. S. Bennett *et al.*, *Phys. Rev. Letters*, **19**, 993 (1967); D. Dorfan *et al.*, *Phys. Rev. Letters*, **19**, 987 (1967); S. Bennett *et al.*, *Phys. Letters*, **27B**, 239 (1968); J. Steinberger, Report at Topical Conference on Weak Interactions, CERN January 1969.
9. S. Bennett *et al.*, *Phys. Rev. Letters*, **19**, 997 (1967); and *Phys. Letters*, **27B**, 244 and 248 (1968).
10. C. Ryan, "Electromagnetic contributions to the charge asymmetry in the semileptonic decays of neutral kaons", I.C.T.P. Trieste preprint IC/69/92. August 1969.
11. D. R. Botterill *et al.*, *Phys. Rev.*, **174**, 1661 (1968).
12. C. Rubbia, Report at Topical Conference on Weak Interactions CERN January 1969.
13. After this work was completed we received two CERN preprints (June 1969) reporting the final results of the X2 collaboration on $K_{\mu_3}^+$. These papers claim to have obtained a consistent picture of all $K_{l_3}^+$ data including Dalitz plot, branching ratio and polarization.

# Author Index